David Capewell Formerly Westfield School, Sheffield
Marguerite Comyns Queen Mary's High School, Walsall
Gillian Flinton All Saints Catholic High School, Sheffield
Paul Flinton Chaucer School, Sheffield
Geoff Fowler Maths Strategy Manager, Birmingham
Derek Huby Mathematics Consultant
Peter Johnson Waitakere College, Auckland, N.Z.
Jayne Kranat Langley Park School for Girls, Bromley
Ian Molyneux St. Bedes RC High School, Ormskirk
Peter Mullarkey Netherhall School, Maryport, Cumbria
Nina Patel Ifield Community College, West Sussex
Claire Turpin Sidney Stringer Community Technology College, Coventry

Great Clarendon Street, Oxford OX2 6DP

Oxford University Press is a department of the University of Oxford.
It furthers the University's objective of excellence in research,
scholarship, and education by publishing worldwide in

Oxford New York

Auckland Cape Town Dar es Salaam Hong Kong Karachi
Kuala Lumpur Madrid Melbourne Mexico City Nairobi
New Delhi Shanghai Taipei Toronto

With offices in

Argentina Austria Brazil Chile Czech Republic France Greece
Guatemala Hungary Italy Japan Poland Portugal Singapore
South Korea Switzerland Thailand Turkey Ukraine Vietnam

Oxford is a registered trade mark of Oxford University Press
in the UK and in certain other countries

First published 2004

British Library Cataloguing in Publication Data

Data available

ISBN-13: 978-0-19-914860-8
ISBN-10: 0-19-914860-0

10 9 8 7 6 5 4 3 2

Typeset by Mathematical Composition Setters Ltd.

Printed in Italy by Rotolito Lombarda

Acknowledgements

The photograph on the cover is reproduced courtesy of Pictor.

The Publisher would like to thank the following for permission to reproduce photographs:

Alamy Images: pp 85, 104 (right), 231; Corbis UK Ltd: p 217; Corbis/Duomo: p 189; Corbis/Bob Krist: p 17; Corbis Royalty Free: p 104 (left); Corel Professional Photos: p 203; Empics/Steve Cuff: p 9; Lonely Planet: p 147; Photodisk: p 1 (all)

Figurative artwork is by Paul Daviz.

About this book

This book has been written specifically for Year 9 of the Framework for Teaching Mathematics. It is aimed at higher ability students who are following the Year 9 teaching programme from the Framework and leads to the 6–8 tier of entry in the NC tests.

The authors are experienced teachers and maths consultants who have been incorporating the Framework approaches into their teaching for many years and so are well qualified to help you successfully meet the Framework objectives.

The book is made up of units based on the sample medium term plans which complement the Framework document, thus maintaining the required pitch, pace and progression.

The units are:

The last five units in this book are designed to consolidate KS3 work and bridge to KS4 work.

Each unit comprises double page spreads that should take a lesson to teach. These are shown on the full contents list.

Problem solving is integrated throughout the material as suggested in the Framework.

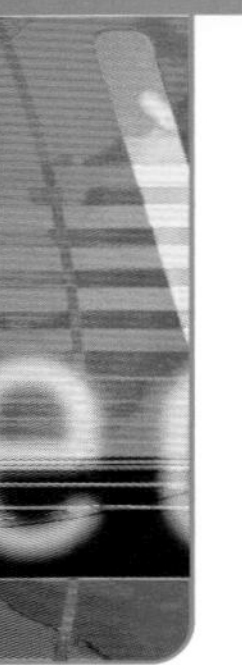

How to use this book

This book is made up of units of work which are colour coded into: Algebra (Blue), Data (Pink), Number (Orange), Shape, space and measures (Green), Problem solving (Light Green) and a Bridging unit (Red).

Each unit of work starts with an overview of the content of the unit, as specified in the Framework document, so that you know exactly what you are expected to learn.

This unit will show you how to:

- Know and use the index laws for multiplication and division of positive integer powers.
- Begin to extend understanding of index notation to negative and fractional powers.
- Multiply and divide by decimals, dividing by transforming to division by an integer.
- Estimate calculations by rounding numbers to one significant figure.

The first page of a unit also highlights the skills and facts you should already know and provides Check in questions to help you revise before you start so that you are ready to apply the knowledge later in the unit:

Before you start

You should know how to ...	Check in
1 Multiply and divide by any integer power of 10.	**1** Multiply each number by 10,100 and 1000: **a** 36.2 **b** 2150 **c** 0.0063
2 Round measurements to a given power of 10.	**2** Round each of these measurements. **a** 0.618 m to the nearest cm **b** 6724 km to the nearest 100 km.

Inside each unit, the content develops in double page spreads which all follow the same structure.

The spreads start with a list of the learning outcomes and a summary of the keywords:

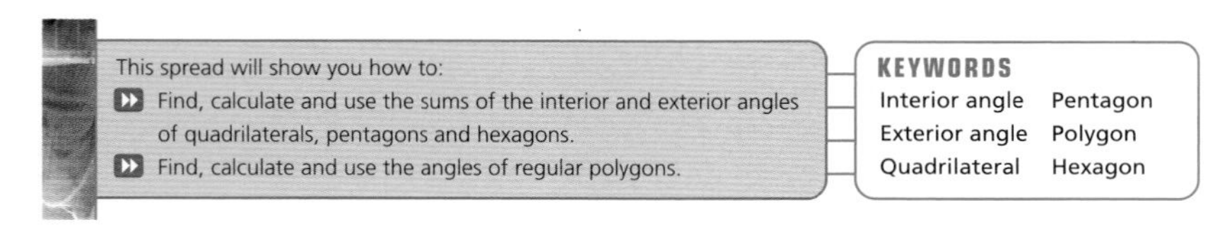
This spread will show you how to:

- Find, calculate and use the sums of the interior and exterior angles of quadrilaterals, pentagons and hexagons.
- Find, calculate and use the angles of regular polygons.

KEYWORDS

Interior angle
Exterior angle
Quadrilateral
Pentagon
Polygon
Hexagon

The keywords are summarised and defined in a Glossary at the end of the book so you can always check what they mean.

Key information is highlighted in the text so you can see the facts you need to learn.

► The interior angle sum of a polygon is $(n - 2) \times 180°$.

Examples showing the key skills and techniques you need to develop are shown in boxes. Also hint boxes show tips and reminders you may find useful:

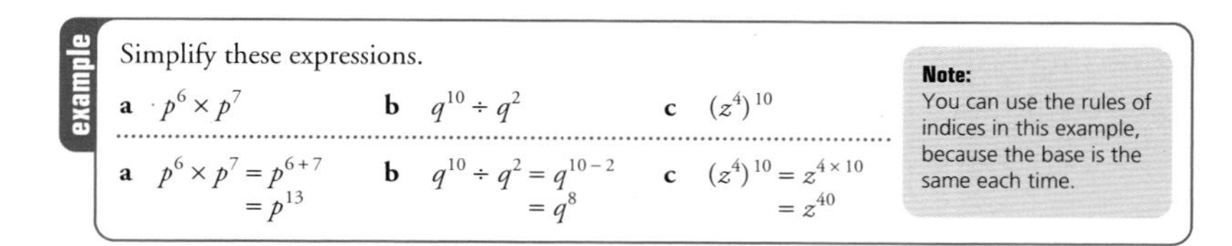

example

Simplify these expressions.

a $p^6 \times p^7$ **b** $q^{10} \div q^2$ **c** $(z^4)^{10}$

a $p^6 \times p^7 = p^{6+7} = p^{13}$

b $q^{10} \div q^2 = q^{10-2} = q^8$

c $(z^4)^{10} = z^{4 \times 10} = z^{40}$

Note:
You can use the rules of indices in this example, because the base is the same each time.

Each exercise is carefully graded, set at three levels of difficulty:

- The first few questions provide lead-in questions, revising previous learning.
- The questions in the middle of the exercise provide the main focus of the material.
- The last few questions are challenging questions that provide a link to further learning objectives.

At the end of each unit is a summary page so that you can revise the learning of the unit before moving on.

Check out questions are provided to help you check your understanding of the key concepts covered and your ability to apply the key techniques. These are all based on actual Key Stage 3 paper questions so they give you practice at the standard required in your examination.

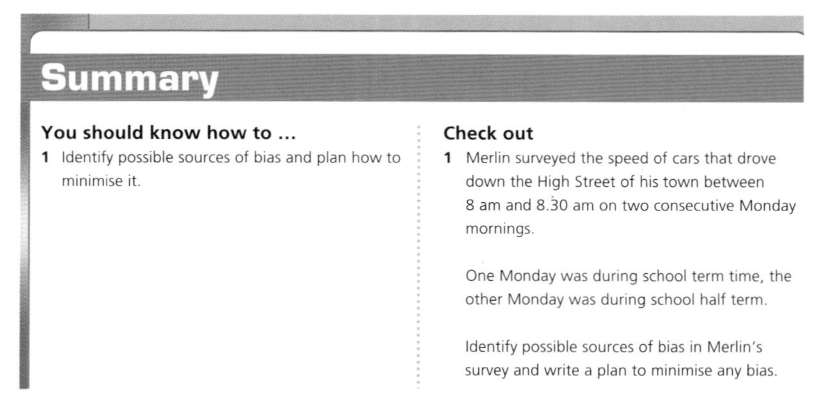

Summary

You should know how to ...

1 Identify possible sources of bias and plan how to minimise it.

Check out

1 Merlin surveyed the speed of cars that drove down the High Street of his town between 8 am and 8.30 am on two consecutive Monday mornings.

One Monday was during school term time, the other Monday was during school half term.

Identify possible sources of bias in Merlin's survey and write a plan to minimise any bias.

The answers to the Check in and Check out questions are produced at the end of the book so that you can check your own progress and identify any areas that need work.

Contents

1 Sequences

This unit will show you how to:

- Generate terms of a sequence using term-to-term and position-to-term definitions.
- Find the next term and the *n*th term of quadratic sequences and functions.
- Generate sequences from practical contexts and write an expression to describe the *n*th term of an arithmetic sequence.
- Deduce properties of triangular and square numbers from spatial patterns.
- Generate fuller solutions to problems.
- Justify generalisations, arguments or solutions.
- Pose extra constraints and investigate whether cases can be generalised further.

Notes in a scale follow a sequence.

Before you start

You should know how to ...

1 Continue a numerical sequence.

2 Generate a sequence from a practical context.

Check in

1 Find the missing numbers in these sequences.

 a 5, 16, 27, 38, __, __, ...

 b 100, 93, 86, __, 72, __, ...

 c 1, __, __, 16, __, 36, 49, ...

2 Playing cards are used to make pyramids:

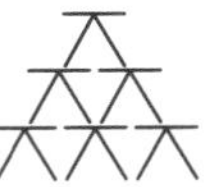

 a Draw the next two diagrams in the pattern.

 b Copy and complete the table.

Number of rows	1	2	3	4	5
Number of cards	3				

 c How many cards will be needed for a pyramid with six rows?

Revising sequences

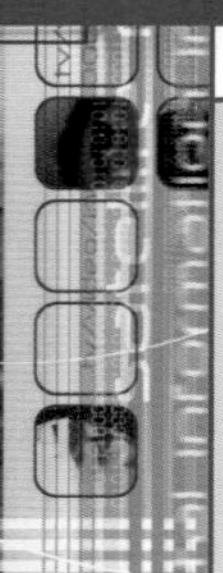

This spread will show you how to:

- Generate terms of a sequence using term-to-term and position-to-term definitions of the sequence.
- Find the next term and the *n*th term of quadratic sequences and functions and explore their properties.
- Generate sequences from practical contexts and write an expression to describe the *n*th term of an arithmetic sequence.

KEYWORDS

Quadratic sequence
Sequence
Linear
Term
Constant
Rule
T(*n*)
*n*th term

Here is a sequence:

You can find a rule for each **term** of the sequence:

The ***n*th term** of the sequence $T(n) = n^3$.

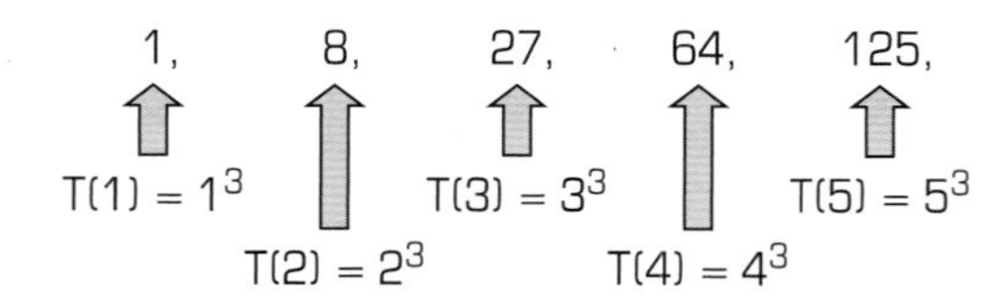

▶ In a **linear sequence**, the difference between successive terms is constant.

example

Find the general term of the sequence: 4, 9, 14, 19, 24, ...

The difference between terms is 5, so compare with the 5 times table:

n	1	2	3	4	5
Multiple of 5	5	10	15	20	25
T(*n*)	4	9	14	19	24

× 5, − 1 ⟹ $T(n) = 5n - 1$

Quadratic sequences are related to the square numbers.

▶ In a quadratic sequence, the second difference is constant.

example

a Show that the sequence 2, 5, 10, 17, 26, ... is quadratic.
b Find a formula for the general term of the sequence.
c Use your formula to find the 25th term of the sequence.

a

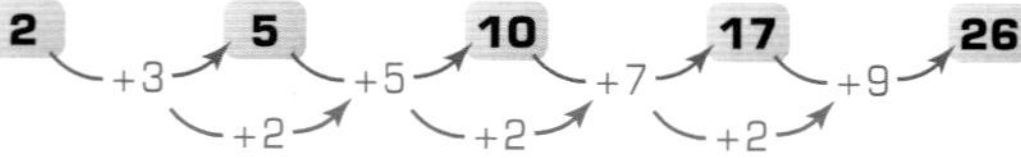

The second difference between terms is always 2.
⟹ The sequence is quadratic.

b Compare with the square numbers:

n	1	2	3	4	5
Square numbers	1	4	9	16	25
T(*n*)	2	5	10	17	26

square, + 1 ⟹ $T(n) = n^2 + 1$

c $T(25) = 25^2 + 1$
$= 626$

Exercise A1.1

1

1, 8, 27, 64, 125, ...	2, 6, 12, 20, 30, ...
5, 8, 11, 14, 17, ...	0, 3, 8, 15, 24, ...
6, 9, 14, 21, 30, ...	4, 8, 12, 16, 20, ...
20, 18, 16, 14, ...	

a Decide if each sequence in the box is linear, quadratic or neither.

b Find a formula for the general term, $T(n)$, of the linear sequences you have identified.

c Match the quadratic sequences with these general term formulae:
$T(n) =$ **i** $n^2 - 1$ **ii** $n^2 + 5$
iii $n^2 + n$

2 Generate the first five terms of the sequences described below.

a Triangular numbers

b $T(n) = 8n + 3$

c $T(n) = 3n^2 + 2$

d Prime numbers

e $T(n) = 50 - n$

f $T(n) = (n + 3)(n + 2)$

3 The tenth term of a linear sequence is 58 and the 12th term is 70.
Write the first five terms of the sequence and find a formula for the general term.

4 Justify the formula given for each pattern.

a 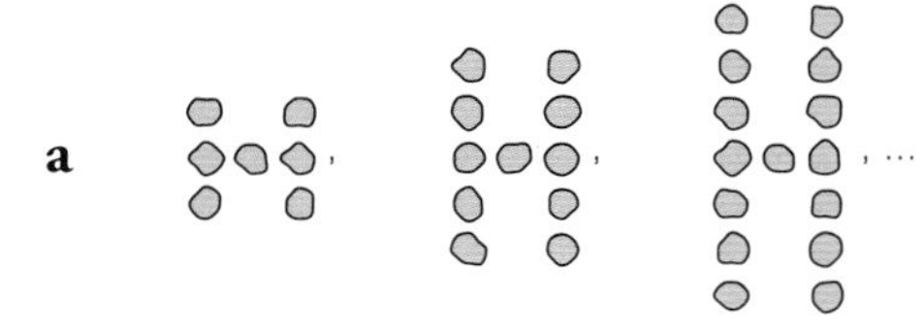

$T(n) = 4n + 3$

b  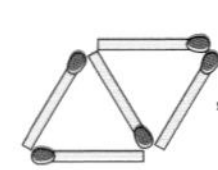 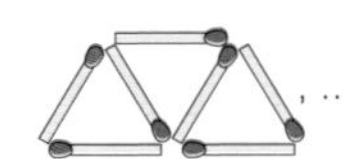

$M = 2T + 1$
M = number of matches
T = number of triangles

5 A plant is 10 cm tall. It grows 2 cm every month. Assuming the plant keeps growing like this, use a sequence formula to work out when the plant will reach 5 m if it is now 1 January 2004.

6

10	11	12	13	14	15	16	← layer 4
	5	6	7	8	9		← layer 3
		2	3	4			← layer 2
			1				← layer 1

a Write a formula for the number on the last brick in each layer of this pyramid.

b A formula for the middle brick in each layer is $T(n) = n^2 - n + 1$. What is the formula for the number on the brick *before* the middle in each layer?

c True or false? 2499 will be in layer 50. Explain your answer.

7 $T(n) = an + b$ is a formula for a linear sequence (a and b are constants) and $T(n) = an^2$ is a formula for a quadratic sequence.
What can you say about a and b if the two sequences have the same first term?

8 Generate the first five terms of $T(n) = 3n + (n - 2)(n + 3) + n^2$.

9 Find a general term formula for:

a 50, 48, 46, 44, ...

b 3, 8, 15, 24, 35, ...

c 53, 56, 61, 68, ...

Hint: For **c**, use **a** and **b**.

10 By considering how each sequence grows, find and justify a general formula for each.

a 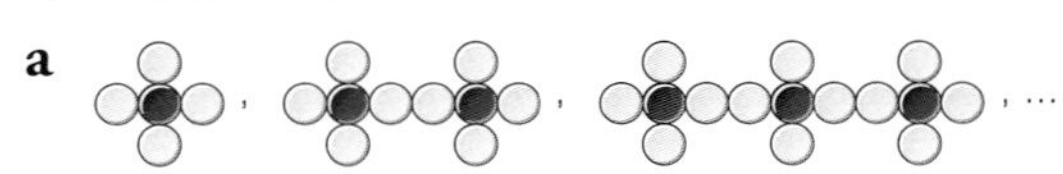

b

A1.2 Exploring sequences

This spread will show you how to:

- Find the next term and the *n*th term of quadratic sequences and functions and explore their properties.

KEYWORDS
Quadratic
Limit
Coefficient
Second difference

▶ **You can write any quadratic sequence in the form:**
$T(n) = an^2 + bn + c$ where a, b and c are constants.

a is always non-zero.
b and c can be zero.

Here are some quadratic sequences:

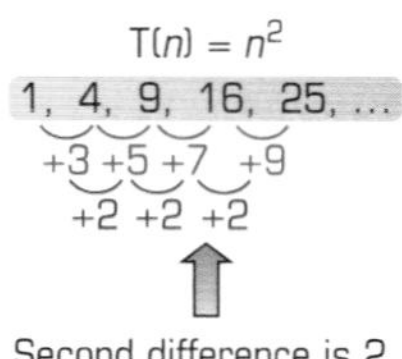

Second difference is 2.
Coefficient of n^2 is 1.

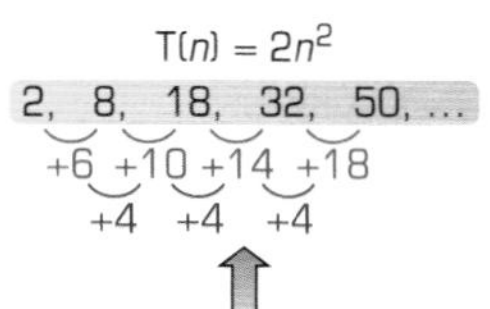

Second difference is 4.
Coefficient of n^2 is 2.

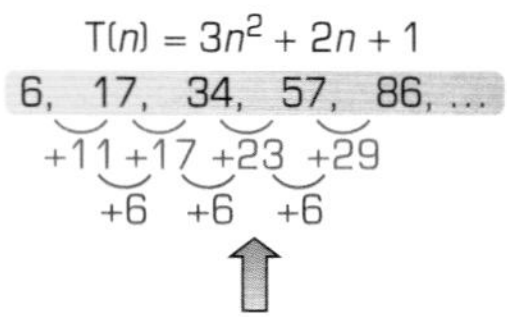

Second difference is 6.
Coefficient of n^2 is 3.

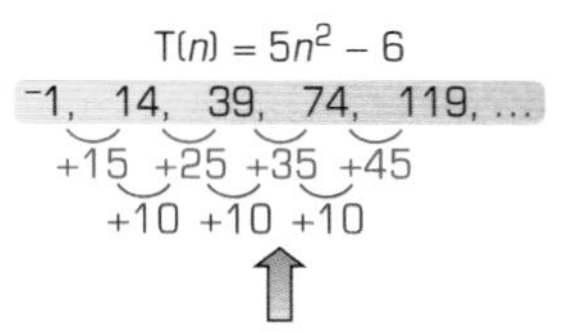

Second difference is 10.
Coefficient of n^2 is 5.

▶ **The second difference in a quadratic sequence is always equal to $2a$.**

A **coefficient** is a number in front of a term.
a is the coefficient of n^2.
b is the coefficient of n.

You can use this fact to find the general term of a quadratic sequence.

example

Here is a quadratic sequence: 3, 12, 25, 42, 63, ...
Find a formula for the general term.

First find the second difference:

3, 12, 25, 42, 63
+9, +13, +17, +21
+4, +4, +4

The second difference is 4, so the formula will contain $2n^2$.

Now draw a table:

Position, n	1	2	3	4	5
Term, $T(n)$	3	12	25	42	63
Quadratic part, $2n^2$	2	8	18	32	50
Remaining part, $T(n) - 2n^2$	1	4	7	10	13
Multiples of 3, $3n$	3	6	9	12	15

When you subtract the quadratic part ($2n^2$), a linear sequence remains ($bn + c$).

It goes up in 3s, so compare with the multiples of 3:
The linear sequence is $3n - 2$

So the general term $T(n) = 2n^2 + 3n - 2$

Check by substituting $n = 5$: $T(5) = 2 \times 5^2 + 3 \times 5 - 2$
$= 63$ This is correct, so the formula works.

Exercise A1.2

1 Copy these quadratic sequences and fill in the missing numbers.

a 8, 20, 38, 62, 92, ...

+12 +18 +24 +30

+6 +6 +6

$T(n) = \square n^2 + 3n + 2$

b 5, 19, 41, 71, 109, ...

+14 +22 +30 +38

+8 +8 +8

$T(n) = \square n^2 + \square n - 1$

2 Find a formula for the general term of each of these quadratic sequences.

a 3, 12, 27, 48, 75, ...

b 1, 7, 17, 31, 49, ...

c 6, 16, 30, 48, 70, ...

d 5, 19, 39, 65, 97, ...

e 2, 2, 4, 8, 14, ...

3 A lucky rabbit has 1000 g of lettuce. He eats half of what he has left each hour, i.e. 500 g in the first hour, 250 g in the next hour and so on.

a Write a sequence to show how much the rabbit eats each hour.

b

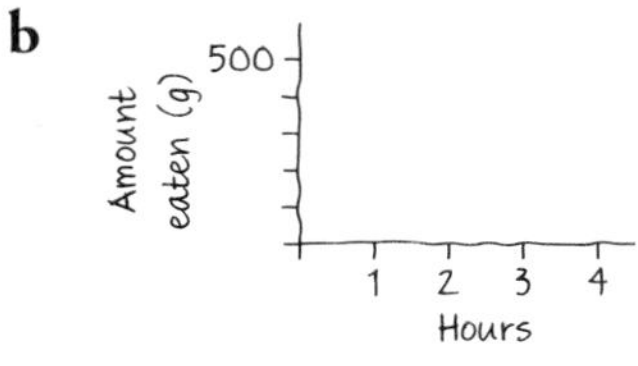

Plot a graph of these results.
Describe what is happening.

c When will all the lettuce be gone?

4 Find a general term formula for these fractional sequences.

a $\frac{5}{12}, \frac{10}{20}, \frac{15}{30}, \frac{20}{42}, \frac{25}{56}, \ldots$ **b** $\frac{5}{1}, \frac{15}{8}, \frac{31}{27}, \frac{53}{64}, \frac{81}{125}, \ldots$

5 The sequence 1, 4, 9, 16, 25, ... has the general term, $T(n) = n^2$.
Find the general term of:

a 4, 9, 16, 25, 36, ...

b 9, 16, 25, 36, 49, ...

6 **a** Copy and complete this table for the diagrams given.

Diagram	1	2	3	4
Number of tiles				

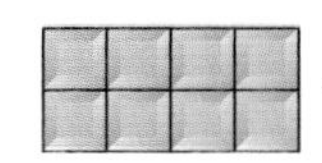

Diagram 1

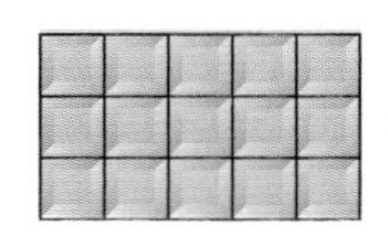

Diagram 2

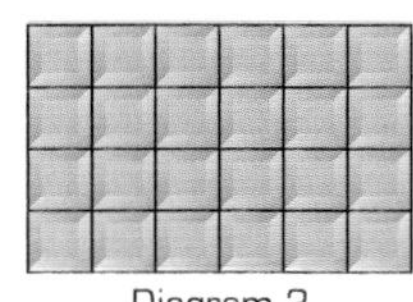

Diagram 3

b Using your results, find a general formula for the sequence.

c Ben found the formula $T(n) = (n + 1)(n + 3)$ ... is he right?

d Can you explain **why** your formula and Ben's formula work?
Which is easier to justify?

7 Here is a sequence: $T(n) = \frac{n}{n+1}$

$T(1) = \frac{1}{2}$, $T(2) = \frac{2}{3}$, $T(3) = \frac{3}{4}$, $T(4) = \frac{4}{5}$, $T(100) = \frac{100}{101}$

The sequence gets closer to 1, but never reaches it. The **limit** of the sequence is 1.
Explore these sequences and decide if they have a limit.

a $T(n) = 2^n$ **b** $T(n) = \frac{2n}{n+1}$

c $T(n) = \frac{n}{2} + 3$

8 **a** The nth term of a sequence is $T(n) = \frac{2n}{n^2 + 1}$. Write the first five terms and the 100th term of the sequence.

b This sequence is infinite.
Which graph best shows how the sequence will continue?

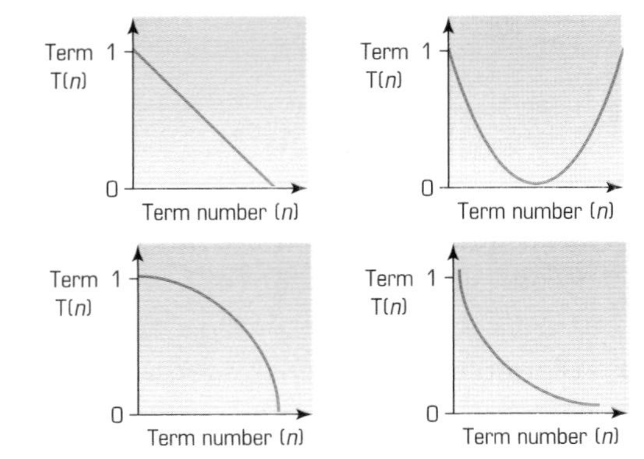

A1.3 The triangular and square numbers

This spread will show you how to:

- Deduce properties of the sequences of triangular and square numbers from spatial patterns.

KEYWORDS
Square
Consecutive
Triangular
Integer

Here is the sequence of **triangular numbers**:

1

3

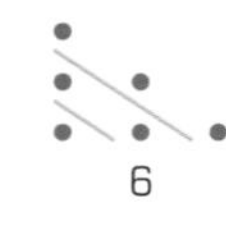

6

10

T(1) = 1   T(2) = 1 + 2   T(3) = 1 + 2 + 3   T(4) = 1 + 2 + 3 + 4

$$T(n) = 1 + 2 + 3 + \cdots + n$$

The tenth triangular number
T(10) = 1 + 2 + 3 + 4 + ⋯ + 9 + 10 = 55

If you wanted the hundredth triangular number, the formula would be very tedious. This sequence of diagrams helps to find a **general** formula:

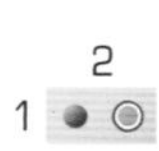

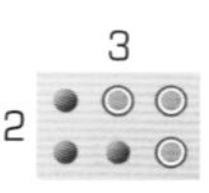

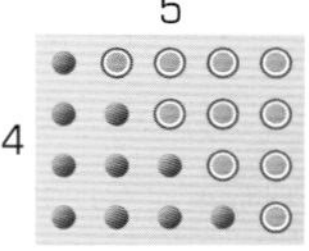

The number of dots in the rectangle is twice the triangular number each time.

So twice the fourth triangular number is 4 × 5.

⟹ Twice the nth triangular number is $n(n + 1)$.

- A formula for the general term of the sequence of triangular numbers is given by: $T(n) = \frac{n(n+1)}{2}$

example

Find the sum of the integers from 1 to 100.

$1 + 2 + 3 + \cdots + 99 + 100$ ← This is the 100th triangular number.

$T(100) = \frac{100 \times 101}{2} = 5050$

You should also know the sequence of **square numbers**.

1

4

9

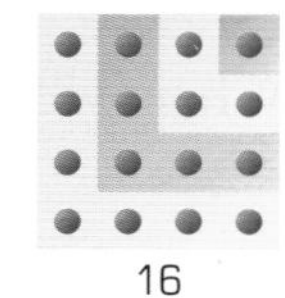

16

$T(1) = 1^2$
$= 1$

$T(2) = 2^2$
$= 1 + 3$

$T(3) = 3^2$
$= 1 + 3 + 5$

$T(4) = 4^2$
$= 1 + 3 + 5 + 7$

- Each square number is the sum of consecutive odd numbers (starting from 1).
  $T(n) = 1 + 3 + 5 + 7 + \cdots + (2n - 1)$
- A formula for the general term of the square numbers is given by: $T(n) = n^2$

So the sixth square number is:
T(6) = 1 + 3 + 5 + 7 + 9 + 11 = 36
or $T(6) = 6^2 = 36$

Exercise A1.3

1 Copy and complete:

a $10^2 = \_\_ \times \_\_$

$= 1 + \cdots$

b Seventh triangular number

$= \frac{\times}{2}$

$= 1 + \cdots$

2 Vicky wanted to know the sum of all the days in August, so she began adding $1 + 2 + 3 + \cdots$

AUGUST						
M	T	W	T	F	S	S
				1	2	3
4	5	6	7	8	9	10
11	12	13	14	15	16	17
18	19	20	21	22	23	24
25	26	27	28	29	30	31

Richard decided he had a quicker way. Use both methods and check they get the same answer.

3 For each set of diagrams, write a formula connecting the height, h, of the staircases with the number of blocks, B, used to make them.

a

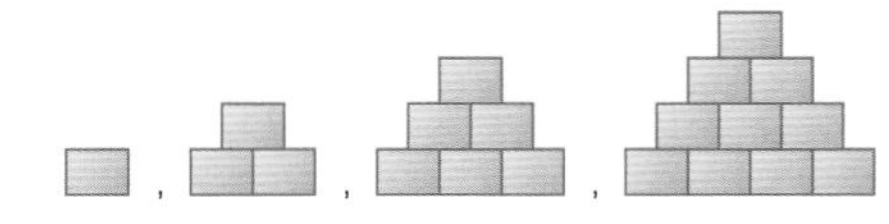

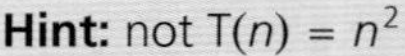

b

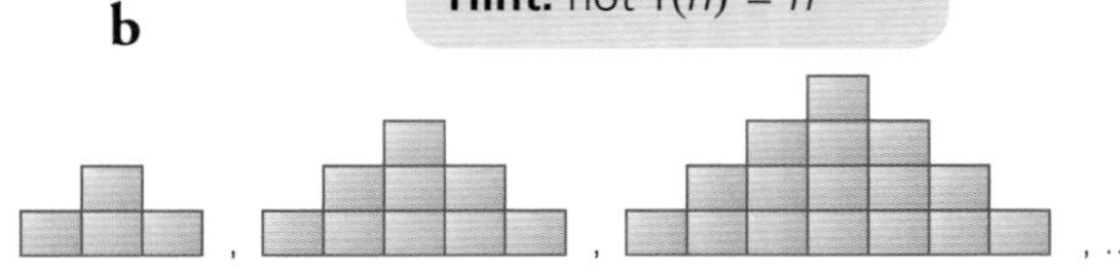

Hint: symmetrical

c

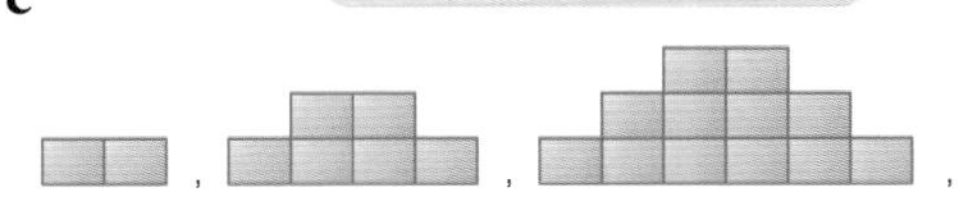

d

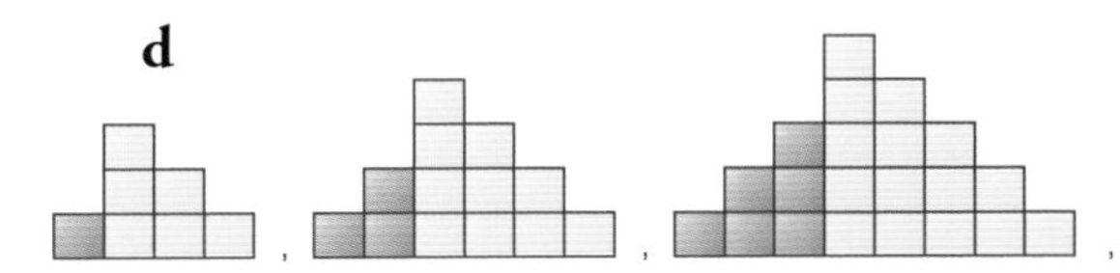

4 On her first birthday, Sarah got one gift. On her second she got two, on her third she got three and so on.

a How many presents did Sarah receive on:

i her 5th birthday

ii her 50th birthday?

b How many presents had Sarah received in total by her 10th birthday?

5 **a** Without using a calculator, find a quick method to sum the numbers from 1 to 1000.

$1 + 2 + 3 + \cdots + 999 + 1000$

b What is the sum of the odd numbers from 1 to 1000 inclusive?

c Hence, find the sum of the even numbers from 1 to 1000 inclusive.

6 A sprout, ①, is placed in the centre slot in a huge crate. After an hour it turns bad. After another hour, the sprouts touching it directly also turn bad (see ②).

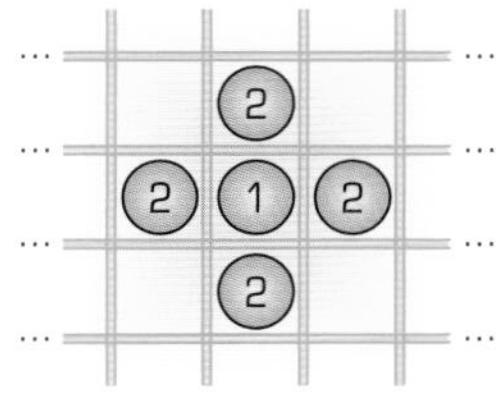

a Investigate the total number of bad sprouts after n hours. Try to write a formula.

b Repeat for the situation where the first sprout is placed in the bottom left-hand corner of the box.

c Repeat for the situation where the first sprout is placed in the centre of the bottom row of the box.

You should know how to ...

1 Find the next term and the nth term of quadratic sequences and functions and explore their properties.

2 Deduce properties of the sequences of triangular and square numbers from spatial patterns.

3 Generate fuller solutions to problems.

Check out

1 **a** Find a general term formula for:

  i 3, 6, 11, 18, 27, ...

  ii 3, 12, 27, 48, 75, ...

  iii 8, 15, 26, 41, 60, ...

 b What is the limit of the sequence

$$T(n) = \frac{n^2 + 1}{3n^2}\ ?$$

2 In each case, find a formula connecting the pattern number (n) to the number of stars used (s).

a

b

c

3 Mr Odd worked for the Odd Ball Company. In his first year, he received £1 salary. In his second year, he received £3, in his third year £5 and so on.

 a If Mr Odd started work at 18 and retired at 65, how much did he receive in his final pay packet?

 b How much did he earn in total over his working life?

A2 Functions and graphs

This unit will show you how to:

- Recognise that equations of the form $y = mx + c$ correspond to straight-line graphs.
- Given values for m and c, find the gradients of lines given by equations of the form $y = mx + c$.
- Find the inverse of a linear function.
- Plot the graph of the inverse of a linear function.
- Know properties of quadratic functions.
- Plot graphs of simple quadratic and cubic functions.
- Generate fuller solutions to problems.
- Justify generalisations, arguments or solutions.
- Pose extra constraints and investigate whether particular cases can be generalised further.

The path of a ball can be described by a quadratic function.

Before you start

You should know how to ...	Check in
1 Substitute values into an expression.	1 If $a = 3$, $b = 4$ and $c = ^-2$: **a** Put these expressions in ascending order.  $3a - 2b$ $2b - 3c$ $2a^2$ abc $\frac{ab}{c}$ $10 - 2a$ **b** True or false? $ab > 3c^2$
2 Plot coordinates in all four quadrants.	2 On a pair of axes labelled from $^-5$ to 5, plot and join these points. What letter is revealed? (1, 1), (3, 1), (3, 3), ($^-3$, 3), ($^-3$, 1), ($^-1$, 1), ($^-1$, $^-3$), (1, $^-3$), (1, 1).
3 Measure the gradient of a straight line.	3 Match each line with its gradient. 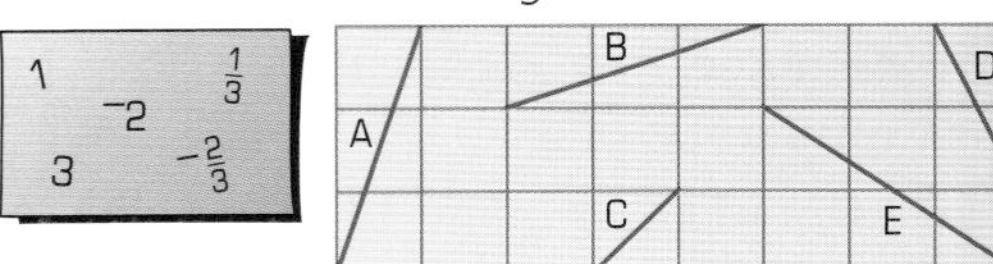 1 $^-2$ $\frac{1}{3}$ 3 $-\frac{2}{3}$ A B C D E

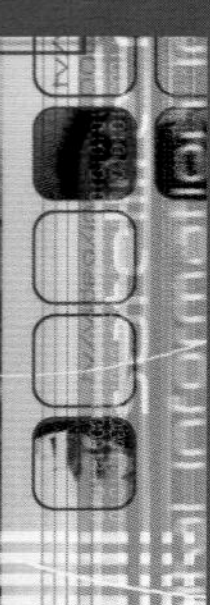

A2.1 Revising straight-line graphs

This spread will show you how to:

- Recognise that equations of the form $y = mx + c$ correspond to straight-line graphs.
- Given values for m and c, find the gradient of lines given by equations of the form $y = mx + c$.
- Investigate the gradients of parallel lines.

KEYWORDS

Linear	Axes
Gradient	Parallel
Intercept	Implicit

▶ **You can describe a straight-line (or linear) graph by the equation $y = mx + c$, where m and c are constants.**

Remember:
Gradient is a measure of steepness.
It is the amount you go up for every unit you go along.

example

Draw the graphs described by these two equations, using suitable axes:

a $y = 3x + 1$ **b** $x + 4y = 8$

For each graph, find the gradient and the point where it crosses the y-axis.

a Draw a table:

x	1	2	3
y	4	7	10

Draw a graph:

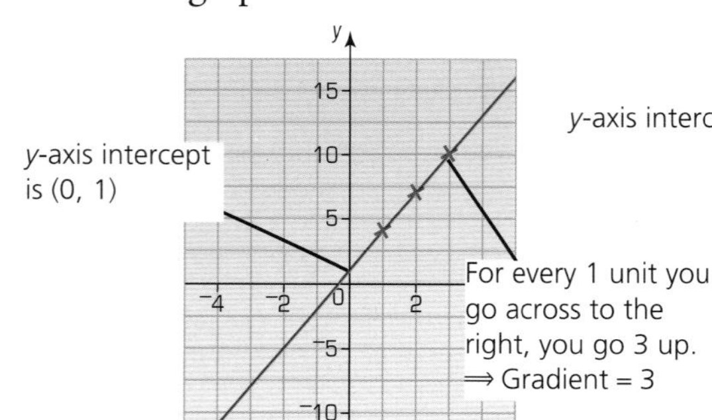

b This is in **implicit** form, so substitute $x = 0$ and $y = 0$ carefully:

x	0	8
y	2	0

There is more about implicit form on page 48.

Draw a graph:

y-axis intercept is (0, 2)

For every unit you go across to the right, you go $\frac{1}{4}$ **down**.
⟹ Gradient = $^{-}\frac{1}{4}$

▶ **In the equation of a straight line, $y = mx + c$, m is the gradient, and c is the y-intercept.**

These facts mean that you can:

- find the gradient and y-intercept of a straight line if you know the equation
- find the equation if you know the line.

example

Find the equation of a line parallel to $y = 10 - 3x$.

Rearrange: $y = {}^{-}3x + 10$

Compare with $y = mx + c$: $m = {}^{-}3$

The gradient is $^{-}3$ so a line parallel to $y = {}^{-}3x + 10$ will have the same gradient.

A possible equation is $y = {}^{-}3x + 2$.

Exercise A2.1

1 The box below contains the equations of various graphs.

$y = x^3$	$2y - x = 10$	$y = 4$
$y = 2x - 1$	$y = \frac{1}{x}$	$y = 2 - x^2$

a Select the equations that represent straight-line graphs.

b Plot each straight-line graph on a separate pair of axes.

2 **a** Does the line with equation $y = 2x - 5$ pass through the point (5, 5)? Explain how you know.

b Give the equation of a line that is parallel to $y = 2x - 5$.

3 Match each graph with an equation.

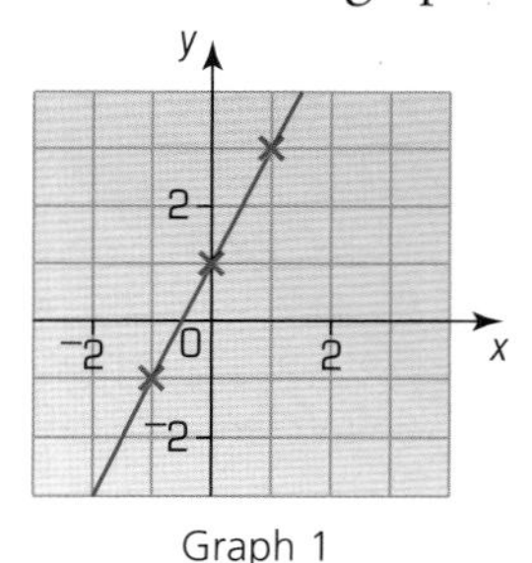

Graph 1

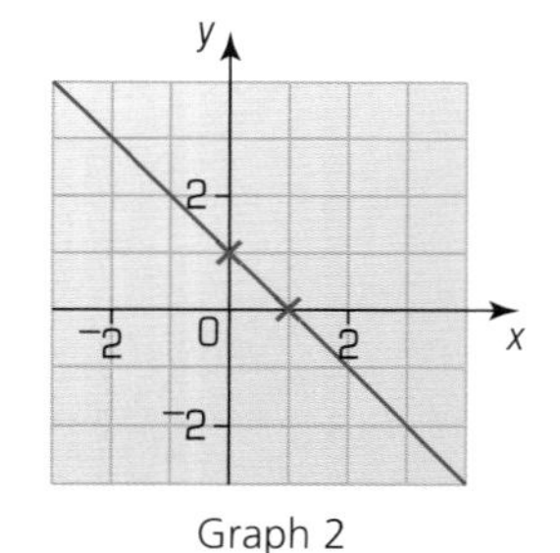

Graph 2

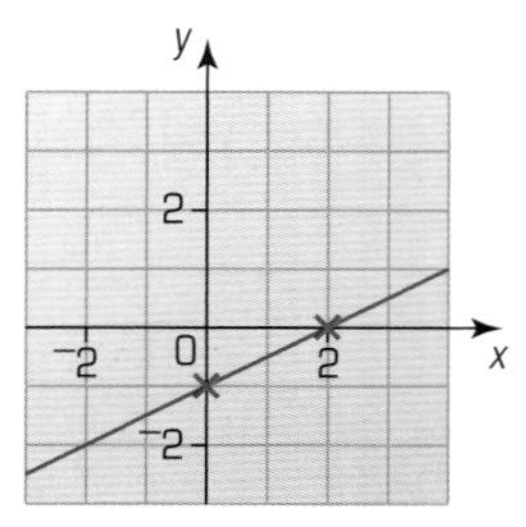

Graph 3

$y = 1 - x$ $\quad$ $2y + 2 = x$ $\quad$ $y = 2x + 1$

4 **a** Explain how you know that there is no point that lies on both $y = \frac{1}{2}x + 3$ and $y = \frac{1}{2}x + 5$.

b True or false? $y = 10 - 2x$ and $2y + 4x = 40$ are parallel. Explain your answer.

5 Copy and complete this table.

Gradient	Positive or negative	y-axis intercept	Equation
			$2y = 3x + 4$
5	Positive	$(0, {}^{-}3)$	
$-\frac{1}{2}$	Negative		$y = \square x + 6$
$\frac{1}{4}$	Positive		$4y = x + 4$

6 The equation of line A is $2x + y = 4$.

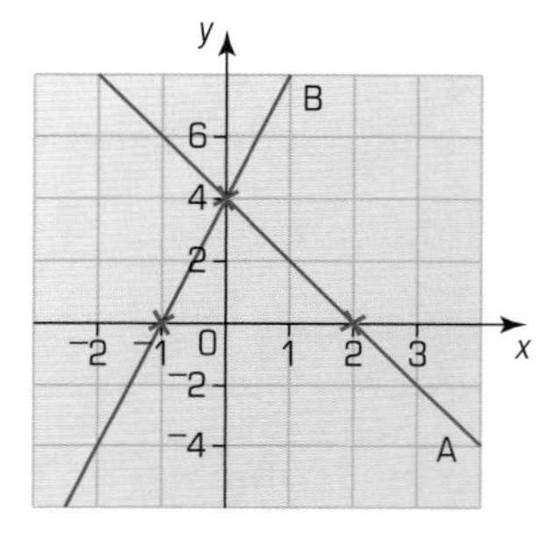

a Write down the equation of line B.

b True or false? $y = 4 - 2x$ intersects both lines A and B. Explain how you know.

c Write down the equation of any line that intersects line B but never intersects line A.

7 Write down the equations of the lines with these characteristics:

> **Hint:** A diagram could be useful, but try first without.

a Gradient = $\frac{1}{2}$, intercept (0, 3)

b Gradient = $^{-}2$, intercept $(0, {}^{-}3)$

c Goes through (0, 2) and (1, 6)

d Goes through (1, 8) and (3, 12).

8 I want a line with equation $y = mx + 6$ to go through (7, 7). What is the value of m?

9 On a pair of axes, construct a rhombus. Find the equation of each side. What do you notice? Investigate for different quadrilaterals.

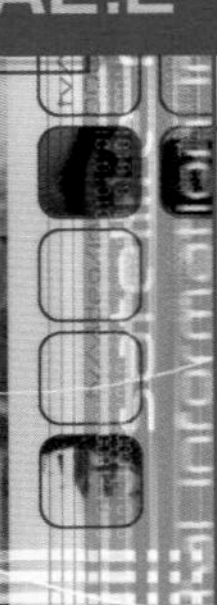

A2.2 Inverse functions

This spread will show you how to:

- Find the inverse of a linear function.
- Plot the graph of the inverse of a linear function.

KEYWORDS
Function
Inverse function

Remember:

- A **function** is a rule that maps a number to one other number only.

$x \to x$ is called the **identity function**. It maps any number on to itself.

$x \longrightarrow 2x + 1$ is a **linear** function.
For any number x, you multiply by 2 and add 1.

You can also write the function as $y = 2x + 1$.

There is an **inverse** function that will reverse the direction of the mapping.

To find the inverse function, you need to 'undo' each operation.

Mapping

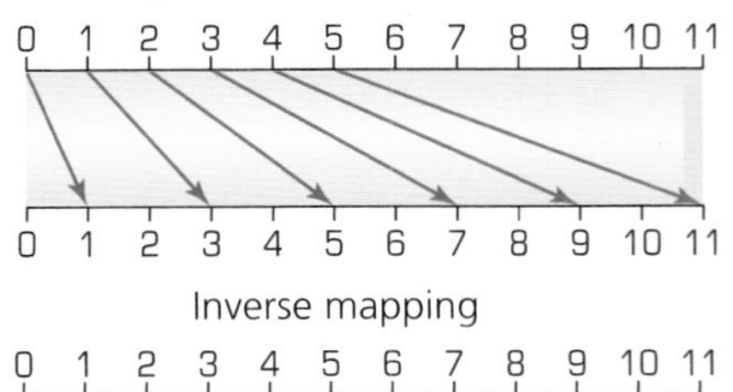

Inverse mapping

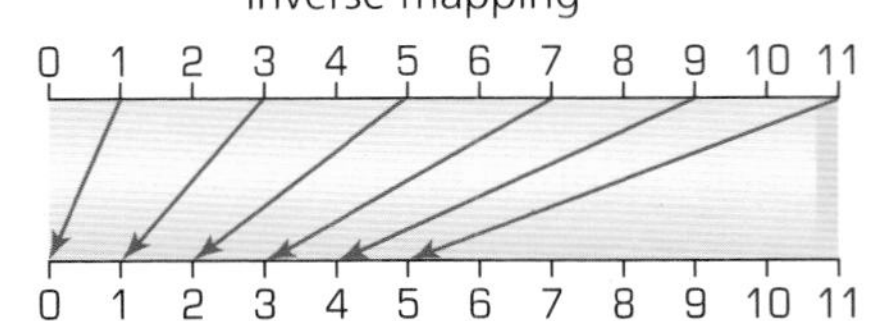

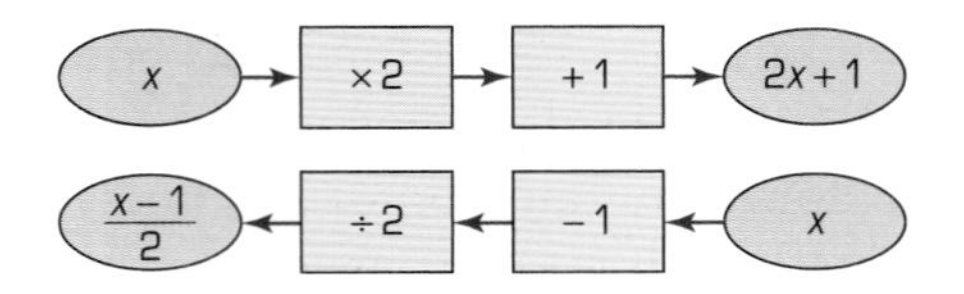

The inverse function is $x \longrightarrow \dfrac{x-1}{2}$ $\left(\text{or } y = \dfrac{x-1}{2}\right)$

Check with $x = 9$:
$x \longrightarrow 2x + 1$
$2 \times 9 + 1 = 19$

Now put $x = 19$ into
$x \longrightarrow \dfrac{x-1}{2}$
$(19 - 1) \div 2 = 9$
This is the number that you started with.

- Every linear function has an inverse function.

You can plot the functions $y = 2x + 1$ and $y = \dfrac{x-1}{2}$ together on a graph.

x	1	2	3
$y = 2x + 1$	3	5	7

x	1	2	3
$y = \dfrac{x-1}{2}$	0	$\frac{1}{2}$	1

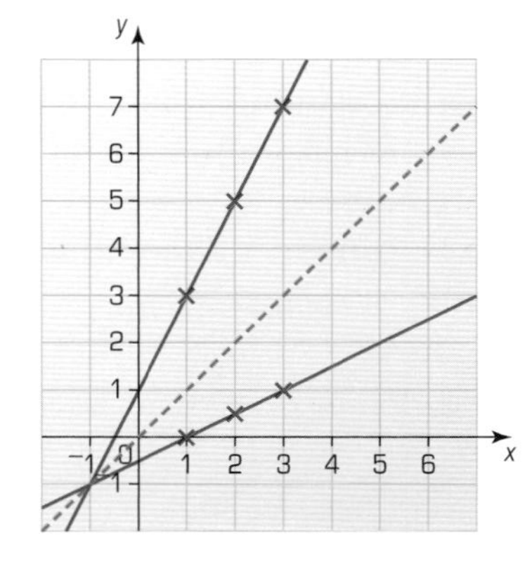

Notice that:

- $y = 2x + 1$ contains the point (1, 3), and $y = \dfrac{x-1}{2}$ contains the point (3, 1).

  The coordinates are swapped around.
- There is a line of symmetry.
  Its equation is $y = x$.

Exercise A2.2

1 Find the inverse of each mapping. For example, $x \longrightarrow \frac{x}{2}$ has inverse $x \longrightarrow 2x$.

a $x \longrightarrow 3x$ **b** $x \longrightarrow \frac{x+1}{2}$

c $x \longrightarrow x^2 - 1$ **d** $x \longrightarrow 3x + 6$

e $x \longrightarrow \sqrt{x}$

2 Check your answers in question 1 by substituting $x = 4$ into each mapping, and then substituting the answer you get into the inverse mapping.

3 True or false? $x \longrightarrow 2(x + 2)$ and $x \longrightarrow 2x + 4$ have the same inverse function.
Explain how you know.

4 **a** Copy and complete this table of inverse functions:

Function	$y = x + 1$	$y = 2x$	$y = \frac{x+2}{2}$	$y = 3x - 2$
Inverse function	$y = x - 1$			

b Plot each function and its inverse on the same set of axes. In each case, add in the line of symmetry.

c Investigate fully the statement: 'All functions and their inverses have the same line of symmetry.'
(If you find it to be true, give the equation of this line.)

5 Sketch copies of these graphs and draw the inverse function on each sketch.

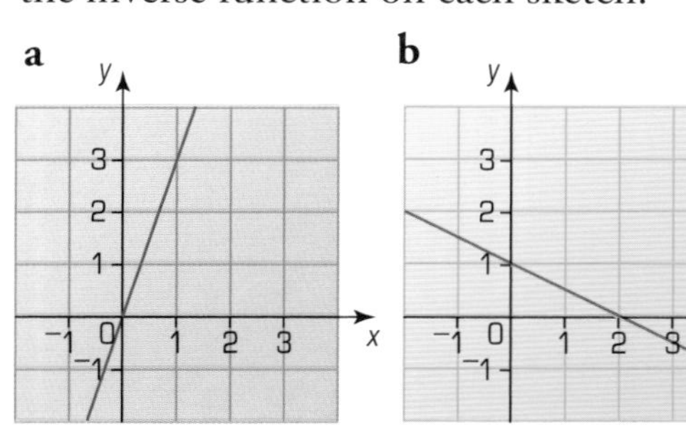

6 Copy the graph and draw the inverse function.

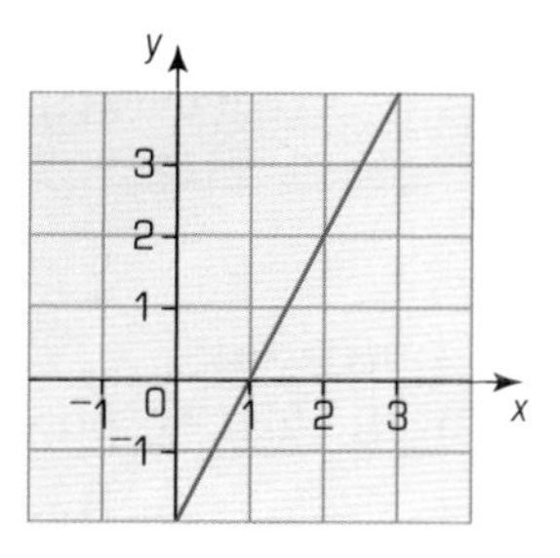

7 **Investigation**

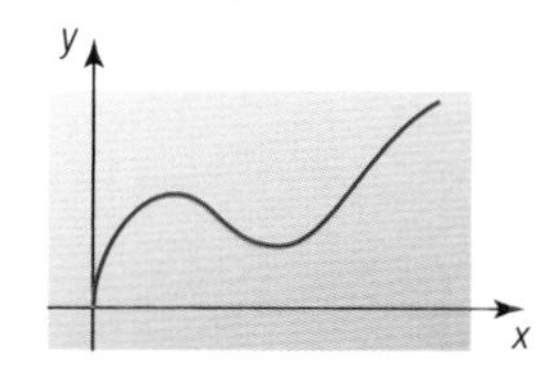

There is no scale given on this pair of axes. Investigate possible shapes for the inverse of the given function.

8 **Investigation**
Investigate the statement: 'No function has an inverse function equivalent to the function itself.'

> If a function had an inverse function equivalent to the function itself, we could say the function was self-inverse.

9 Use the graphs given to help you sketch the functions asked for:

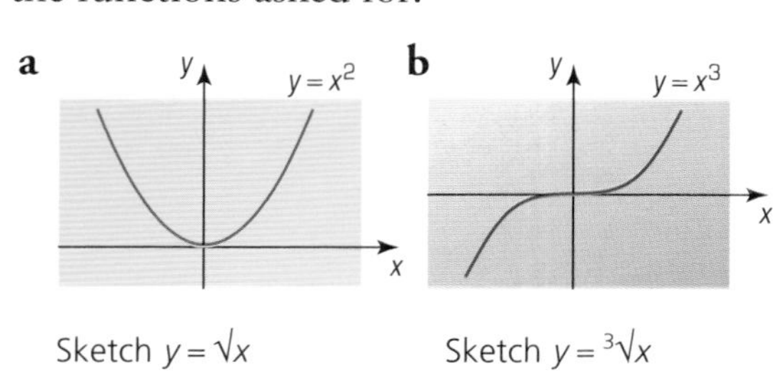

a Sketch $y = \sqrt{x}$ **b** Sketch $y = \sqrt[3]{x}$

Quadratic functions and curved graphs

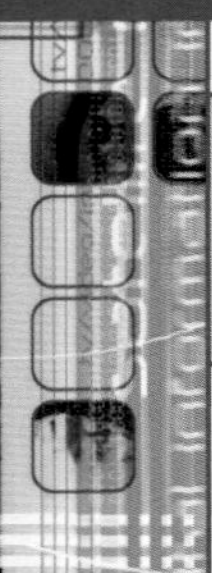

This spread will show you how to:

- Know simple properties of quadratic functions.
- Plot graphs of simple quadratic and cubic functions.

KEYWORDS

Quadratic, Minimum, Parabola, Cubic, Maximum, Curve

Here are three **quadratic functions**: $y = x^2$ $\quad y = 3x^2 + 4$ $\quad y = 2x^2 + 3x - 2$
They all have an x^2 term as the highest power.

The graph of a quadratic function is a curve so you should calculate at least seven points.

example

Plot the graph of the function $y = x^2 + x$.

First draw a table of values:

x	$^-4$	$^-3$	$^-2$	$^-1$	0	1	2	3
y	12	6	2	0	0	2	6	12

For example, $(^-4)^2 + (^-4) = 16 + (^-4) = 12$

Then plot a graph:

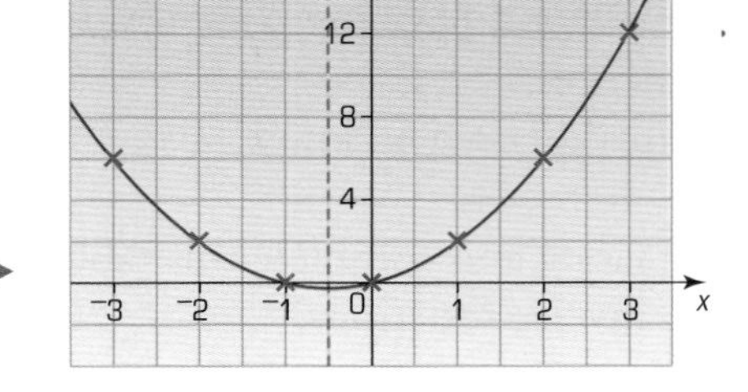

- The shape of a quadratic graph is called a **parabola**.

Every parabola has a vertical **line of symmetry**.

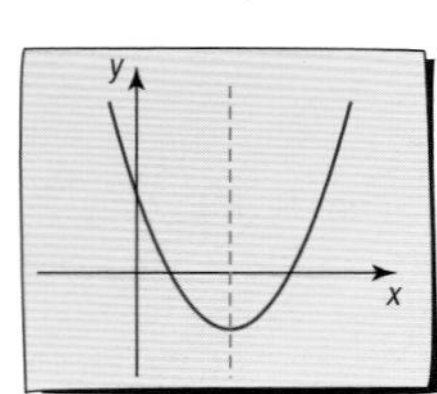

It either has a **maximum** point ...

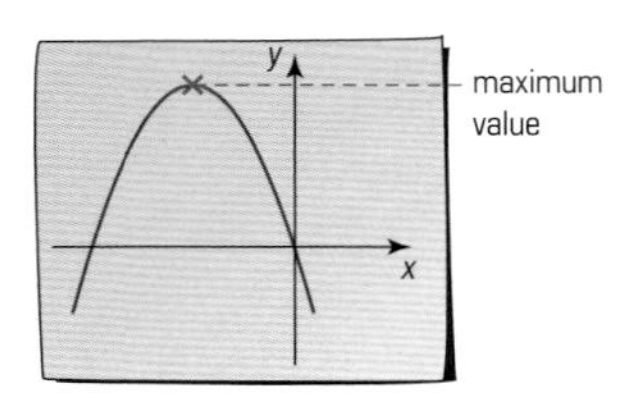

... or a **minimum** point.

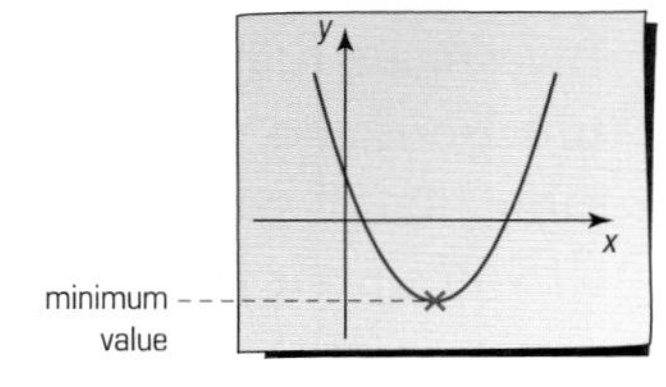

- **Cubic functions** contain a term in x^3 as the highest power.

Like quadratics, cubic graphs have a distinctive shape.

example

Accurately construct the graph of the function $y = x^3 + 2x^2 - x - 3$, for $^-3 \leqslant x \leqslant 2$.

It may help to have a row for each term ...

x	$^-3$	$^-2$	$^-1$	0	1	2
x^3	$^-27$	$^-8$	$^-1$	0	1	8
$2x^2$	18	8	2	0	2	8
$-x$	3	2	1	0	$^-1$	$^-2$
$^-3$	$^-3$	$^-3$	$^-3$	$^-3$	$^-3$	$^-3$
y	$^-9$	$^-1$	$^-1$	$^-3$	$^-1$	11

... then add the terms at the end.

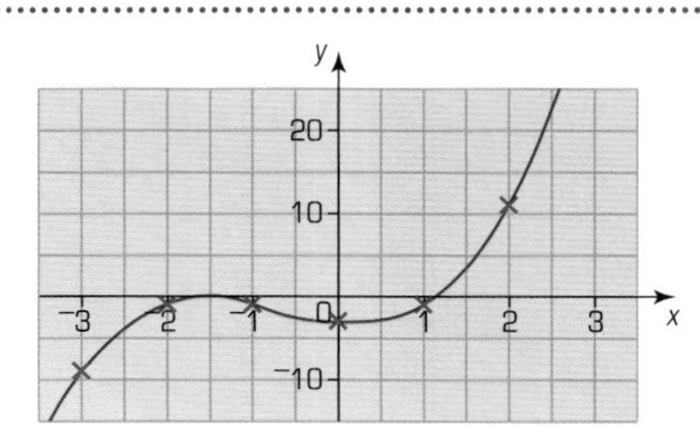

Exercise A2.3

1 Copy and complete the table of coordinates for each function.
Use the tables to plot the parabolas.

a $y = x^2 + 2$

x	$^-3$	$^-2$	$^-1$	0	1	2	3
x^2							
2	2	2	2	2	2	2	2
y							

b $y = x^2 + 2x - 1$

x	$^-4$	$^-3$	$^-2$	$^-1$	0	1	2
x^2							
$2x$							
$^-1$							
y							

c $y = x^2 - 3x$

x	$^-2$	$^-1$	0	1	2	3	4	5
x^2								
^-3x								
y								

2 For each graph in question 1:

a Give the equation of the line of symmetry.

b Identify the coordinates of the maximum or minimum point on each parabola.

3 The sketch graph shows the curve $y = 9 - x^2$.

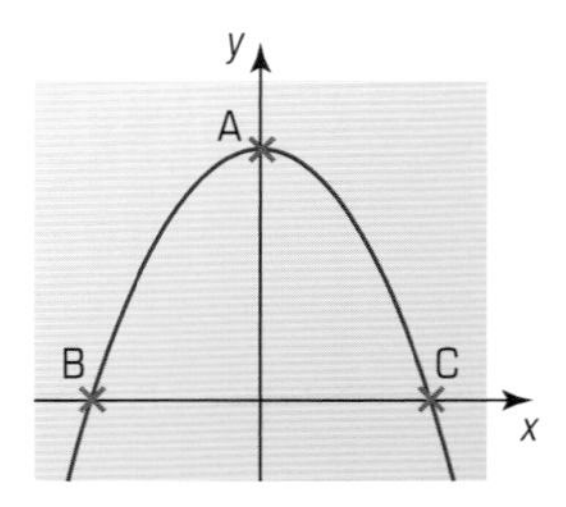

a Explain how you can tell from the equation that the graph is an 'upside-down U-shape'.

b Write down the coordinates of A, B and C. (Use a table of values if you find this difficult.)

4 **Investigation**

a We know that equations with an x^2 term produce a quadratic curve or parabola. But what effect does changing the other numbers in the equation have?
In each case below, draw enough graphs until you are satisfied what effect the constant has. Write down what you notice.

i $y = x^2 + a$ (a is a constant)
For example, $y = x^2$, $y = x^2 + 1$, $y = x^2 + 2$, …, $y = x^2 - 3$

ii $y = bx^2$ (b is a constant)
For example, $y = x^2$, $y = 2x^2$, $y = 3x^2$, …, $y = {}^-4x^2$

b Using part **a**, predict what these graphs look like by drawing a sketch graph.

i $y = x^2 + 10$

ii $y = 6x^2$

iii $y = x^2 - \frac{1}{2}$

iv $y = {}^-10x^2$

v $y = 2x^2 + 3$

5 True or false? $y = x^2 + 100$ and $y = 100x^2$ have the same line of symmetry.

6

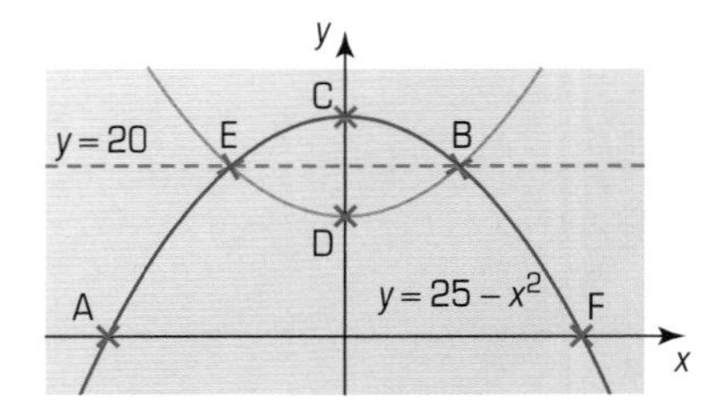

The diagram shows the graph $y = 25 - x^2$ together with its reflection in the line $y = 20$.

a What are the coordinates of the points A, B, C, D, E and F?

b What is the equation of the new curve obtained after reflection in $y = 20$?

A2 **Summary**

You should know how to ...	Check out
1 Plot the graph of the inverse of a linear function.	**1 a** Write down the inverse of these functions: **i** $y = 3x + 1$ **ii** $y = \frac{x}{2} - 1$ **iii** $y = 4(x - 3)$ **b** Copy this graph. Draw on the graph of the inverse function.

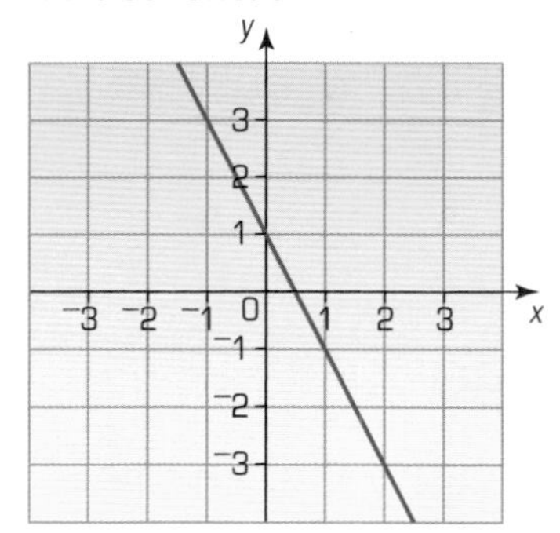

2 Know simple properties of quadratic functions.	**2 a** Will $y = x^2 + 3$ have a maximum or minimum? Check by plotting the graph. Write down the equation of the line of symmetry. **b** By completing this table of values, plot the graph $y = x^2 - 3x + 8$. Use your graph to estimate the value of y when $x = 1.5$.

x	$^-3$	$^-2$	$^-1$	0	1	2	3
x^2							
^-3x							
8							
y							

3 Generate fuller solutions to problems.	**3 a** Write down the gradient, the y-axis intercept, and hence the equation of this line.

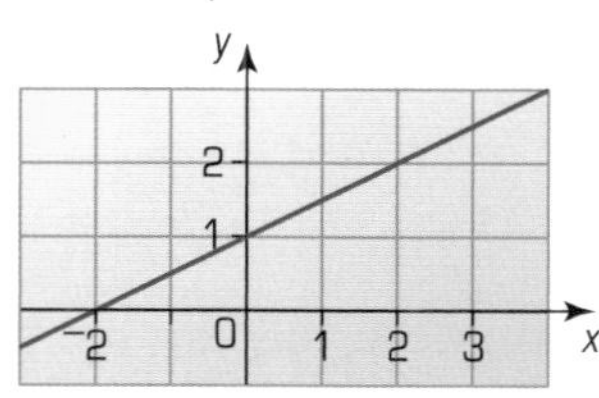

b If x represents the number of people in a tent, and y the cost of a campsite (in pounds), discuss what the gradient and y-intercept tell you about staying at this campsite.

1 Proportional reasoning

This unit will show you how to:

- Use efficient methods to add, subtract, multiply and divide fractions.
- Recognise and use reciprocals.
- Solve problems involving percentage changes.
- Recognise when fractions or percentages are needed to compare proportions.
- Use proportional reasoning to solve a problem, choosing the correct numbers to take as 100%.
- Use laws of arithmetic and inverse operations.
- Understand the effects of multiplying and dividing by numbers between 0 and 1.
- Interpret and use ratio in a range of contexts.
- Understand the implications of enlargement for area and volume.
- Understand proportionality and calculate proportional change.
- Round numbers to a given number of significant figures.
- Estimate calculations by rounding numbers to one significant figure.
- Understand the order of precedence and the effect of powers.
- Generate fuller solutions to problems.

The lengths in a scale model are in proportion to the real lengths.

Before you start

You should know how to ...	Check in
1 Find the HCF and LCM of two numbers.	1 a Find the HCF of 36 and 48. b Find the LCM of 14 and 6.
2 Calculate a percentage of an amount.	2 Calculate 36% of £416.
3 Simplify a ratio.	3 Write each of these ratios in its simplest form. a 26 : 38 b 18 : 54 c 155 : 50
4 Round a number to a given number of decimal places, or a given power of 10.	4 Round each number to the accuracy stated. a 32 149 (to the nearest 10) b 2178 (to the nearest 100) c 6.712 (to 1 dp) d 3.0684 (to 2 dp)

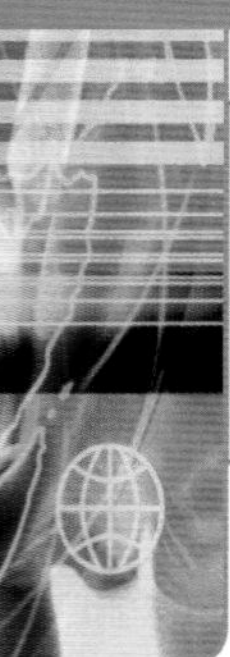

N1.1 Adding and subtracting fractions

This spread will show you how to:

- Use efficient methods to add and subtract fractions.
- Add simple algebraic fractions.
- Understand the equivalence of simple algebraic fractions.

KEYWORDS
Algebraic fraction
Equivalent fraction
Lowest common multiple (LCM)
Highest common factor (HCF)
Common denominator
Cancel

You can use equivalent fractions to ...

- compare fractions

Which is bigger, $\frac{13}{15}$ or $\frac{4}{5}$?
$\frac{4}{5} = \frac{12}{15}$, so $\frac{13}{15}$ is bigger.

- add or subtract fractions

Work out $\frac{15}{32} - \frac{13}{48}$

The LCM of 32 and 48 is 96.
96 is the **common denominator**.

$$\frac{15}{32} - \frac{13}{48} = \frac{45}{96} - \frac{26}{96}$$
$$= \frac{19}{96}$$

Use prime factors:
$32 = 2^5$
$48 = 2^4 \times 3$
LCM $= 2^5 \times 3$
$= 96$

You can simplify fractions by dividing by the HCF.

example

Simplify the fraction $\frac{102}{255}$.

The HCF of 102 and 255 is 51.

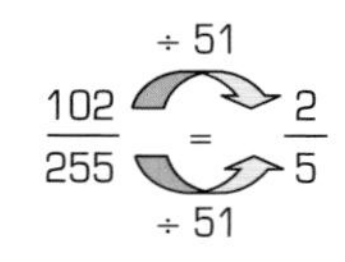

Use prime factors:
$102 = 17 \times 3 \times 2$
$255 = 17 \times 3 \times 5$
$\text{HCF} = 17 \times 3 = 51$

You can add or subtract algebraic fractions as well.

example

Write $\frac{a}{b} + \frac{c}{d}$ as a single fraction.

The LCM of b and d is bd.

$$\frac{a}{b} + \frac{c}{d} = \frac{a \times d}{b \times d} + \frac{c \times b}{b \times d}$$
$$= \frac{ad + cb}{bd}$$

Multiples of b are b, $2b$, $3b$, ..., $\mathbf{db}$, ...
Multiples of d are d, $2d$, $3d$, ..., $\mathbf{bd}$, ...
So bd is the LCM of b and d.

You can simplify an algebraic fraction by cancelling to find an equivalent fraction.

example

Simplify these algebraic fractions:

a $\frac{2a}{20}$ **b** $\frac{abc}{c^2}$ **c** $\frac{15p}{3p^2qr}$

a $\frac{2a}{20} = \frac{a}{10}$ **b** $\frac{abc}{c^2} = \frac{ab}{c}$ **c** $\frac{15p}{3p^2qr} = \frac{5p}{p^2qr} = \frac{5}{pqr}$

÷ by 3, then ÷ by p.
Or just ÷ by $3p$ in one go.

Exercise N1.1

1 Work out these answers as mixed numbers where appropriate.

a $\frac{5}{9}+\frac{3}{4}$ **b** $\frac{14}{15}-\frac{7}{10}$ **c** $\frac{11}{14}+\frac{6}{7}$

d $\frac{5}{7}-\frac{2}{21}$ **e** $\frac{6}{11}+\frac{^-2}{15}$ **f** $5\frac{2}{3}-\frac{4}{7}$

g $5\frac{2}{9}+4\frac{3}{4}$ **h** $\frac{16}{9}-\frac{^-5}{11}$

2 Write each of these fractions in its simplest form.

a $\frac{36}{52}$ **b** $\frac{^-12}{20}$ **c** $\frac{81}{150}$ **d** $\frac{17}{102}$

e $\frac{15}{^-135}$ **f** $\frac{^-14}{63}$ **g** $\frac{125}{1800}$ **h** $\frac{232}{1392}$

3 For each of the following, insert the correct sign, >, < or =, in between the fractions.
Show clearly how you worked out your answer.

a $\frac{9}{13}$ $\frac{6}{17}$ **b** $\frac{1}{18}$ $\frac{2}{31}$ **c** $\frac{^-3}{16}$ $\frac{^-5}{18}$

d $6\frac{3}{7}$ $\frac{59}{8}$ **e** $\frac{335}{132}$ $2\frac{3}{5}$

4 Puzzle

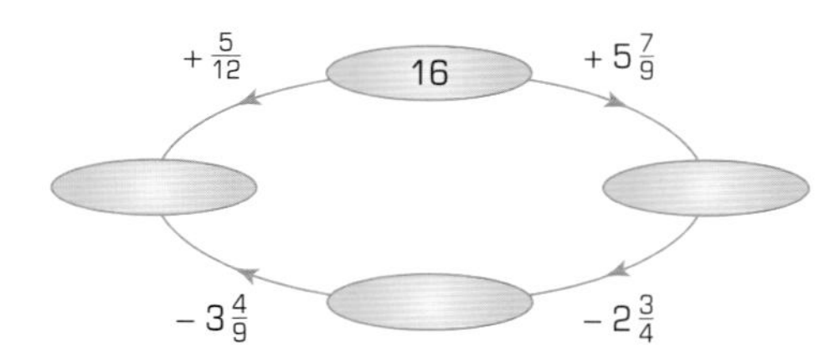

In this fraction loop, some of the fractions are missing. Copy the diagram and fill in the blanks.

5 a A container holds $9\frac{3}{4}$ litres of liquid. Liquid is poured from this container to fill two glasses with capacities of 0.75 litres to the brim.
How much liquid remains in the container?

b A trapezium has sides of length $6\frac{1}{4}$ m, $2\frac{2}{5}$ m, $3\frac{1}{2}$ m and $4\frac{3}{8}$ m.
What is its perimeter?

6 Write each of these as a single fraction.

a $\frac{2}{a}+\frac{1}{a}$ **b** $\frac{5}{t}-\frac{2}{t}$ **c** $\frac{1}{t}+\frac{1}{u}$

d $\frac{2}{x}-\frac{2}{y}$ **e** $\frac{1}{ab}+\frac{1}{c}$ **f** $\frac{2}{xy}-\frac{3}{m}$

g $\frac{5a}{t}+\frac{3a}{ty}$ **h** $\frac{6m}{n^2}-\frac{5}{n}$

7 James's maths teacher was having a bad day. Whilst teaching his class how to do algebraic fractions he wrote:
$\frac{1}{a}+\frac{1}{b}=\frac{2}{ab}$
Use your understanding of adding algebraic fractions to show that the **correct** answer is $\frac{a+b}{ab}$.

8 The sum of two algebraic fractions is:
$\frac{2m+3t}{tm}$
One of the fractions is $\frac{2}{t}$.
Work out the other fraction.

9 Choose simple numbers for w, x, y and z to show that these expressions are true.

a $\frac{w}{x}+\frac{y}{z}=\frac{wz+xy}{xz}$

b $\frac{w}{x}-\frac{y}{z}=\frac{wz-xy}{xz}$

10 Use your knowledge of the equivalence of algebraic fractions to cancel each of the following to its simplest form.
For example, $\frac{3a}{9ab}=\frac{1}{3b}$

a $\frac{xy}{xz}$ **b** $\frac{nm^2}{2nm}$ **c** $\frac{2ab}{6b^2}$ **d** $\frac{7pr}{pst}$ **e** $\frac{15st^3m}{24t^4}$

11 Investigation

Consider the sequence:

$1, \frac{1}{2}, \frac{1}{4}, \frac{1}{8}$

a Write the next three terms in the sequence.

b Write down the rule for finding the next term in the sequence.

c By calculating the sum of the series: $1+\frac{1}{2}+\frac{1}{4}+\frac{1}{8}+\ldots$ suggest a number that the sum of the series would tend towards.
Investigate for other series, for example $1+\frac{1}{3}+\frac{1}{9}+\frac{1}{27}+\ldots$

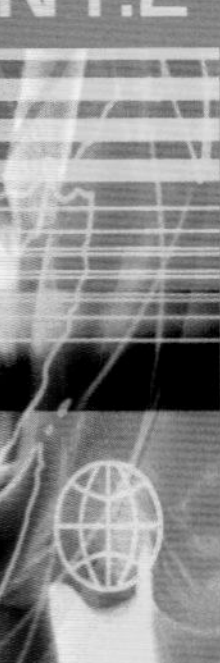

N1.2 Multiplying by fractions

This spread will show you how to:
- Use efficient methods to multiply fractions.
- Cancel common factors before multiplying.
- Simplify algebraic fractions by finding common factors.

KEYWORDS
Numerator
Denominator
Cancelling
Algebraic fraction
Improper fraction
Mixed number
Simplify

You multiply two fractions by: $\frac{1}{2} \times \frac{3}{4}$
- multiplying the numerators
- multiplying the denominators.

$\frac{1 \times 3}{2 \times 4} = \frac{3}{8}$

You can often simplify the answer by cancelling.
You can either do this ...

- before the multiplication or

$$\frac{7}{24} \times \frac{4}{5} = \frac{7 \times 4}{24 \times 5} = \frac{7}{6} \times \frac{1}{5} = \frac{7}{30}$$

Cancel here.

- after the multiplication.

$$\frac{7}{24} \times \frac{4}{5} = \frac{7 \times 4}{24 \times 5} = \frac{28}{120} = \frac{7}{30}$$

Cancel here.

Cancelling before multiplying leads to smaller numbers in your working.
You should convert mixed numbers to improper fractions before multiplying.

example

Calculate $4\frac{2}{3} \times \frac{7}{8}$.

$4\frac{2}{3} = \frac{14}{3}$

$$4\frac{2}{3} \times \frac{7}{8} = \frac{14}{3} \times \frac{7}{8} = \frac{7}{3} \times \frac{7}{4} = \frac{49}{12} = 4\frac{1}{12}$$

Remember to convert back to a mixed number at the end.

- You multiply algebraic fractions in the same way as numerical fractions.
  - $\frac{a}{b} \times \frac{c}{d} = \frac{ac}{bd}$

example

Simplify $\frac{4cd}{5ef} \times \frac{15c}{bd}$.

$$\frac{4cd}{5ef} \times \frac{15c}{bd} = \frac{4c}{ef} \times \frac{3c}{b} = \frac{12c^2}{bef}$$

This is simplified because it is written as a single fraction.

Exercise N1.2

1 Calculate, giving your answers in their simplest form:

a $7 \times \frac{3}{4}$ **b** $\frac{14}{15} \times 8$ **c** $3 \times \frac{9}{11}$

d $\frac{7}{16} \times 5$ **e** $\frac{4}{5} \times \frac{1}{3}$ **f** $\frac{2}{7} \times \frac{5}{8}$

g $\frac{9}{10} \times \frac{4}{7}$ **h** $\frac{5}{14} \times \frac{4}{15}$

2 Calculate, giving your answers in their simplest form:

a $3\frac{1}{14} \times \frac{2}{3}$ **b** $\frac{5}{7} \times 2\frac{2}{21}$ **c** $3\frac{6}{7} \times \frac{3}{4}$

d $3\frac{14}{15} \times \frac{13}{8}$ **e** $3\frac{6}{11} \times 3\frac{2}{15}$ **f** $5\frac{2}{3} \times 1\frac{4}{7}$

g $^{-}3\frac{2}{9} \times 5\frac{3}{7}$ **h** $\frac{^{-}16}{9} \times \frac{^{-}4}{20}$

3 Work out each of these using the values given.

$x = \frac{^{-}2}{3}$, $y = \frac{4}{5}$, $z = \frac{^{-}2}{31}$

a x^2 **b** xy **c** xyz

d z^3 **e** xy^2z **f** x^2zy

4 In these fraction pyramids, each fraction is the product of the two fractions below. Copy and complete the pyramids.

a

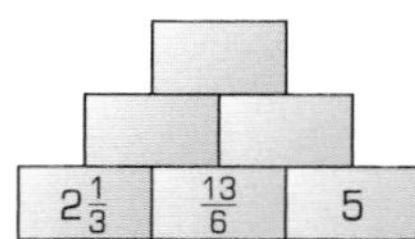

b

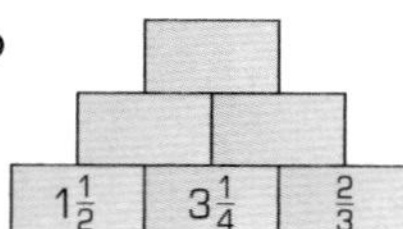

c

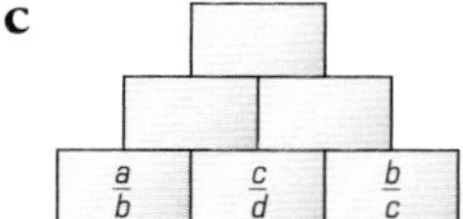

5 Simplify these algebraic fractions.

a $\frac{3}{c} \times \frac{1}{b}$ **b** $\frac{6}{m} \times \frac{2}{t}$ **c** $\frac{4}{d} \times \frac{1}{h}$

d $\frac{5}{e} \times \frac{4}{y}$ **e** $\frac{6}{ab} \times \frac{a}{c}$ **f** $\frac{2m}{xy} \times \frac{3x}{m}$

g $\frac{5ya}{3t} \times \frac{3a}{ty}$ **h** $\frac{12p}{q^2} \times \frac{5p^2}{q}$

6 A trapezium has the dimensions shown. Calculate its area.

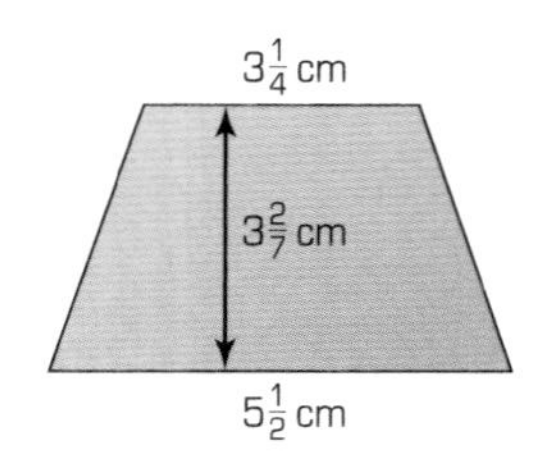

7 Use your knowledge of multiplying fractions to work out the area and perimeter of each of these shapes. Let $\pi = \frac{22}{7}$ and give your answers as mixed numbers.

a

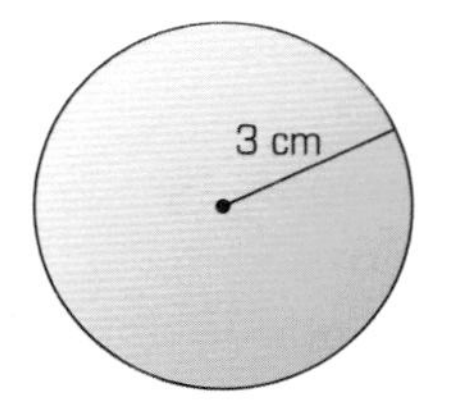

b

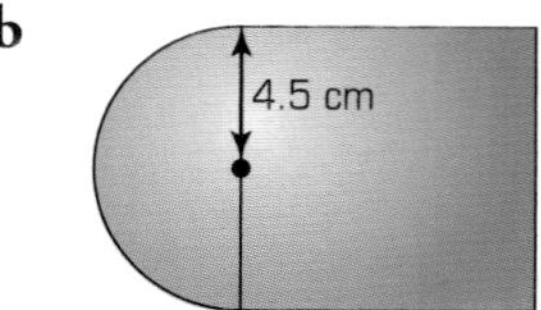

8 A table at a wedding reception has 42 glasses, each with a capacity of $\frac{4}{11}$ of a litre. 20 of them are full, 14 are one-quarter full and the rest are half full.

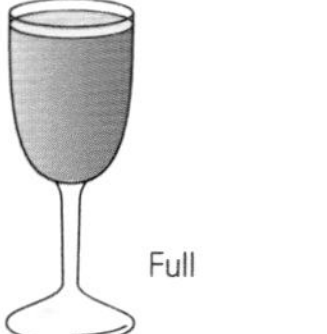

Show that the total volume of wine in the glasses is equal to 10 litres.

9 Swinfield Town Football Club won $\frac{5}{9}$ of their matches last season.
Of these matches, they won $\frac{3}{7}$ of them by two clear goals.

a What fraction of Swinfield Town's matches did they win by two clear goals?

b Swinfield Town played 45 matches in total last season.
How many matches did they not win?

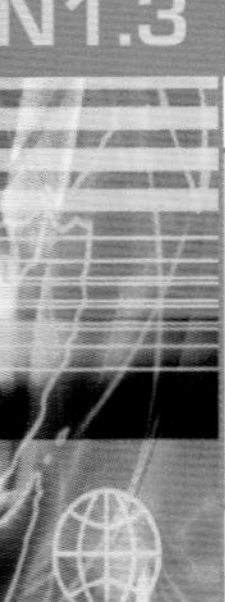

N1.3 Dividing by fractions

This spread will show you how to:

- Use efficient methods to divide fractions.
- Simplify algebraic fractions by finding common factors.
- Recognise and use reciprocals.

KEYWORDS

Divisor
Cancel
Reciprocal
Multiplicative inverse

- You can divide any amount by a fraction by:
  - replacing the ÷ with a ×
  - turning the **divisor** upside-down:

$$4 \div \tfrac{2}{5} = 4 \times \tfrac{5}{2} = \tfrac{4}{1} \times \tfrac{5}{2} = \tfrac{2}{1} \times \tfrac{5}{1} = \tfrac{10}{1} = 10$$

Remember:
The divisor is the number that you are dividing by.

$\frac{5}{2}$ is the **reciprocal** of $\frac{2}{5}$. Similarly, $\frac{2}{5}$ is the reciprocal of $\frac{5}{2}$.

- The reciprocal of a number is 1 divided by that number.
- $\frac{a}{b}$ is the **multiplicative inverse** of $\frac{b}{a}$.

You should convert any mixed numbers to improper fractions during your working.

example

Calculate $\frac{2}{3} \div 3\frac{2}{7}$

$$\frac{2}{3} \div 3\frac{2}{7} = \frac{2}{3} \div \frac{23}{7} = \frac{2}{3} \times \frac{7}{23} = \frac{2 \times 7}{3 \times 23} = \frac{14}{69}$$

Nothing will cancel because there are no common factors.

- You can divide algebraic fractions in the same way as numerical fractions.

$$\frac{a}{b} \div \frac{c}{d} = \frac{a}{b} \times \frac{d}{c}$$

example

Simplify $\frac{x^2y}{z} \div \frac{x}{yz}$

$$\frac{x^2y}{z} \div \frac{x}{yz} = \frac{x^2y}{z} \times \frac{yz}{x} = \frac{xy}{1} \times \frac{y}{1} = \frac{xy^2}{1} = xy^2$$

Exercise N1.3

1 Calculate, giving your answers in their simplest form:

a $9 \div \frac{1}{4}$ **b** $16 \div \frac{1}{8}$ **c** $\frac{1}{2} \div \frac{1}{5}$

d $\frac{1}{7} \div \frac{2}{21}$ **e** $\frac{1}{8} \div 6$ **f** $9 \div \frac{4}{11}$

g $\frac{9}{13} \div \frac{4}{5}$ **h** $\frac{7}{16} \div \frac{28}{18}$

2 Calculate, giving your answers in their simplest form:

a $5\frac{1}{8} \div 6\frac{2}{5}$ **b** $6\frac{2}{3} \div 2\frac{1}{7}$ **c** $5\frac{2}{7} \div 1\frac{3}{7}$

d $\frac{14}{9} \div 3$ **e** $9 \div \frac{16}{5}$ **f** $^{-}4 \div \frac{3}{2}$

g $\frac{-16}{5} \div 3\frac{1}{4}$ **h** $^{-}4\frac{3}{8} \div {}^{-}2\frac{1}{3}$

3 Work out each of these using the values given.

$x = \frac{1}{3}$, $y = \frac{-2}{5}$, $z = \frac{-1}{16}$

a $x \div y$ **b** $xzy \div y$ **c** xz

d Explain why your answers to **b** and **c** are the same.

4 Find the reciprocal of:

a $\frac{2}{11}$ **b** $3\frac{2}{5}$ **c** $\frac{-5}{6}$ **d** 5.6

e Find the square root of $\frac{1}{4}$.

f Find the cube root of $\frac{1}{125}$.

5 Fill in these 'fraction snakes' by copying the diagrams and writing the missing values.

a

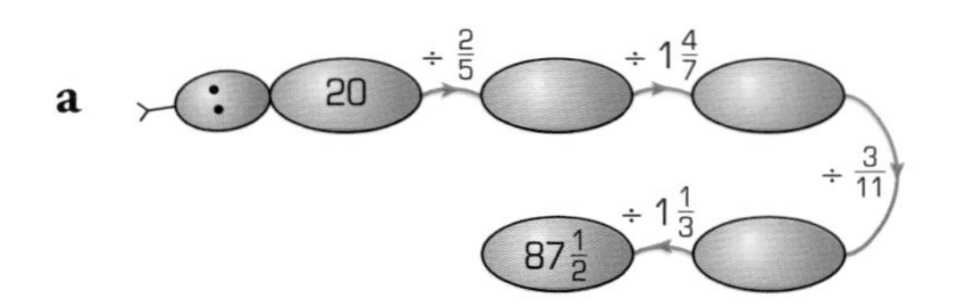

b

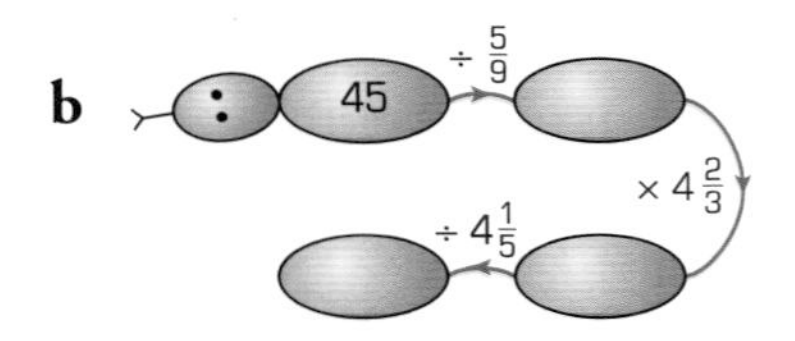

6 Use your knowledge of dividing fractions to work out the diameter of each of the shapes. Let $\pi = \frac{22}{7}$, and give your answer correct as a fraction or mixed number.

a

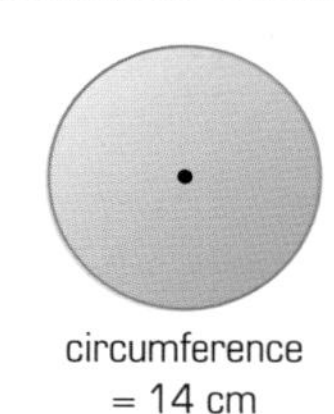

b

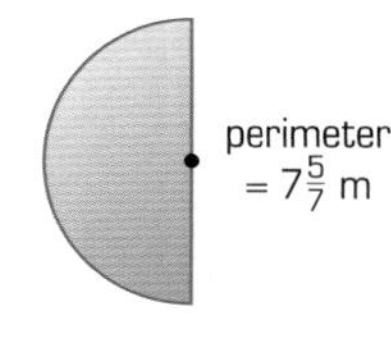

7 Simplify:

a $\frac{5}{a} \div \frac{1}{m}$ **b** $\frac{2}{p} \div \frac{5}{x}$ **c** $\frac{3}{a} \div \frac{a}{f}$

d $\frac{4}{d} \div \frac{d}{7}$ **e** $\frac{12}{ab} \div \frac{6}{abc}$ **f** $\frac{2a}{bc} \div \frac{3b}{a}$

g $\frac{5ma}{3r} \div \frac{30a}{y}$ **h** $\frac{14u^2}{g} \div \frac{g^2}{2k}$

8 **a** A wooden container will hold $43\frac{2}{3}$ kg. How many tins each weighing $2\frac{6}{7}$ kg could the container hold?

b If the tins were made from a lighter material and only weighed 2 kg, how many *extra* tins could be put into the container?

9 A path made of flagstones laid end to end is $13\frac{4}{15}$ m long. Each of the flagstones has a length of $1\frac{1}{8}$ m. The last flagstone has to be cut to fit.

How long will the last flagstone be?

10 Work out the missing expression in each of the following:

a $p^2 \div \square = \frac{p^3}{6}$ **b** $\frac{m}{4} \div \square = \frac{5}{4m}$

c $\frac{6t}{s^2} \div \square = \frac{6p}{s^2 t}$ **d** $1 \div \square = \frac{t}{p^2}$

e $3 \div \square = \frac{3x}{t^2}$ **f** $\frac{m^2 tv}{5} \div \square = \frac{m^3}{5x^2}$

11 Show that $\frac{2x+3}{5} \div \frac{6x+9}{10}$ can be simplified to $\frac{2}{3}$.

Percentage change

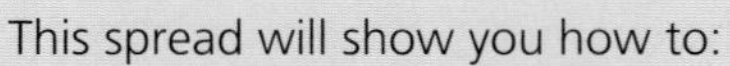

This spread will show you how to:

- Solve problems involving percentage changes.

KEYWORDS
Percentage change
Compound interest

Percentage change problems involve a quantity that increases or decreases.

example

A cornflower has a height of 7.5 cm.
It grows by 3% in height in one month.
Calculate its new height after one month.

Here are two ways of solving this problem.

- Work out the percentage change:

  3% of 7.5

  $= \frac{3}{100} \times 7.5$

  $= 0.225$

- Add the percentage change:
  $7.5 + 0.225 = 7.725$
  The new height is 7.725 cm.

- Work out 103% of the original amount:

  This method is quicker because it involves a single calculation.

  The new height is 7.725 cm.

When you save money in a bank you earn **interest**.

example

Jon opens a savings account and puts £200 into it.
Each year 4% interest is added to his account.
He does not withdraw any of his savings, or add any extra money.

Best Bank
4% INTEREST

a How much does Jon have in his account after three years?
b What is the interest that Jon has earned over three years?

a After 1 year: $200 \times \frac{104}{100} = 208$ (Work out 104% of 200.)
After 2 years: $208 \times \frac{104}{100} = 216.32$
After 3 years: $216.32 \times \frac{104}{100} = 224.9728$
After three years, Jon will have £224.97 (to the nearest penny) in his account.

b Interest earned $= 224.9728 - 200$
$= 24.9728$
Jon will have earned £24.97 interest.

In the example, you needed to work out 4% of a larger amount each time.
This type of interest is called **compound interest**.

Exercise N1.4

1 Calculate each of these **without** using a calculator.

a 9% of 150 kg **b** 16% of €300
c 22% of 250 people **d** 5% of £314
e 35% of 720 cm **f** 41% of $425
g 17.5% of £27 **h** 12.5% of 14 m

2 Value Added Tax, or VAT, is a government tax on goods and services.
It is currently 17.5% of the value of the item.
Do-it-Right DIY specialists price their items 'pre-VAT'. This means that VAT is added on to the ticket price.
Work out the VAT to be added to each of these 'pre-VAT' prices.

3 **a** Show, using a written method, two different ways in which a percentage increase of 7% can be calculated. Use £340 as the amount to be increased.
b Show, using a written method, two different ways in which a percentage decrease of 12% can be calculated. Use $260 as the amount to be decreased.

4 An increase of 50% means that you are adding on half as much again.
Explain, in simple terms, the meaning of these percentage changes.

a An increase of 100%
b A decrease of 50%
c An increase of 1000%
d A decrease of 100%.

5 Copy and complete this 'percentage change' snake.

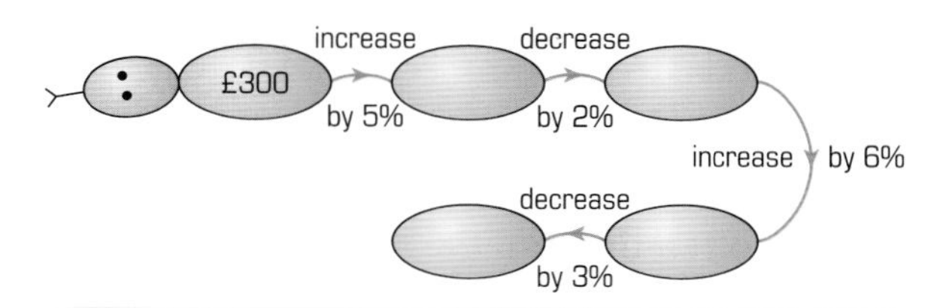

Note: Give your answers to the nearest penny but use the calculator answer for the next calculation.

6 Use your calculator to work out these percentage changes, giving your answers correct to two decimal places.

a Increase 231 kg by 16.3%.
b Increase €300 by 3.2%.
c Decrease 250 mm by 43.2%.
d Increase 6 ml by 5.4%.
e Decrease 710 cm by 19.3%.
f Decrease $325 by 54.8%.
g Increase $56 by 121%.
h Decrease 14 litres by 25.5%.

7 A factory produces terracotta plant pots.
The cost price of each plant pot is how much it costs to produce. It includes materials and labour.
The sales manager calculates the selling price by increasing the cost price by 35%.
Copy and complete the table:

	Cost price	Selling price
15 cm pot	25p	
20 cm pot		81p
35 cm pot	£1.10	
30 cm pot	95p	

Hint: Round your answers up to the nearest penny.

8 A car salesperson estimates that a new car with a value of £9400 will lose value, or **depreciate**, at a rate of 17% per year for the first 3 years and at a rate of 9% thereafter.
Use this to estimate the value of the car after 5 years.

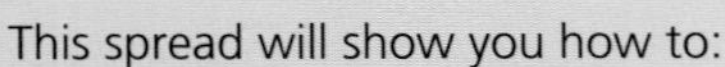

N1.5 Comparing proportions

This spread will show you how to:

- Recognise when fractions or percentages are needed to compare proportions.
- Solve problems involving percentage changes.

KEYWORDS
Percentage change
Proportion

You will often find it useful to work out a percentage change.

example

Nina used to earn £4 per hour in her job in a supermarket.
Now she earns £6 per hour.
By what percentage has Nina's pay increased?

- Work out the actual increase: $6 - 4 = 2$
- Express the increase as a fraction of the original: $\frac{2}{4} = \frac{1}{2}$
- Convert the fraction into a percentage: $\frac{1}{2} \times 100\% = 50\%$

- **Percentage increase/decrease** $= \frac{\text{actual increase/decrease}}{\text{original amount}} \times$ **100%**

You can use percentage change to compare proportions.

example

The pie charts show the ages of members of a youth club in 1997 and 2003.

a What was the proportion of 11–12 year-olds:
i in 1997 **ii** in 2003?

b By how much has the proportion of 11–12 year-olds changed?

c Find the percentage change in the **actual** number of 13–14 year-olds between 1997 and 2003.

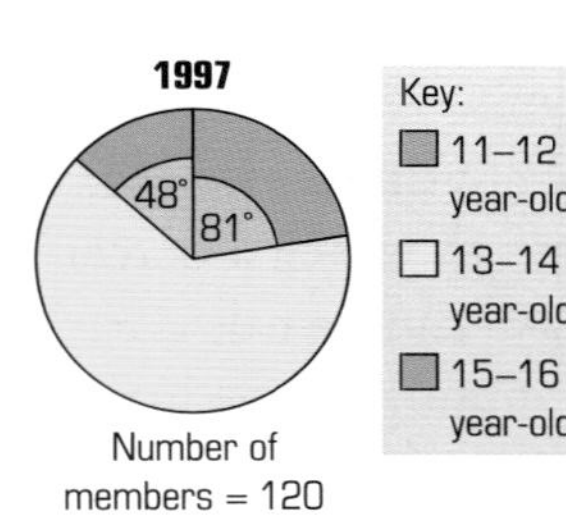

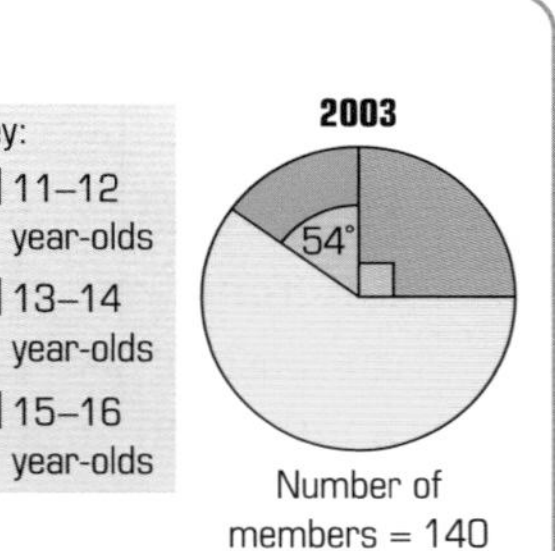

a i Proportion of 11–12 year-olds in 1997 $= \frac{81}{360} \times 100\% = 22.5\%$

ii Proportion of 11–12 year-olds in 2003 $= \frac{90}{360} \times 100\% = 25\%$

b $25 - 22.5 = 2.5$
The proportion of 11–12 year-olds has increased by 2.5%.

c Angle corresponding to 13–14 year-olds in 1997: $360° - (48° + 81°) = 231°$
Total number of 13–14 year-olds in 1997: $= \frac{231}{360} \times 120 = 77$

Angle corresponding to 13–14 year-olds in 2003: $360° - (54° + 90°) = 216°$
Total number of 13–14 year-olds in 2003: $= \frac{216}{360} \times 140 = 84$

Percentage change $= \frac{84 - 77}{77} \times 100\% = 9.1\%$
The number of 13–14 year-olds increased by 9.1% between 1997 and 2003.

Exercise N1.5

1 Calculate the percentage decrease or increase that has occurred between each of these pairs of amounts. Round your answer to one decimal place where appropriate.

a £20 to £25
b 18 kg to 30 kg
c 24 people to 21 people
d 4 eggs to 7 eggs
e 135 cm to 170 cm
f \$51 to \$42
g £12.70 to £10
h 13.1 litres to 9.7 litres

2 The hourly pay of staff on grades A, B and C at a company increase from £4, £5.50 and £6.40 an hour to £4.45, £5.87 and £6.81 an hour respectively.
Calculate the grade of staff that received the highest percentage increase.
Show your working.

3 Mr Green is told by the local council that he must give 5 m^2 of his land to his neighbour, Mr Plant. Mr Green's garden is 62 m^2 and Mr Plant's is 51 m^2.
Work out the percentage increase or decrease that occurs to their gardens.

4 Five children are asked to guess the number of marbles in a jar.
The actual number was 437.
The results are shown below.

	Guess
Mariah	520
Jenny	452
James	379
Femi	360
Anne	420

Work out the percentage error for each child's guess.

Hint: 'Percentage error' is the error as a percentage of the actual number.

5 The table shows the number of girls and boys at Learn-a-lot Comprehensive School in 1997, 2001 and 2003.

	Boys	Girls
1997	671	590
2001	620	631
2003	625	650

a Create a table, similar to the one above, showing the proportion of boys and girls as a percentage of the total for each year.
b In what year was the proportion of girls the highest?
c By how many would the number of boys need to increase in 2003 to match the proportion of boys in 1997?

6 The radius of the small circle is 6.3 cm and that of the larger circle 8.4 cm.
Calculate the area of the shaded part as a proportion of:
a the small circle
b the large circle.
Give your answers as percentages.

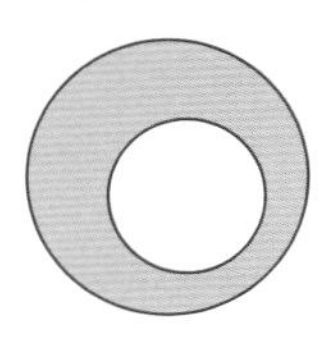

7 After playing a game in class Jenny calls out:

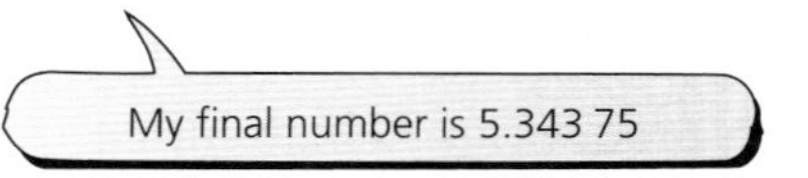

The teacher had asked her to think of a number, work out 25% of it, add on 5 and find $\frac{3}{4}$ of the result.
What number did Jenny think of originally?

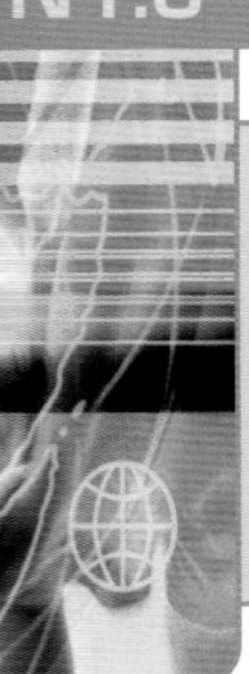

N1.6 Reverse percentages

This spread will show you how to:

- Solve problems involving percentage changes.
- Use proportional reasoning to solve a problem, choosing the correct numbers to take as 100%, or as a whole.
- Use the laws of arithmetic and inverse operations.
- Understand the effects of × and ÷ by numbers between 0 and 1.

KEYWORDS
Unitary method
Inverse
Compound interest
Decimal multiplier

You can use your knowledge of percentages to work problems backwards.

example

A CD player is in a sale priced at £66.
It has been reduced by 20%.
What was the original price of the CD player?

You could use the **unitary method**:
£66 is 80% of the original price.
So £66 ÷ 80 = £0.825
is 1% of the original price.
£0.825 × 100 = £82.50
£82.50 is the original price.

You could use **inverse relationships**:

original price → reduced price: $\times 0.8$ or $\times \frac{80}{100}$ — 80% = 0.8

reduced price → original price: $\div 0.8$ or $\times \frac{100}{80}$ — $\frac{100}{80}$ is the **reciprocal** of $\frac{80}{100}$.

So original price = £66 ÷ 0.8 = £82.50

- **When you divide by a number between 0 and 1, the answer is larger than the number you started with.**

You can extend these techniques to compound interest problems.

example

Zeinab's bank statement shows that she has £4544.92 in her savings account. She opened her account four years ago and she has not added or withdrawn any money since.
Her account has an annual interest rate of 6%.
How much money did Zeinab initially put into her savings account?

Amount after 1 year = 106% of original amount
= $\frac{106}{100}$ × original amount
= 1.06 × original amount

So amount after 4 years = 1.06 × 1.06 × 1.06 × 1.06 × original amount
£4544.92 = 1.06^4 × original amount
⟹ Original amount = £4544.92 ÷ 1.06^4 = £3600.00

Note:
1.06 is the **decimal multiplier** in this problem.

Exercise N1.6

1 Find the reciprocal of:

a $\frac{5}{9}$ **b** $\frac{14}{15}$ **c** $\frac{11}{14}$ **d** $\frac{5}{7}$ **e** $\frac{6}{11}$ **f** $\frac{4}{7}$ **g** $4\frac{3}{4}$ **h** $\frac{^-5}{11}$

2 Copy and complete each of these statements.

a If an item is reduced by 24%, the sale price is __% of the original price.

b If an item increases in price by 12%, the new price is __% of the original price.

3 Mary says that her exam results have improved this year. Her results are shown in the table below. Copy and complete the table, giving last year's results to the nearest whole number.

Subject	Last year's result	This year's result	Percentage increase
Maths		78%	7%
English		100%	5%
Science		67%	2%
History		89%	17%
Drama		69%	34%
French		71%	20%

4 **a** Hayley managed to get a 40% reduction on the price of a holiday by agreeing to travel at an unpopular time. She paid £430. How much was the original price?

b Kevin is to get a 3.5% pay rise, which will increase his annual salary by £560. How much will he earn annually after the increase?

5 Anna's bank account has a balance of £4118.40. She had invested £3900 exactly one year ago. What annual rate of interest, as a percentage, has the bank paid?

6 The table below shows the heights of four plants, the number of years they have been growing, and the rate, as a percentage, that they grow each year.
Copy and complete the table, finding the original height for each plant.

	Rate	Number of years	Height	Original height
Plant A	18%	2	32 cm	
Plant B	9%	3	51 cm	
Plant C	20%	1	26 cm	
Plant D	34%	3	16 cm	

7 A house has increased in value by a rate of 14% each year for the previous three years. It is now valued at £120 000. Work out its value at the end of each of the previous years.

8 **Investigation**

a Multiply £200 by 0.83^2.
Final amount =

b Multiply £200 by 1.07^3.
Final amount =

Suggest a context that **a** and **b** could represent in terms of percentage increase or decrease.

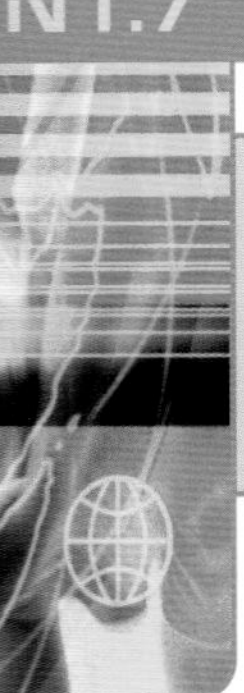

N1.7 Using ratio

This spread will show you how to:
- Compare two ratios.
- Interpret and use ratio in a range of contexts.
- Understand the implications of enlargement for area and volume.

KEYWORDS
Ratio
Enlargement
Unitary form
Proportion

A ratio compares the sizes of parts with each other.

Robert the builder mixes four buckets full of sand with $2\frac{2}{5}$ buckets of cement.
You can write the ratio of sand to cement in whole-number form or in unitary form $1 : m$.

Unitary ratios are easier to compare than whole-number ratios.

Whole-number form: $4 : \frac{12}{5}$
$= 20 : 12$
$= 5 : 3$

Unitary form: $4 : \frac{12}{5}$
$= 1 : \frac{12}{20}$
$= 1 : \frac{3}{5}$

example

The ratio of boys to girls in class 9Y is $5 : 7$ and in class 9Z is $11 : 14$.
Which class has the greater proportion of girls?

Class 9Y
$5 : 7 = 1 : 1.4$

Class 9Z
$11 : 14 = 1 : 1.27$

Class 9Y has a greater **proportion** of girls than class 9Z.

Note:
This does not mean that class 9Y has more girls than class 9Z.

You can use ratio to describe enlargements.

example

The red cube has edges of length 4 cm.
The ratio of lengths between the red cube and the green cube is $2 : 3$.

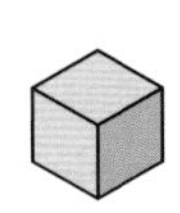
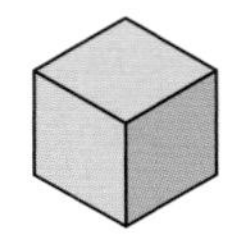

a Find the ratio of areas between a red face and a green face.
b Find the ratio of volumes.

a Ratio of lengths is $2 : 3$, or $1 : 1.5$
Green length $= 4 \text{ cm} \times 1.5 = 6 \text{ cm}$
Area of red face $= 4 \text{ cm} \times 4 \text{ cm} = 16 \text{ cm}^2$
Area of green face
$= 6 \text{ cm} \times 6 \text{ cm} = 36 \text{ cm}^2$
Ratio red area : green area
$= 16 : 36 = 1 : 2.25$ (or $1 : 1.5^2$)

b Red volume
$= 4 \text{ cm} \times 4 \text{ cm} \times 4 \text{ cm} = 64 \text{ cm}^3$
Green volume
$= 6 \text{ cm} \times 6 \text{ cm} \times 6 \text{ cm} = 216 \text{ cm}^3$
Ratio red volume : green volume
$= 64 : 216 = 1 : 3.375$ (or $1 : 1.5^3$)

When you enlarge a 3-D shape, if the ratio of corresponding lengths is $1 : k$
- The ratio of corresponding areas is $1 : k^2$
- The ratio of volumes is $1 : k^3$

Exercise N1.7

1 Express each of these ratios in whole-number form as simply as possible.

a $0.4 : 6$ **b** $1.4 : 1.5$
c $1\frac{1}{3} : 4$ **d** $5 : \frac{1}{7}$
e $6 : 1.1$ **f** $3.4 : \frac{2}{5}$
g $4\frac{3}{4} : 4$ **h** $\frac{5}{2} : 3.4$

2 Write each of these ratios in the form $1 : n$.

a $3 : 8$ **b** $4 : 5$
c $2\frac{1}{3} : 4$ **d** $3 : \frac{1}{8}$
e $3.2 : 1.2$ **f** $6.4 : \frac{1}{5}$
g $4\frac{3}{4} : 4.2$ **h** $5.2 : 3\frac{1}{4}$

3 Write each of these ratios in the form $n : 1$.

a $5 : 9$ **b** $2 : 7$
c $6\frac{2}{3} : 4$ **d** $5 : \frac{7}{8}$
e $7.4 : 2.3$ **f** $7.4 : \frac{2}{5}$
g $3\frac{1}{2} : 5.2$ **h** $2.9 : 8\frac{3}{5}$

4 James asked 18 students to name their favourite fruit juice.
Five liked blackcurrant, nine liked orange and the rest liked lemon.
There are 600 students in James's school.
Estimate how many students in the school like each of the three drinks.

5 When a tree was planted in 2002 it was 15 cm tall.
One year later it had grown by 32%.

a Calculate the height of the tree after one year.
b Write the ratio of the tree's height in 2002 to its height in 2003:
 i in whole-number form
 ii in the form $1 : n$
 iii in the form $n : 1$.

6 **Puzzle**
The price of an adult ticket for Adventure Land is $2\frac{1}{8}$ times that for a child and 1.6 times that for a senior citizen. A senior citizen ticket costs £3.20.
What is the total cost of admission for two adults, three senior citizens and one child?
Show your working.

7 Two shapes are similar if one is a 'scaled' version of the other.
The length of a model car is $\frac{1}{50}$ that of the real car.
Use your knowledge of area and volume to write down the ratios of the areas and volumes of the model car compared with the real one.

8 **Investigation**
Two maps, A and B, representing the same location, have scales of 1 : 30 000 and 1 : 60 000 respectively.
The length of a line on map A is twice that of a line on map B.
Investigate the relationship between the areas covered by objects on each map.

9 A model train is $\frac{3}{121}$ the size of the real train.

a Write as a ratio in whole-number form
length of model train : length of the real train
b The length of a window on the model train is 4.9 cm.
What scale factor would you need to multiply by to find the length of the window on the real train? Express the scale factor as a decimal to 2 dp.
c What is the length of the window on the real train?

Proportionality

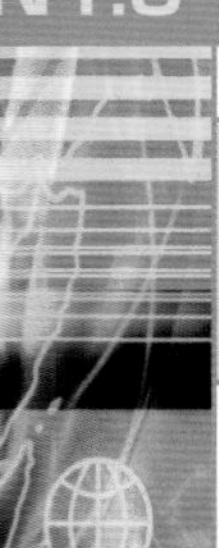

This spread will show you how to:

- Understand and use proportionality and calculate the result of any proportional change using multiplicative methods.

KEYWORDS
Proportional
Proportionality
Ratio

The table shows some conversions between litres and gallons.

Litres	Gallons
3	0.660
5	1.100
9	1.980

→ Ratio litres : gallons = 3 : 0.660 = 1 : 0.220
→ Ratio litres : gallons = 5 : 1.100 = 1 : 0.220
→ Ratio litres : gallons = 9 : 1.980 = 1 : 0.220

The number of litres is **proportional** to the number of gallons.

- **Two sets of numbers, X and Y, are proportional when the ratio of corresponding numbers is the same.**

You can write $X \propto Y$.

$\propto$ is a symbol that means 'is proportional to'.

You can use proportionality to solve problems.

example

£1 is equivalent to €1.75. £1 is also equivalent to $1.55.
How many euros are equivalent to $85?

Convert dollars to pounds: $85 \div 1.55 = 54.8387$
Convert pounds to euros: $54.8387 \times 1.75 = 95.9677$
$\Rightarrow$ $85 is equivalent to €95.97.

You could do this in a single step:
$85 \div 1.55 \times 1.75 = 95.9677$
This is quicker and eliminates rounding errors.

In real life, quantities may only be proportional over a limited range of values.

example

The number of ice creams bought is thought to be proportional to the maximum temperature.
On Monday when the temperature reached 32 °C, 296 ice creams were sold. On Tuesday the temperature only reached 25 °C.

a Give an estimate for the number of ice creams sold on Tuesday.
b Do you think the proportionality would still hold at **i** 45 °C **ii** $^{-}5$ °C?
Justify your answers.

a Number of ice creams $\propto$ temperature
$296 \div 32 = 9.25$ $25 \times 9.25 = 231.25$
$\Rightarrow$ Roughly 231 ice creams would have been sold on Tuesday.
b At 45 °C, ice cream would melt and it would be too hot to be out.
At $^{-}5$ °C, the proportionality suggests that you would have a negative number of ice creams, which is impossible.

Exercise N1.8

1 Each of the ratios below represents two quantities A and B respectively. For each ratio, write the scale factor for changing quantity A into quantity B.

a 3 : 4 **b** 6 : 11 **c** 2 : 5 **d** 5 : 8
e 7 : 2 **f** 10 : 3 **g** 25 : 15 **h** 21 : 7

2 Mandy is told that she can currently get €4.00 for £3.00.

a Write down the decimal number that Mandy needs to multiply by to change:
  i euros into pounds
  ii pounds into euros.

b Copy and complete the table.

Pounds	Euros
£3.12	
	€23.42
	€320.15
	€45.70
£245.16	
£298.54	

3 Calculate the size of the angle, to 1 dp where appropriate, that is needed in a pie chart to represent each of these.
Show your method clearly.

a 3 people out of a total of 10
b 4 kg out of a total of 25 kg
c $2\frac{1}{2}$ cm out of a total of 15 cm
d 46 hens out of a total of 90 animals
e 2060 voters out of a total of 2300 voters.

4 The cost of milk delivered to a school is directly proportional to its volume.
The cost of 83 litres of milk is £15.04.

a How much would 100 litres of milk cost?
b What volume was delivered when the cost was £23.20?

5 A car travels on a motorway at a constant speed for 6 hours and covers a distance of 346 km.
Calculate how far the car travels in:

a 3 hours 40 minutes
b 38 minutes.

6 Mr William Hugh, who works for an American company, is responsible for sending out invoices to Europe.
Using the exchange rates \$1.55 = £1.00 and €1 = 66p, copy and complete the table to allow him to convert the bill in dollars to euros to the nearest cent.

Dollars	Euros
\$57.97	
\$789.45	
\$37.42	
\$376.00	
\$52.56	
\$86.96	

7 To change kilograms to pounds there is a multiplying factor (scale factor) of 2.2.
This means that to change an amount in kilograms to pounds you multiply by 2.2.

a Write down, to 2 dp, the multiplying factor for changing pounds (lb) to kilograms (kg).

b Use your answer to **a** to decide whether these lorries that each have a weight of 8000 kg are over-loaded:
  i Weight including load = 19 480 kg
  maximum permissible
  load = 25 210 lb
  ii Weight including load = 30 000 lb
  maximum permissible
  load = 22 000 kg

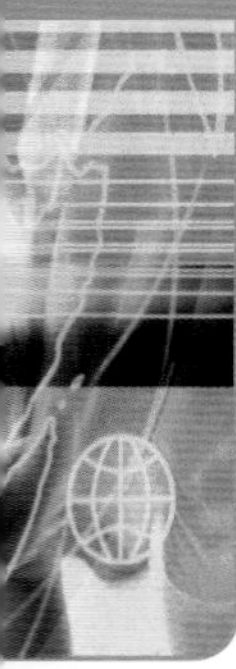

N1.9 Significant figures

This spread will show you how to:
- Round numbers to a given number of significant figures.
- Estimate calculations by rounding numbers to one significant figure and multiplying or dividing mentally.
- Understand the order of precedence and the effect of powers.

KEYWORDS
Significant figures
Estimate
Power
Order of operations
Root

The first non-zero digit of a number is called its first **significant figure**.

The first significant figure of 8062 is 8.
The second significant figure is 0.

The first significant figure of 0.005803 is 5.
The second significant figure is 8.

You can round numbers to:
- a power of 10
- a number of decimal places
- a number of significant figures.

Exact number		Rounded number
36 452	Rounded to the nearest 100	36 500
3.141 592	Rounded to 2 decimal places	3.14
591	Rounded to 1 significant figure	600 Replace the remaining two digits with zero.

example

Round:

a 498 320 to 3 significant figures

b 0.030 529 to 2 significant figures.

a Count the digits: 498 320
498 320 is 498 000 to 3 significant figures (sf).

b Count the digits: 0.030 529
Round up because of the 5: 0.031
0.030 529 is 0.031 to 2 sf.

You don't need to add extra zeros after the decimal point.

You can estimate a calculation by rounding to one significant figure.

example

Calculate: $3.2 + \frac{4.2\sqrt{(5.1^2 - (^-1.9)^2)}}{^-2.3}$ Give your answer to 3 sf.

First **estimate**: $3.2 + \frac{4.2\sqrt{(5.1^2 - (^-1.9)^2)}}{^-2.3} \approx 3 + \frac{4\sqrt{(5^2 - (^-2)^2)}}{^-2}$

$= 3 + \frac{4\sqrt{21}}{^-2} \approx 3 - 2\sqrt{25}$

$= 3 - 2 \times 5 = 3 - 10 = {}^-7$

Now calculate: $3.2 + \frac{4.2\sqrt{(5.1^2 - (^-1.9)^2)}}{^-2.3}$

$= 3.2 + \frac{4.2\sqrt{22.4}}{^-2.3}$

$= 3.2 + \frac{4.2 \times 4.73286 \ldots}{^-2.3}$

$= 3.2 - 8.64262 \ldots$

$= {}^-5.44262 \ldots$

$= {}^-5.44$ to 3 sf

Remember the correct order of operations:
- Brackets
- Powers or roots
- × or ÷
- × or −

$^-5.44$ is fairly close to the estimate of $^-7$.

Exercise N1.9

1 Round each of these numbers to the accuracy stated.

a 21 960 (2 sf) **b** 3146 (2 sf)
c 5967 (1 sf) **d** 0.3106 (3 sf)
e 0.003 21 (2 sf) **f** 0.2018 (2 sf)
g 1.3092 (3 sf) **h** 0.000 481 3 (2 sf)

2 Calculate each of these, giving your final answer to 3 significant figures where appropriate.

a $(^{-}5)^2 + 16$ **b** $(7 - 15)(5 - 7)$
c $(6 \div 7)^2$ **d** $\frac{(2 - 3.1)(4 - 3.6)^2}{43}$
e $\frac{(5 - 8)^3}{\sqrt{6} \times 4}$ **f** $\frac{(^{-}5)^2 - 81}{(3 \times 0.5)^2}$
g $\frac{(12 - 4)^3(5 - 2.3)}{(5 - 6.5)^2}$
h $^{-}(405 \times 6 + 3) + 5 \times 344 - (^{-}597)$

3 Insert the correct operations and brackets to make each of these calculations correct.

a $\frac{1}{2} \square 4 \square\ 3 = {}^{-}11.5$
b $\sqrt{2^2 \square 3 \square 11} = 14.7$ (1 dp)

4 What are the least and greatest possible numbers of people living in these towns?

Town	Population
a Croyfield	8100 (2 sf)
b Murton	20 000 (1 sf)
c Harcote	82 100 (3 sf)

5 On Tom's calculator, the key sequence for working out:

$$(3 + 4.1)^2$$

is     

Write the key sequence for **your** calculator for each of the following calculations.

a $(\frac{5}{8} + 1)^2$ **b** $\frac{(3 + 2.1)(14 - 1.2)^3}{12}$

6 Use the value $x = 4$ to work out the values of:

a $(3x)^2$
b $3x^2$
c $9x^2$
d Write down anything you notice about your answers.

7 Find the values of m and n when $p = 6$ and $b = {}^{-}5$.

$m = \frac{p^2}{b} + (pb - 7)$

$n = \frac{4b^2(p - 4)}{8p}$

8 By rounding the numbers within the calculation to 1 sf estimate the answer to:

$$\frac{32.97 \times 41.8}{62 - (4.9 \times 0.034)}$$

Use your calculator to check how close to the exact answer your estimate is.

9 The volume of a sphere is given by:

$$V = \frac{4}{3}\pi r^3$$

By rounding the figures to 1 sf, estimate the volume of a football with a radius of 11.6 cm.

10 Henry works out how many 2.31 litre bottles he can fill with 2324.9 litres of water.
He rounds each of the values to 2 significant figures before starting the calculation.
What is the error this causes?

N1 Summary

You should know how to ...

1 Understand and use proportionality and calculate the result of any proportional change using multiplicative methods.

2 Generate fuller solutions to problems.

Check out

1 The recipe for blackcurrant and apple squash is:

3.5 litres of water
40 cl of blackcurrant juice
0.6 litres of apple juice

a Write the ratio of water : blackcurrant : apple in integer form.

b Write the proportions of each liquid in the drink.

c Use your answer to **b** to calculate:
  i the amount of blackcurrant juice in 13.2 litres of the drink
  ii the amount of apple juice in a drink that contains 2 litres of water.

2 a Two cuboids have the same cross-sectional area.
Cuboid X has a height of 2.5 cm and cuboid Y has a height of 7.2 cm.
The volume of cuboid Y is 37.9 cm^3.
What is the volume of cuboid X?

b Cuboids X and Z have volumes in the ratio 1 : 2.5. The cross-sectional area of cuboid Z is 20 cm^2.
What is the height of cuboid Z?

3 Solving equations

This unit will show you how to:

- Construct and solve linear equations, using an appropriate method.
- Use systematic trial and improvement methods to find approximate solutions to equations.
- Solve a pair of simultaneous linear equations by eliminating one variable.
- Link a graphical representation of a pair of equations to the algebraic solution.
- Consider cases that have no solution or an infinite number of solutions.
- Generate fuller solutions to problems.
- Justify generalisations, arguments or solutions.
- Pose extra constraints and investigate whether particular cases can be generalised further.

You can use algebra to solve problems.

Before you start

You should know how to ...	Check in
1 Solve a simple linear equation.	**1** Three of these equations have the same solution. Which is the odd one out? $3x - 4 = 23$ $2(x + 1) = 20$ $\frac{2x+1}{3} = 5$ $\frac{x}{3} + 4 = 7$
2 Derive a formula.	**2** Norman has a £50 note. He buys n items, costing £3 each. Write a formula for C, the change received (in pounds).
3 Substitute decimals into expressions involving powers.	**3** If $x = 0.2$ and $y = 0.03$, put these expressions in descending order of size:  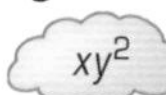   x^2 xy^2 $(xy)^2$ y^3 $(x - 2y)^2$
4 Plot a linear graph, given in explicit or implicit form.	**4** Plot the graphs $y = 2x - 5$ and $x + 2y = 14$ on one pair of axes. At which point do they intersect?

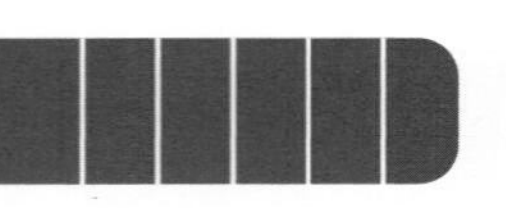

A3.1 Revising equations

This spread will show you how to:

- Construct and solve linear equations, using an appropriate method.

KEYWORDS

Equation	Solve
Linear	Term
Unknown	Brackets

You should know how to solve different types of linear equations.

In a two-sided equation, the unknown appears on both sides.

example

Solve the equation $5x + 9 = 3x + 2$.

Subtract the **s**mallest algebra term:

Solve the one-sided equation:

$$5x + 9 = 3x + 2 \quad (-3x)$$
$$2x + 9 = 2 \quad (-9)$$
$$2x = {}^{-}7 \quad (\div 2)$$
$$x = {}^{-}3\tfrac{1}{2}$$

An equation might contain negatives or brackets.

example

Solve the equations:

a $10 - 2x = 7x + 11$ **b** $3(2x - 4) = 4(3x + 1)$

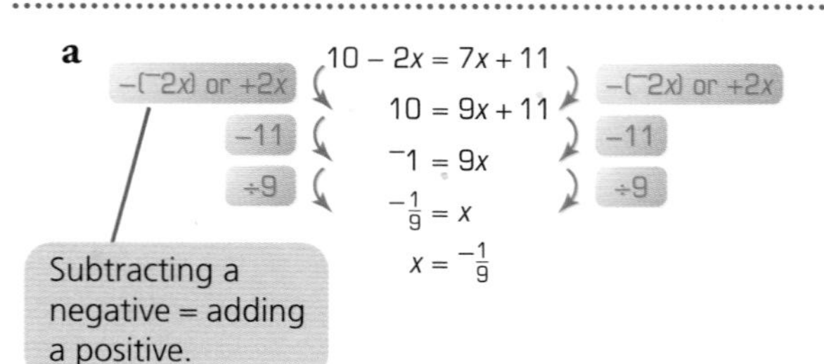

a

$$10 - 2x = 7x + 11 \quad (-({}^{-}2x) \text{ or } +2x)$$
$$10 = 9x + 11 \quad (-11)$$
$$^{-}1 = 9x \quad (\div 9)$$
$$-\tfrac{1}{9} = x$$
$$x = -\tfrac{1}{9}$$

Subtracting a negative = adding a positive.

b

$$3(2x - 4) = 4(3x + 1)$$
$$6x - 12 = 12x + 4 \quad (-6x)$$
$$^{-}12 = 6x + 4 \quad (-4)$$
$$^{-}16 = 6x \quad (\div 6)$$
$$^{-}\tfrac{16}{6} = x$$
$$x = -\tfrac{16}{6} = {}^{-}2\tfrac{2}{3}$$

First expand the brackets.

An equation might contain fractions.

example

Solve the equations: **a** $\frac{5}{x+3} = \frac{7}{x+2}$ **b** $\frac{t}{3} + 8 = {}^{-}4$

a

$$\frac{5}{x+3} = \frac{7}{x+2}$$
$$5(x + 2) = 7(x + 3)$$
$$5x + 10 = 7x + 21$$
$$10 = 2x + 21$$
$$^{-}11 = 2x$$
$$^{-}5\tfrac{1}{2} = x \text{ or } x = {}^{-}5\tfrac{1}{2}$$

Cross-multiply to remove the fractions.

b

$$\frac{t}{3} + 8 = {}^{-}4$$
$$\frac{t}{3} = {}^{-}12$$
$$t = {}^{-}36$$

Only cross-multiply when there is a single term on each side.

Exercise A3.1

1 Solve these equations. Copy the number grid and circle your answer: you will know you are wrong if it is not in the grid.

7	3	15
18	52	9
4	21	11

a $2t - 4 = 10$
b $2a + 15 = 4a - 15$
c $70 - 2b = b + 7$
d $2(x + 4) = 5(x - 5)$
e $\frac{y+2}{4} = \frac{y-8}{2}$
f $100 - x = 48$
g $\frac{3(x+2)-3}{5} = 6$

2 $2x + y = 22$
If x and y represent two consecutive numbers, use an algebraic method to find their values.

3 Solve these equations.
Their solutions are not integers.

a $2(x - 4) = 11$ **b** $5y + 8 = 2y + 3$
c $11 - 3x = 6$ **d** $\frac{5}{p+3} = 10$

4 In each case, form an equation to represent the information and solve it to find the unknown.

a

b

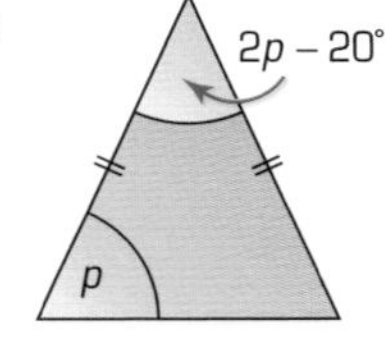

c

7, 11, 15, 19, 23, …

nth term is 203

5 Students started to solve the equation $8x + 10 = 6x + 13$ in different ways and the teacher marked their attempts.

Amy
8x + 10 = 6x + 13
So 18x = 19x ✗

Bruno
8x + 10 = 6x + 13
So 8x = 6x + 3 ✓

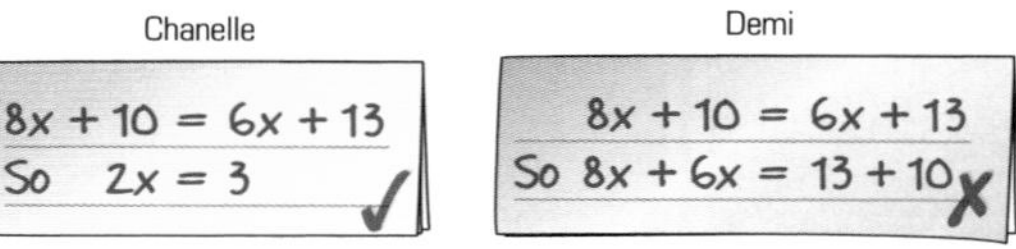

Erroll
8x + 10 = 6x + 13
So 10 = ⁻2x + 13 ✓

a Explain what each student has done and why this makes their work right or wrong.
b Alison solved $8x + 10 = 6x + 13$ using trial and improvement.
Was this a good method to use?

6 A rectangle is cut in two different ways.
What value of m makes the perimeters of the two new rectangles equal?

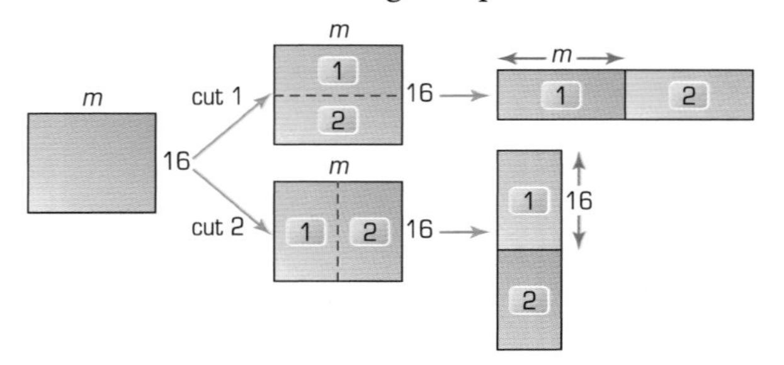

7 The graphs $y = 2x + 7$ and $y = 3x + 2$ intersect at a point.
Form an equation and solve it to find this point.

A3.2 Trial and improvement

This spread will show you how to:

- Use systematic trial and improvement methods to find approximate solutions to equations.

KEYWORDS
Expression
Formula
Equation
Trial and improvement

You should be able to derive a formula.

example

Derive a formula for the volume of this cuboid.

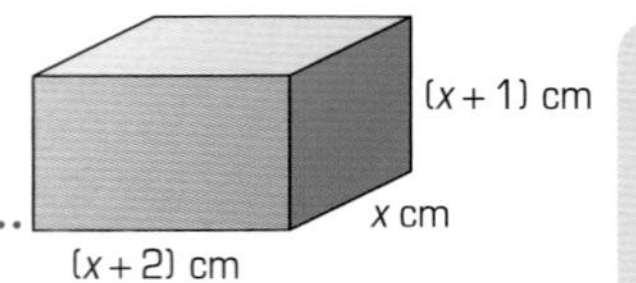

Volume of a cuboid = length × width × height

$V = x(x+1)(x+2)$

$x + 2$ is an **expression** because it does not contain an equals sign.

$V = x(x+1)(x+2)$ is a **formula** because if you know the value of one of the variables, you can work out the other.

In the example, let $V = 1716\ \text{cm}^3$.
Then $x(x+1)(x+2) = 1716$ is an equation.

$x(x+1)(x+2) = 1716$ is an **equation** because it contains an = sign and you can solve it to find x.

You can use a **trial and improvement** method to solve difficult equations like this.

Construct a table:

Decide on possible values for x, but be systematic.

x	$x+1$	$x+2$	$x(x+1)(x+2)$	
7	8	9	504	too small
8	9	10	720	too small
9	10	11	990	too small
10	11	12	1320	too small
11	12	13	**1716**	just right

$\Rightarrow x = 11$

The dimensions of the cuboid are 11 cm, 12 cm and 13 cm.

example

Use trial and improvement to solve the equation $x^3 + 2x = 700$, giving your answer to 1 dp.

Start with integers before moving on to 1 dp.

This is just to decide whether x is closer to 8.8 or 8.9.

x	x^3	$2x$	$x^3 + 2x$	
8	512	16	528	too small
9	729	18	747	too big
$8 < x < 9$				
8.5	614.125	17	631.125	too small
8.8	681.472	17.6	699.072	too small
8.9	704.969	17.8	722.769	too big
$8.8 < x < 8.9$				
8.85	693.154125	17.7	710.854125	too big
$8.8 < x < 8.85$				

8 — 8.5 — 9

8.8 — 8.85 — 8.9

So $x = 8.8$ to 1 dp.

▸ When you use trial and improvement, always go to one more decimal place than you need.

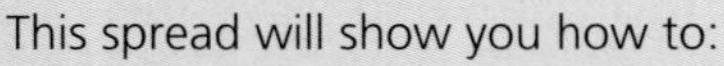

Exercise A3.2

1 **a** Copy the table and write these items under the correct headings.

Identity	Equation	Formula	Function	Inequality

$V = lbh$ $\quad$ $x(x-1) = x^2 - x$ $\quad$ $10 - 2x > 6$

$2x + x^2 - 3(x-1) = x^2 - x - 1$ $\quad$ $y = 3x^2 - 1$ $\quad$ $2(x+4) = 3(x-2)$

b Write an example of your own in each column.

2 Equations and identities are easy to confuse.

a Copy the table below and tick (✓) the correct box for each algebraic statement.

	Correct for no y values	Correct for one y value	Correct for two y values	Correct for all y values
$4y + 6 = 8$				
$y + 5 = y - 5$				
$2(y-4) = 2y - 8$				
$6 + y = 6 - y$				
$y^2 = 16$				

b Hence, list any identities in the table.

3 Use trial and improvement to solve these equations.

a $x^2 + x = 380$ $\quad$ **b** $x^3 - 2x^2 = 1440$

c $x^4 - x = 11.1321$

4 Solve the following equations by trial and improvement, giving your solutions to the stated degree of accuracy.

a $x^2 - x = 71$ (1 dp)

b $2x^3 + x^2 = 99$ (1 dp)

c $3^x = 41$ (2 dp)

d $\sqrt[3]{x} + 2x = 10$ (2 dp)

5 Construct equations to represent the information given below. Use trial and improvement to find the value of the unknown in each case.

a x

Length is 2 cm more than width.
Area is 575 cm^2.

b

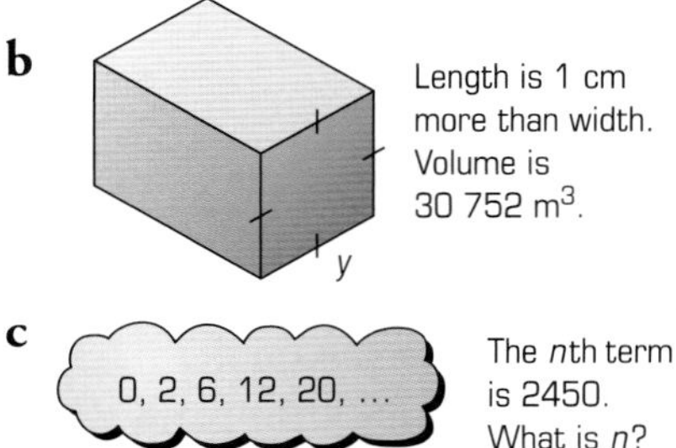

Length is 1 cm more than width.
Volume is 30 752 m^3.

c 0, 2, 6, 12, 20, …

The nth term is 2450.
What is n?

In questions 6–9, you need to construct an equation and solve it, by trial and improvement, to 1 dp if the solution is not an integer.

6 The product of three consecutive numbers is 83 140. What are the numbers?

7 The product of three consecutive odd numbers is 15 525.
What are the numbers?

8 The square of a number is 30 times as much as its cube root. What is the number?

9 This net is folded to produce a box with a volume of 500 cubic units. What is x?

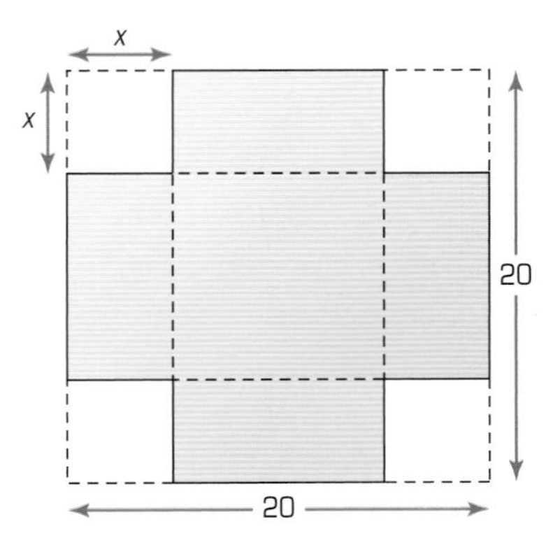

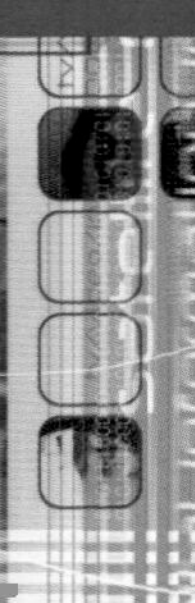

A3.3 Simultaneous equations

This spread will show you how to:

- Solve a pair of simultaneous linear equations by eliminating one variable.

KEYWORDS
Simultaneous equations
Elimination
Substitute

Clare and Jason are playing a jumping game.
Jason does three long jumps and five short jumps, covering 11 m.
Clare does three long jumps and two short jumps, covering 8 m.

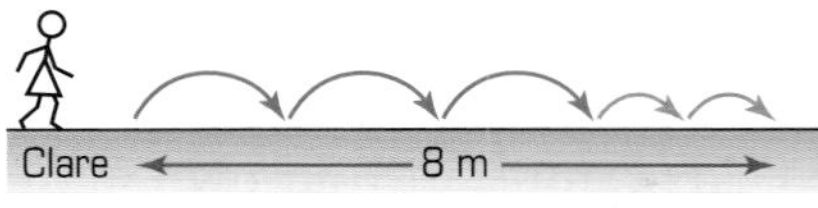

Clare ← 8 m →

How big are their jumps?

You can use algebra to solve this problem.

Assume that Clare and Jason have equal-length jumps.

Use letters to describe the variables ...
Let x = length of long jump, and
y = length of short jump.

... then write two equations:
Jason: $3x + 5y = 11$ (1)
Clare: $3x + 2y = 8$ (2)

Label your equations.

Subtract equation (2) from equation (1): $0 + 3y = 3$
$\Rightarrow y = 1$

x has been eliminated.

Substitute $y = 1$ into equation (2): $3x + 2 = 8$
$\Rightarrow 3x = 6$ and $x = 2$

So each long jump is 2 m, and each short jump is 1 m.

Check using equation (1):
$3 \times 2 + 5 \times 1 = 6 + 5$
$= 11$
as required.

- **Simultaneous equations** are true at the same time.

- You can solve simultaneous equations using **elimination**.
Add or subtract the equations to eliminate one variable.

example

Solve these simultaneous equations using the method of elimination.

a $2x + 4y = 18$
$2x + y = 12$

b $4p + 3q = 24$
$p + 3q = 15$

---

a Label and subtract:
(1) $2x + 4y = 18$
(2) $2x + y = 12$
(1)–(2) $3y = 6 \Rightarrow y = 2$
Substitute in (2): $2x + 2 = 12$
$2x = 10 \Rightarrow x = 5$
Check with (1): $2 \times 5 + 4 \times 2 = 18$

So the solution is $x = 5$, $y = 2$.

b Label and subtract:
(1) $4p + 3q = 24$
(2) $p + 3q = 15$
(1)–(2) $3p = 9 \Rightarrow p = 3$
Substitute in (2): $3 + 3q = 15$
$3q = 12 \Rightarrow q = 4$
Check with (1): $4 \times 3 + 3 \times 4 = 24$

So the solution is $p = 3$, $q = 4$.

Exercise A3.3

1 Solve these simultaneous equations using the method of elimination.
Remember to find the value of both unknowns and to check your answers.

a $5x + 2y = 16$
$x + 2y = 4$

b $5a + b = 22$
$2a + b = 10$

c $m + 3n = 11$
$m + 2n = 9$

d $4x + 5y = 9$
$4x - 2y = 2$

e $6p + q = 31$
$3p + q = 16$

f $4x + 3y = {}^{-}5$
$7x + 3y = {}^{-}11$

Note:
Take extra care with negatives.

2 Norman and Aura are having a jumping competition. Their short jumps and long jumps are the same size.
Norman does 2 long jumps and 4 short jumps:

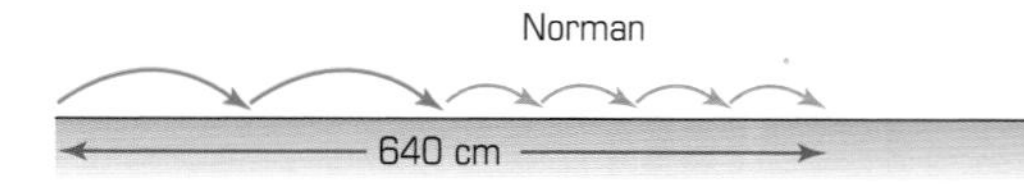

Aura does 3 long jumps and 5 short jumps:

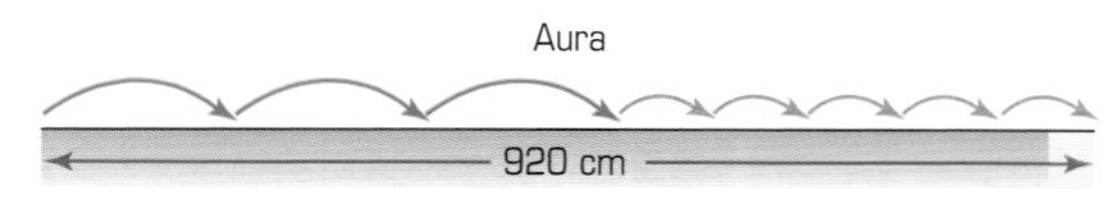

a Write a pair of simultaneous equations to show the jumping competition.

b Using elimination, calculate how long each jump is.

3 2 cups of tea and a cola cost £2.80 and 3 cups of tea and 2 colas cost £4.70.
I have £5. Do I have enough for 6 cups of tea?

4 Some simultaneous equations are best rearranged before solving.
For example,

$2x = 7 - y \xrightarrow{+y} 2x + y = 7$

$0 = 5 - x - y \xrightarrow{+x+y} x + y = 5$

Rearrange these equations and solve them.

a $3x = 14 - y$
$x + 2y - 3 = 0$

b $2y + 5x = 23$
$2y = 10 - 3x$

5 It is possible to use a method called 'substitution' to solve simultaneous equations, rather than elimination.
Follow this example:

$2x + y = 10 \quad (1)$
$y = x + 1 \quad (2)$

Substitute equation (2) into (1):

$2x + (x + 1) = 10$
$3x + 1 = 10$
$3x = 9$
$x = 3$

Hence, $2 \times 3 + y = 10$
$y = 4$

Try using substitution to solve these:

a $3x + y = 18$
$y = 3x$

b $2x - y = 7$
$y = x - 7$

c $x + 2y = 14$
$y + 2 = x$

6 **Challenge**
If we have three equations, we can find three unknowns.
Can you use elimination to find x, y and z?

$$x + y + z = 14$$
$$2x + y + 3z = 30$$
$$3x + y + z = 18$$

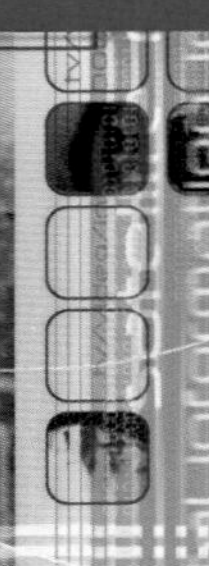

A3.4 Further simultaneous equations

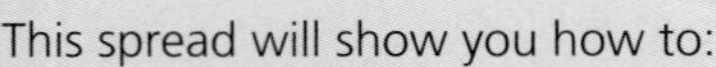

This spread will show you how to:

- Solve a pair of simultaneous linear equations by eliminating one variable.

KEYWORDS
Simultaneous equations
Elimination Substitute
Variable

Clare and Jason are playing their jumping game.

Clare does five long jumps and two short jumps, covering 5 m.

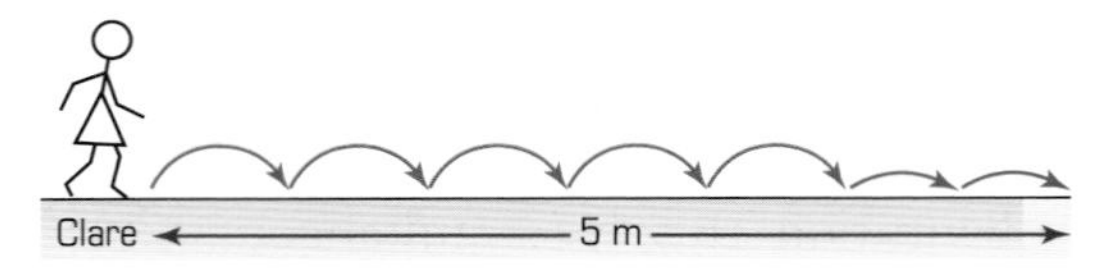

Using algebra, $5x + 2y = 500$

Jason does three long jumps, but jumps back two short jumps, covering 1.4 m.

Jason 1.4 m

Using algebra, $3x - 2y = 140$

x = length of long jump (in cm), y = length of short jump (in cm)

Hint:
Use cm rather than m in the equations to avoid decimals.

You can work out the lengths of their jumps by solving the equations simultaneously.

(1) $5x + 2y = 500$

(2) $3x - 2y = 140$ +

If you **subtract** (1) – (2), you get $2x + 4y = 360$ (check). You do not eliminate either of the variables.

Now **add** (1) + (2): $8x + 0 = 640 \Rightarrow x = 80$

Substitute in (2): $3 \times 80 - 2y = 140$

$240 - 2y = 140$

$2y = 100 \Rightarrow y = 50$

So $x = 80$ cm, $y = 50$ cm

Check by substituting in (1): $5 \times 80 + 2 \times 50 = 400 + 100 = 500$

- When you eliminate a variable, look at the signs.
  - If the signs are opposite, then add.
  - If they are the same, then subtract.

Remember:
Same **s**igns **s**ubtract.

Sometimes, neither of the variables matches up.
You may need to multiply one or both of the equations.

example

Solve these simultaneous equations.

a $5x - y = 8$
$7x + 4y = 22$

b $2x + 3y = 5$
$5x - 2y = {}^{-}16$

a (1) $5x - y = 8 \xrightarrow{\times 4} 20x - 4y = 32$

(2) $7x + 4y = 22 \rightarrow 7x + 4y = 22$

$27x = 54$

$\Rightarrow x = 2$

Substitute in (1): $5 \times 2 - y = 8$

$\Rightarrow 10 - y = 8 \Rightarrow y = 2$

Check in (2): $7 \times 2 + 4 \times 2 = 14 + 8 = 22$

b (1) $2x + 3y = 5 \xrightarrow{\times 5} 10x + 15y = 25$

(2) $5x - 2y = {}^{-}16 \xrightarrow{\times 2} 10x - 4y = {}^{-}32$

$19y = 57$

$\Rightarrow y = 3$

Substitute in (1): $2x + 9 = 5$

$2x = {}^{-}4 \Rightarrow x = {}^{-}2$

Check in (2): $5 \times {}^{-}2 - 2 \times 3 = {}^{-}10 - 6 = {}^{-}16$

Exercise A3.4

1 Solve these simultaneous equations by adding to eliminate either x or y.

a $3x + 2y = 19$
$8x - 2y = 58$

b $5a + 2b = 16$
$3a - 2b = 8$

c $7m - 3n = 24$
$2m + 3n = 3$

d $2x + 3y = 19$
$^{-}2x + y = 1$

2 Solve these simultaneous equations by deciding whether to add or subtract in order to eliminate one unknown.

a $x + y = 3$
$3x - y = 17$

b $5m - 2n = 4$
$3m + 2n = 12$

c $5p + q = {}^{-}7$
$5p - 2q = {}^{-}16$

d $3x + y = 14$
$3x - y = 10$

3 Solve these simultaneous equations by multiplying one or both equations by a suitable constant. Decide whether to add or subtract in order to eliminate one unknown.

a $2x + y = 8$
$5x + 3y = 12$

b $3x + 2y = 19$
$4x - y = 29$

c $2a + 3b = 20$
$a + 5b = 31$

d $5p - q = 8$
$7p + 4q = 22$

e $8x - 3y = 30$
$3x + y = 7$

f $2x + 3y = 12$
$5x + 4y = 23$

g $9m + 5n = 15$
$3m - 2n = {}^{-}6$

h $3x - 2y = 11$
$2x - y = 8$

4

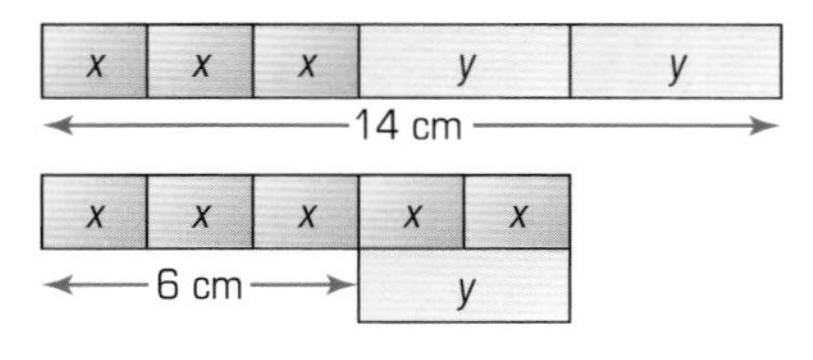

Different blocks of wood are put together to make the arrangements shown.

a Devise a pair of simultaneous equations to show this information.

b Solve the equations, by elimination, to find the length of a block of wood of each type.

5 Make as many pairs of simultaneous equations as you can from A, B and C.

A $3x + y = 14$

B $x - 2y = 7$

C $2x + 3y = 21$

Which pair gives the largest product xy?

6 Solve these simultaneous equations. Rearrange them first.

a $y = 2x + 7$
$x + y - 1 = 10$

b $3a = 19 - b$
$5a = 6 + 2b$

c $3p = q - 17$
$3q = 1 - p$

7 Solve these simultaneous equations which involve fractions.

Hint: How could you get rid of the fractions to make the equations easier to deal with?

a $\frac{x}{2} + 3y = 1$
$5x - 7y = 47$

b $\frac{m}{3} - \frac{n}{4} = 1\frac{1}{2}$
$2m + n = 14$

c $\frac{a}{4} + 3b = {}^{-}1$
$a - \frac{2}{3}b = 8\frac{2}{3}$

8 There are two cafés in small town.

What is the price of:

a an orange juice

b a coffee

in each café?

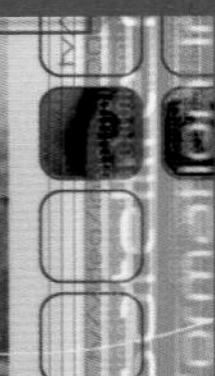

A3.5 Using simultaneous equations to solve problems

This spread will show you how to:
- Solve a pair of simultaneous linear equations by eliminating one variable.
- Represent problems and synthesise information in algebraic form.

KEYWORDS
Simultaneous equations
Elimination Substitute

You can use simultaneous equations to solve problems containing two unknowns.

example

Ahmed buys two CDs and three DVDs for £55.
Sarah buys three CDs and four DVDs for £76.50.

Dominic goes into the same shop.
Can he afford one of each for £20?

Define the unknowns:
x = price of a CD (in pence)
y = price of a DVD (in pence)

Write a pair of **simultaneous equations**:
Ahmed: $2x + 3y = 5500$
Sarah: $3x + 4y = 7650$

Hint: It is easier to work in pence to avoid decimals.

Solve using elimination:
(1) $2x + 3y = 5500$
(2) $3x + 4y = 7650$
Subtract because the x's are the **s**ame:

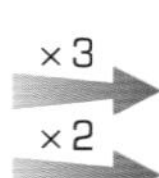

$$6x + 9y = 16\,500$$
$$6x + 8y = 15\,300$$
$$y = 1200$$

Substitute in (1) to find x:
$$2x + 3600 = 5500$$
$$2x = 1900$$
$$x = 950$$

Answer the problem:
A CD costs £9.50, and a DVD costs £12.
Dominic cannot afford to buy a CD and a DVD, as the total cost is £21.50.

Some equations may require other mathematical knowledge.

example

In the isosceles triangle shown, angle b is 9° more than angle a.
Find the angles.

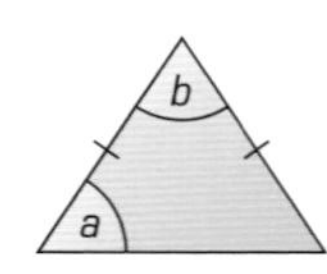

Angles of a triangle $a + b + c = 180$
Isosceles $\Rightarrow c = a$
So $2a + b = 180$ (equation 1)

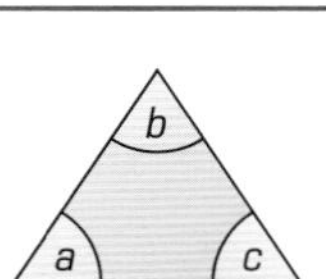

b is 9° more than a $\Rightarrow b = a + 9$
Rearrange: $a - b = {}^{-}9$ (equation 2)
$$2a + b = 180$$
Add: $$3a = 171$$
$$a = 57$$

$b = 57 + 9 = 66$ $c = a = 57$
So the angles of the triangle are 57°, 57° and 66°.

Exercise A3.5

In questions 1–4, form a pair of simultaneous equations and solve them to find the solution to the problem.

1 Elaine buys two bars of chocolate and a packet of crisps for 88p, while Ben buys a bar of chocolate and four packets of crisps for £1.28.
How much does each item cost?

2 I am thinking of two numbers. Twice the larger number plus the smaller number is 6. Their difference is also 6.
What is their product?

3 Find two numbers with a sum of 25 and a difference of 9.

4 The mean of two numbers is 45.
The larger number is 18 more than the smaller number.
What are the numbers?

5 **a** Solve these equations.

$$x + 4y = 58$$
$$5x + y = 62$$

b Invent a problem that could give rise to these two simultaneous equations.

6 **a** A puzzle appeared in a newspaper. The number at the end of the row is the sum of the values of the symbols in that row.

Use algebra to find the value of the symbols.

b Invent a similar problem for your neighbour to try.

7 Ben commented to his father:

> In five years' time, you will be twice my age.

His father replied:

> Yes, and in 13 years the sum of our ages will be 100.

How old is Ben now?

8 For a school trip, cars and minibuses are used. Each car can take four pupils and each minibus twelve. If nine vehicles are used and 60 pupils go on the trip, how many of each vehicle are taken?

9 If the perimeter of this triangle is 91 cm, how long is each side?

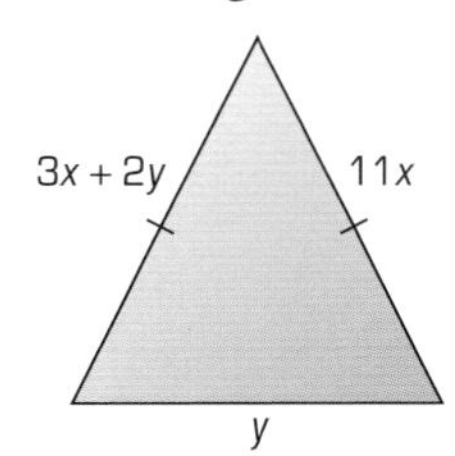

10 The line $y = mx + c$ passes through (2, 9) and (6, 29). Use simultaneous equations to find the equation of the line.

11 **a** A quadratic sequence has formula $T(n) = an^2 + b$. If the third term is 29 and the ninth term is 245, find the formula and, hence, the first five terms of the sequence.

b Why wouldn't we be able to complete part **a** if the formula was $T(n) = an^2 + bn + c$? How could we find the exact formula – what extra information would be needed?

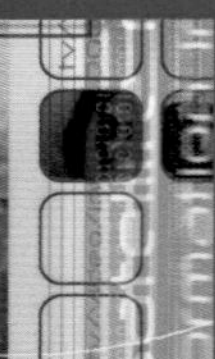

A3.6 Solving simultaneous equations by graphs

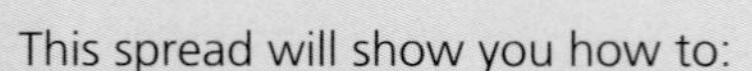

This spread will show you how to:

- Solve simultaneous linear equations by eliminating one variable.
- Link graphical representations of equations to the algebraic solutions.
- Consider cases that have no solution or an infinite number of solutions.

KEYWORDS

Simultaneous equations
Intersection
Implicit form

You can plot simultaneous equations on a graph.

example

Here are two simultaneous equations:

$2x - y = 1$
$x + 2y = 8$

a Draw their graphs on the same pair of axes.
b Find the point of intersection. What does this represent?

a $2x - y = 1 \quad \Rightarrow y = 2x - 1$

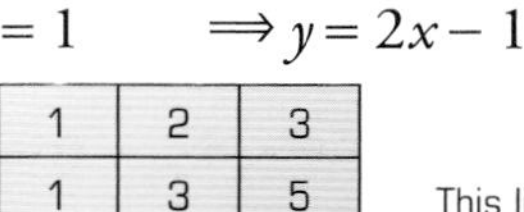

x	1	2	3
y	1	3	5

$x + 2y = 8$

x	0	8
y	4	0

When an equation is in **implicit** form, just let both y and x equal 0.

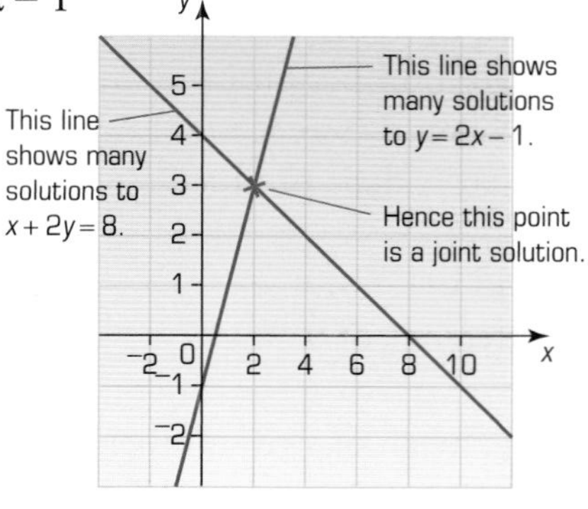

b The point of intersection is (2, 3).
At this point, the two equations are true simultaneously.
So the point represents the solution to the equations.
$x = 2$, $y = 3$ is the solution.

▶ **You can solve a pair of simultaneous equations by representing them graphically and finding their point of intersection.**

example

On Reena's mobile phone, three text messages and two local calls cost 36p.
Two text messages and four local calls cost 40p. What is the cost of each type of call?

Let x = cost of text message (in pence) and
y = cost of local call (in pence).

(1) $3x + 2y = 36$

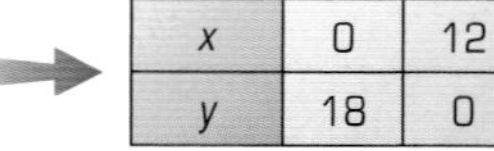

x	0	12
y	18	0

(2) $2x + 4y = 40$

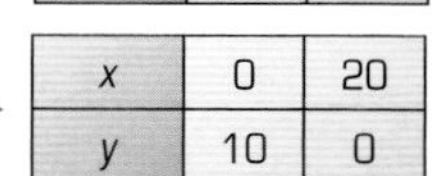

x	0	20
y	10	0

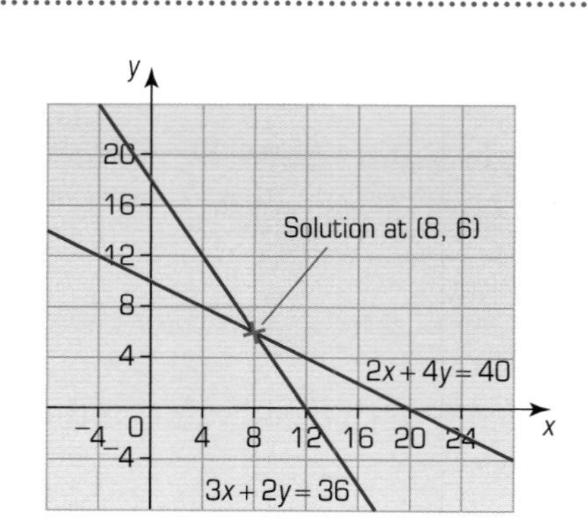

From the graph, the solution is (8, 6).
$x = 8 \Rightarrow$ A text message costs 8p.
$y = 6 \Rightarrow$ A local call costs 6p.

Exercise A3.6

1 **a** Use the diagram to solve these pairs of simultaneous equations.

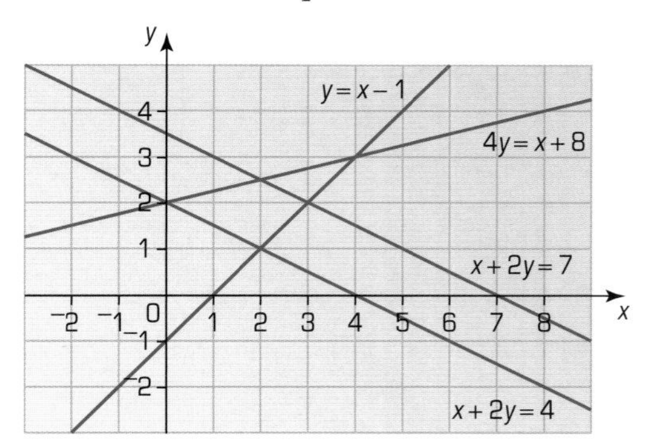

i $y = x - 1$
$x + 2y = 4$

ii $y = x - 1$
$x + 2y = 7$

iii $x + 2y = 4$
$4y = x + 8$

iv $4y = x + 8$
$y = x - 1$

b Using the diagram, explain why the simultaneous equations $x + 2y = 4$ and $x + 2y = 7$ have no solution.

2 Plot your own graphs to solve these simultaneous equations.

a $y = 3x - 2$
$x + y = 6$

b $y = 2x + 1$
$x + y = 10$

c $y = x - 1$
$2x + 3y = 12$

3 I am thinking of two numbers.
Here are some clues about them.

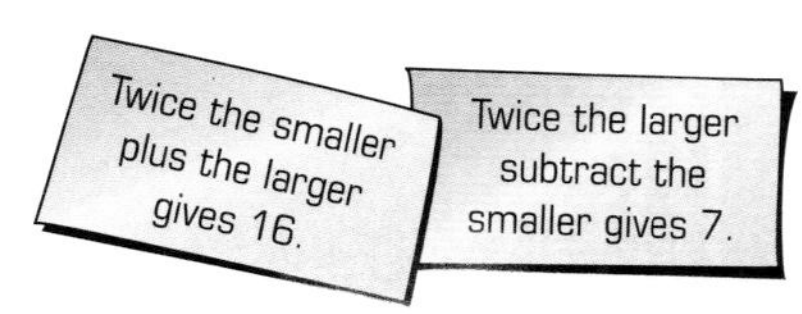

Use a graphical method to find the values of my numbers. Confirm your answer by using an algebraic method.

4 The diagram shows a line, A.

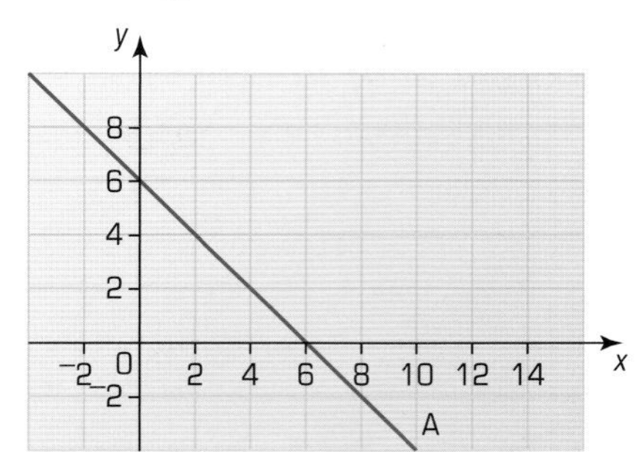

a Write down the equation of line A.

b Copy the diagram and add the graph $y = 3x - 2$. Label this line B.

c Solve the simultaneous equations given by the equations of line A and line B.

5 Use graphs to explain if it is possible to have a pair of simultaneous linear equations with:

a no solutions **b** exactly two solutions.

6 By considering their graphs, explain why these equations have no simultaneous solution.

a $y = 3x + 2$
$y = 3x + 5$

b $2x + y = 7$
$4x + 2y = 14$

7 In each of these diagrams, find the simultaneous equations that have been solved.
Use an algebraic method to check that your equations give the correct solution.

a

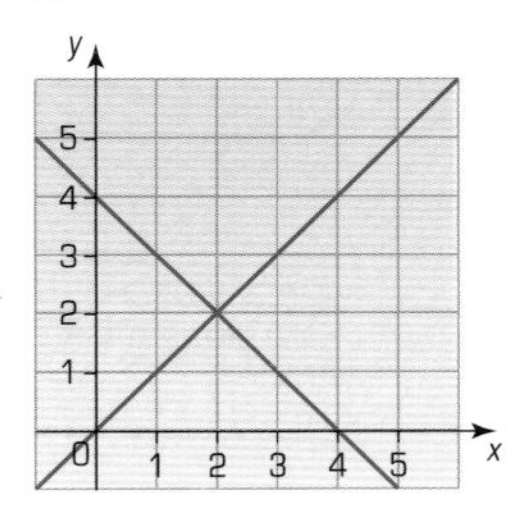

b

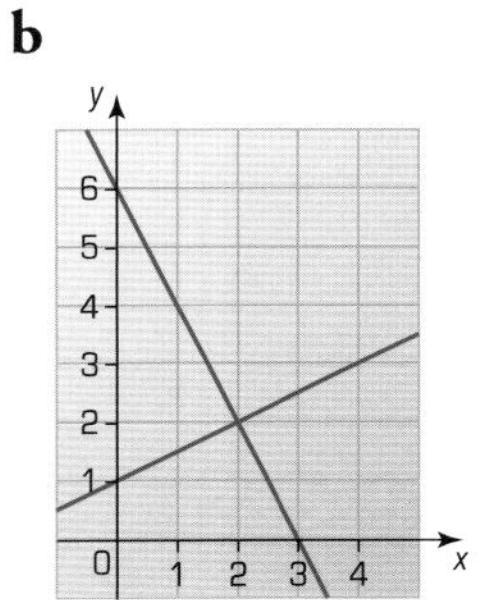

A3 Summary

You should know how to ...

1 Solve a pair of simultaneous linear equations by eliminating one variable.

2 Link a graphical representation of a pair of equations to the algebraic solution.

3 Generate fuller solutions to problems.

Check out

1 Solve these pairs of equations.

a $x + 2y = 10$
$3x + 2y = 14$

b $2x - y = 5$
$x + 3y = {}^{-}8$

c $3x + 2y = 17$
$4x + 3y = 23$

2 **a** Solve the equations:
$y = 2x - 2$
$2x + 3y = 6$
using the graph shown.

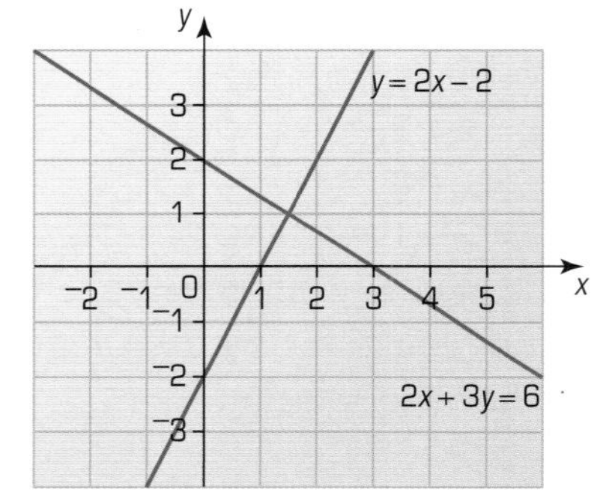

b Solve the equations algebraically to confirm your solution.

3 Here are two adverts for different mobile phone companies.

Write a report detailing which of these mobile phone companies gives the best value for money for different phone users. Use graphs to support your findings.

1 Geometrical reasoning and construction

This unit will show you how to:

- Distinguish between conventions, definitions and derived properties.
- Find, calculate and use the interior and exterior angles of regular polygons.
- Solve problems using properties of angles, of parallel and intersecting lines, and of polygons.
- Understand and apply Pythagoras' theorem.
- Explain why inscribed regular polygons can be constructed by equal divisions of a circle.
- Know that the tangent at any point on a circle is perpendicular to the radius at that point.
- Generate fuller solutions to problems.
- Explain why the perpendicular from the centre to a chord bisects the chord.
- Use straight edge and compasses to construct a triangle, given RHS.
- Know that triangles given SSS, SAS, ASA or RHS are unique.
- Find the locus of a point that moves according to a simple rule.
- Extend to more complex rules involving loci and simple constructions.
- Distinguish between practical demonstration and proof.

The origins of geometry are over two thousand years old.

Before you start

You should know how to ...

1 Calculate angles in triangles and quadrilaterals.

2 Recognise angles in parallel and intersecting lines.

3 Calculate square roots.

4 Use compasses to construct triangles and bisectors.

Check in

1 Calculate the unknown angles:

a

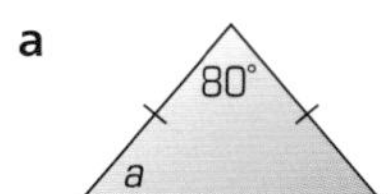

b

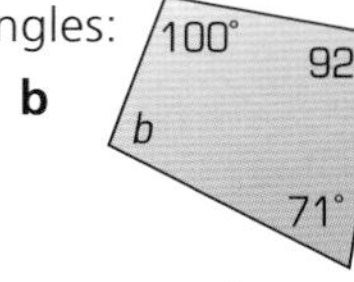

2 Copy the diagram and label the angles equal to x.

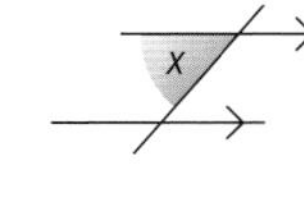

3 Find, to 2 dp:

a $\sqrt{50}$ **b** $\sqrt{102}$ **c** $\sqrt{950}$

4 **a** Construct triangle ABC, where AB = 5.4 cm, BC = 6.3 cm, CA = 7.9 cm.

b Construct the bisector of: **i** $\angle A$ **ii** AB

Geometrical language

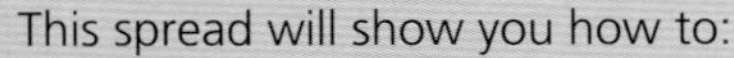

This spread will show you how to:

- Distinguish between conventions, definitions and derived properties.
- Distinguish between a practical demonstration and a proof.

KEYWORDS

Convention Definition
Derived property
Practical demonstration
Proof

► A **convention** is an agreed way of describing a situation.

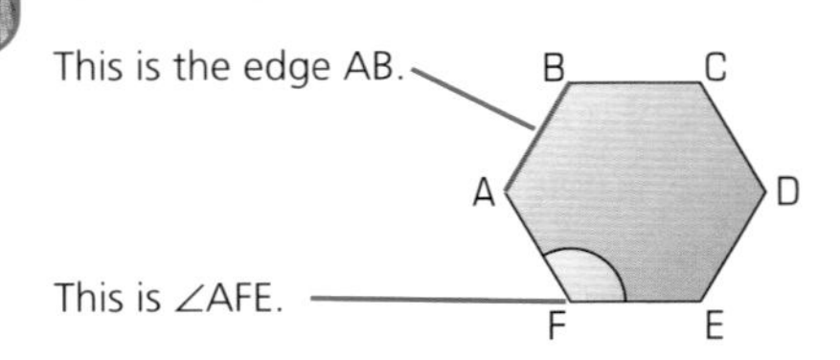

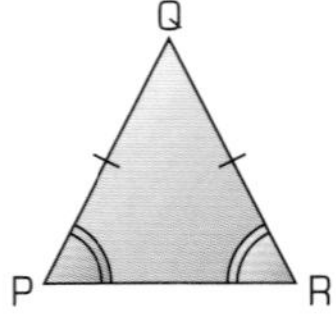

PQ = QR
∠P = ∠R

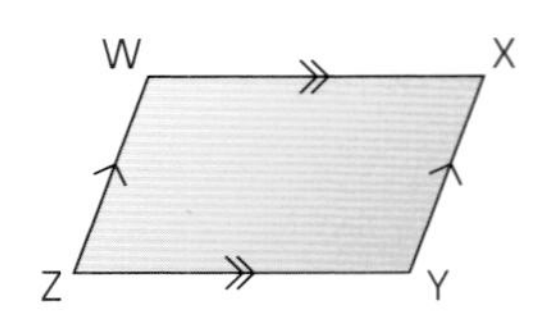

WX and ZY are parallel.
WZ and XY are parallel.

► A **definition** is a set of conditions needed to specify an object or quantity.

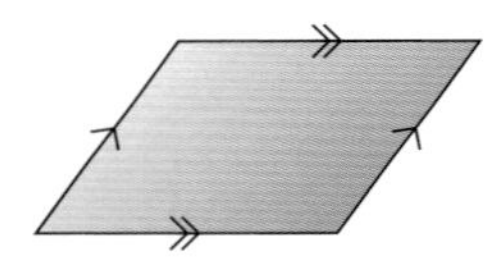

A parallelogram is a quadrilateral with opposite sides parallel.

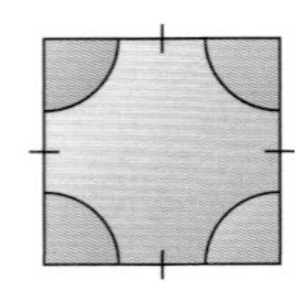

A square is a quadrilateral with all sides equal and all angles equal.

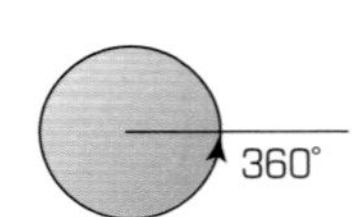

A degree (°) is a unit for measuring angles, in which one complete rotation is divided into 360°.

► A **derived property** is a consequence of a definition.

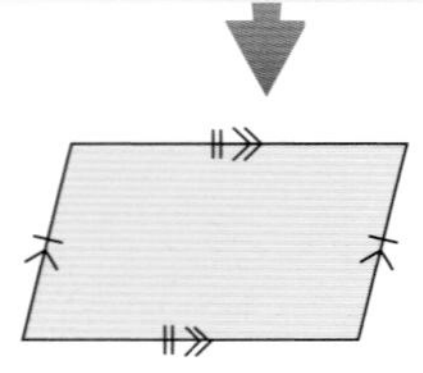

The opposite sides of a parallelogram have equal lengths.

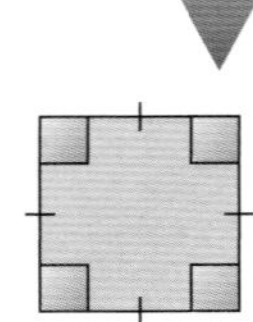

Each angle in a square is 90°.

Half a complete turn is 180°.

► A **practical demonstration** can show that something is true by observation.

If you fit the corners of a quadrilateral together they make 360°.
⟹ The angles of a quadrilateral add up to 360°.

► A **proof** requires a deductive argument.

In the diagram,

$a + d + e = 180°$ (angles on a straight line)
$d = b$ (alternate angles)
$e = c$ (alternate angles)
$\Rightarrow$ $a + b + c = 180°$
$\Rightarrow$ Angles in a triangle add up to 180°.

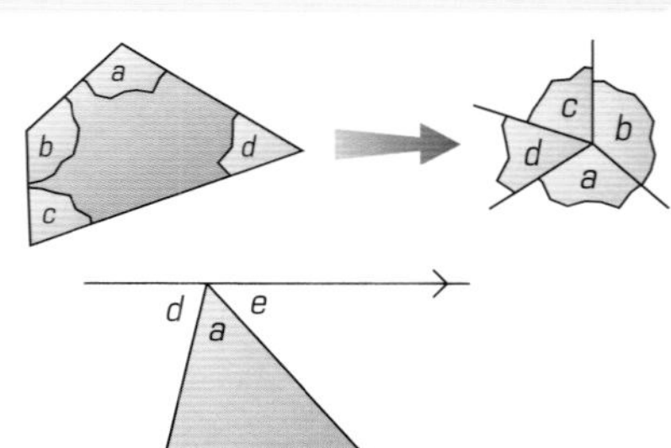

Exercise S1.1

1 Use conventions to sketch these diagrams.

a the line CD

b the angle CAB

c the triangle ABC

d the parallelogram PQRS

e the equilateral triangle XYZ

f the isosceles triangle EFG where EF = EG

g the trapezium KLMN where KL and MN are parallel and $\angle M = \angle N$

h the kite EFGH where EF = FG and GH = HE

i the triangle ABC where AB = BC and AC has been extended to D.

2 Write your own definitions for these terms.

a edge

b tessellation

c kite

d trapezium

e perpendicular

Give examples to help explain your definitions. You may find it helps to draw a diagram.

Read your definitions to a partner and see if they know what you're describing.

3 Copy and complete the following derived properties:

a The angles of a triangle add up to __.

b The diagonals of a square are __ in length.

c The interior angle of a square is __ degrees.

d The sides of a rhombus are __.

e Opposite angles of a parallelogram are __.

f The diagonals of a square bisect each other at __ degrees.

g The diagonals of a kite are __ to one another.

4 Take a square piece of paper.

Fold the corners of your square to a common point in the centre.

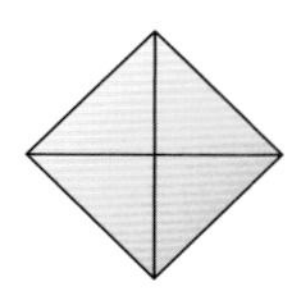

Does this always work?

Prove that the angles in a square add up to 360°.

Prove that the angles in a quadrilateral add up to 360°.

Convince your partner that this is true.

5 Copy and complete this proof.

The sum of the interior angles of a triangle is 180°.

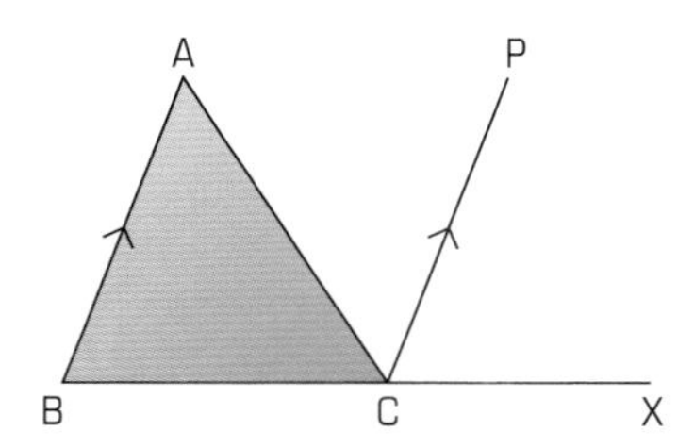

Given	Any triangle ABC with BC produced to X.
To prove	$\angle CAB + \angle ABC + \angle BCA = 180°$.
Construction	CP is parallel to AB through C.

6 Prove that the angles of a quadrilateral add up to 360°.

S1.2 Angles in polygons

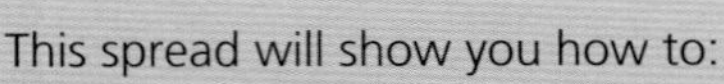

This spread will show you how to:

- Find, calculate and use the sums of the interior and exterior angles of quadrilaterals, pentagons and hexagons.
- Find, calculate and use the angles of regular polygons.

KEYWORDS

Interior angle, Exterior angle, Quadrilateral, Pentagon, Polygon, Hexagon

Remember: A polygon has **interior** and **exterior** angles.

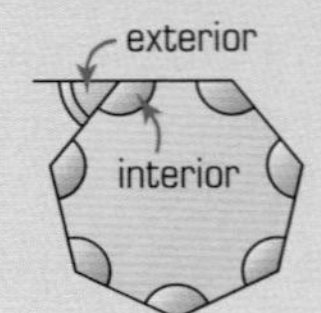

Interior angles

► **The sum of the interior angles of a quadrilateral is 360°.**

You can show this by sticking two triangles together:

The angles in each triangle add to 180°.
⟹ The angles in a quadrilateral equal 360°.

You can extend this to the interior angles of different polygons.

A pentagon (5 sides) is made from 3 triangles.

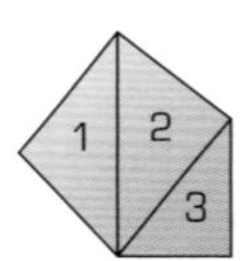

The interior angle sum is $3 \times 180° = 540°$.

A hexagon (6 sides) is made from 4 triangles.

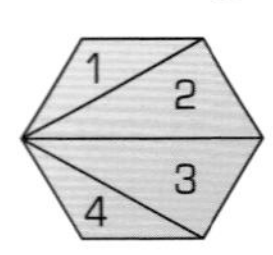

The interior angle sum is $4 \times 180° = 720°$.

An n-sided polygon is made from $(n - 2)$ triangles.

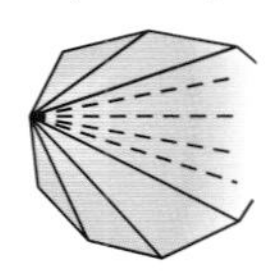

The interior angle sum is $(n - 2) \times 180°$.

► **The interior angle sum of a polygon is $(n - 2) \times 180°$.**

example

Find the value of x in this pentagon.

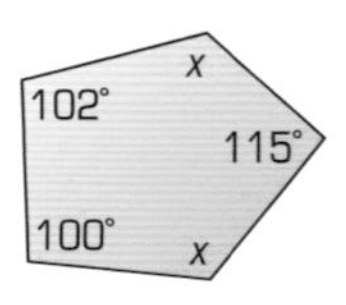

$$100° + 102° + 115° + x + x = (5 - 2) \times 180° = 540°$$
$$317° + 2x = 540°$$
$$2x = 223°$$
$$x = 111.5°$$

Exterior angles

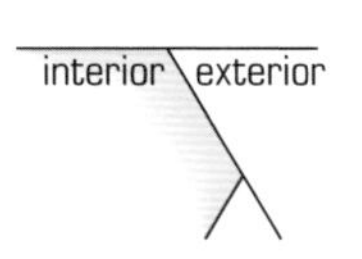

At each vertex of a polygon:

interior angle + exterior angle = 180°

For a polygon with n vertices:

sum of interior angles + exterior angles is $n \times 180°$

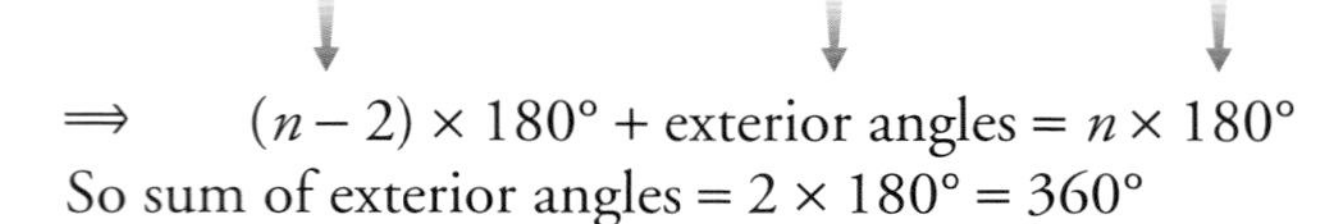

⟹ $(n - 2) \times 180°$ + exterior angles = $n \times 180°$

So sum of exterior angles = $2 \times 180° = 360°$

► **The exterior angles of a polygon always add up to 360°.**

Exercise S1.2

1 Find the unknown angles.

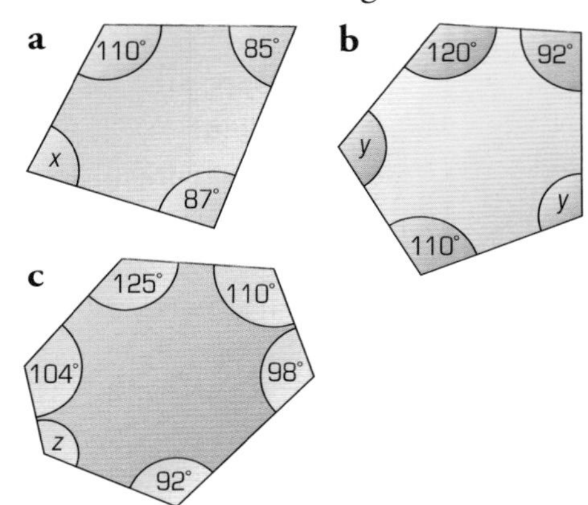

2 Here is a regular decagon.
It has 10 sides of equal length.

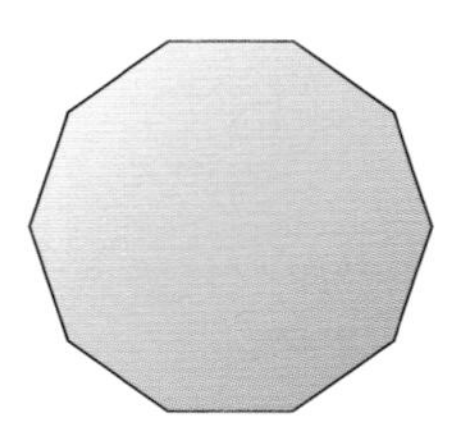

This formula gives you the sum of the interior angles of a regular polygon:

$s = (n - 2) \times 180°$

where s = sum of interior angles
n = number of sides.

a Work out the sum of the interior angles of a regular decagon.

b Find the size of an interior angle of a regular decagon.

3 Find the size of an exterior angle of a regular nonagon.

4 A regular polygon has exterior angles of 20°. How many sides does it have?

5 A regular polygon has interior angles of 140°. How many sides does it have?

6 Find angle x. Show all your working.

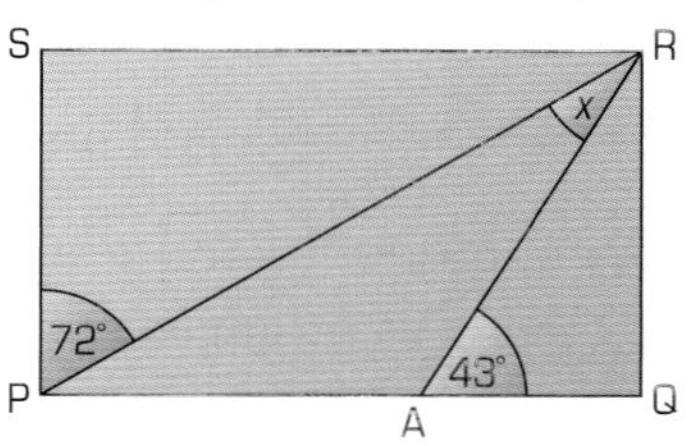

7 This diagram shows how C and D fit together to make a right-angled triangle.

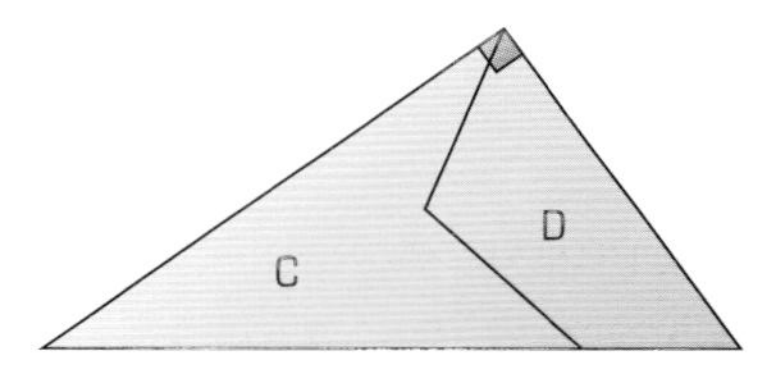

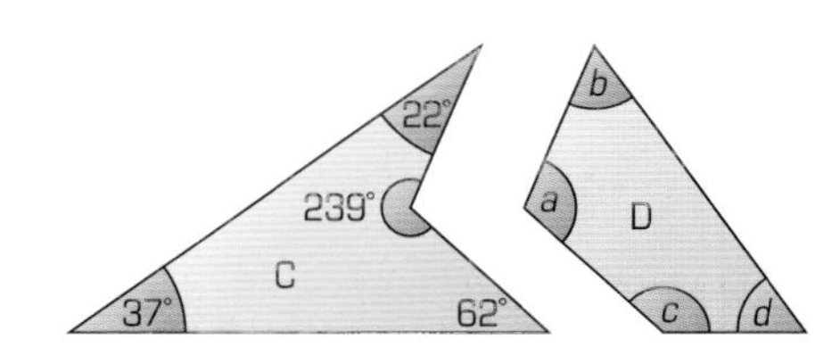

Work out the size of each of the angles in D.

8 Could a regular polygon have interior angles of 165.6°?
Explain your answer.

9 Could a regular polygon have interior angles of 152°?
Explain your answer.

S1.3 Harder angle problems

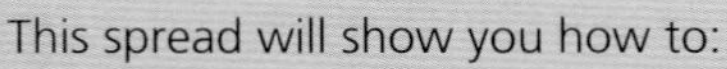

This spread will show you how to:

- Solve problems using properties of angles, of parallel and intersecting lines, and of triangles and other polygons.

KEYWORDS

Vertically opposite
Corresponding
Alternate
Parallel
Intersecting
Symmetry

Here are some properties of angles and lines:

a and b are vertically opposite angles.

$a = b$

c and d are corresponding angles.

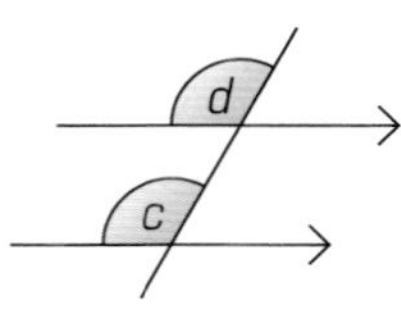

$c = d$

e and f are alternate angles.

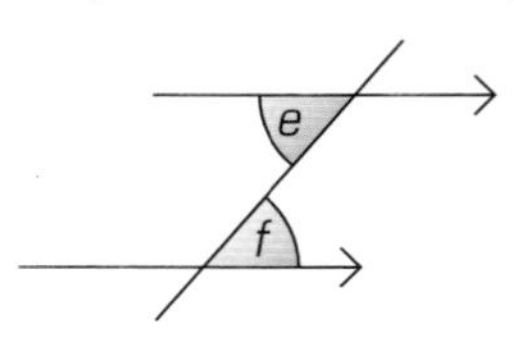

$e = f$

You can use these properties to solve problems in parallel and intersecting lines.

example

Find x, y and z.

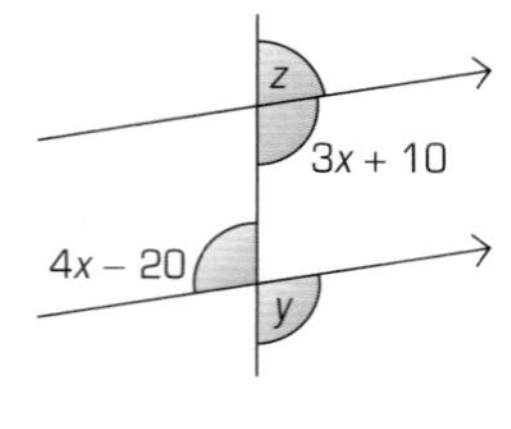

$3x + 10 = 4x - 20$ (alternate angles)

$10 = x - 20 \quad \Rightarrow \quad x = 30°$

$y = 4x - 20$ (opposite angles)
$= 120° - 20° = 100°$

$z = 180° - (3x + 10)$ (angles on a straight line)
$= 180° - 100° = 80°$

When solving angle problems, you should justify your answers.

You can also use your knowledge of angles and symmetry to solve problems in shapes.

example

a Find angles a, b and c in this trapezium.

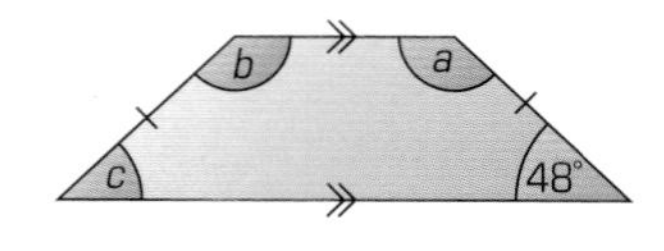

b Find angles d, e and f in this regular octagon.

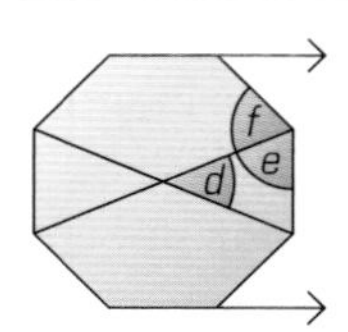

a $a = 180° - 48° = 132°$ (interior angles)

$b = 132°$ (symmetry)

$c = 48°$ (symmetry)

b $d = 360° \div 8 = 45°$ (regular octagon)

$e = \frac{180 - 45}{2} = 67\frac{1}{2}°$ (isosceles △)

$f = 67\frac{1}{2}°$ (symmetry)

Exercise S1.3

1 Find the unknown angles.
Give reasons for your answers.

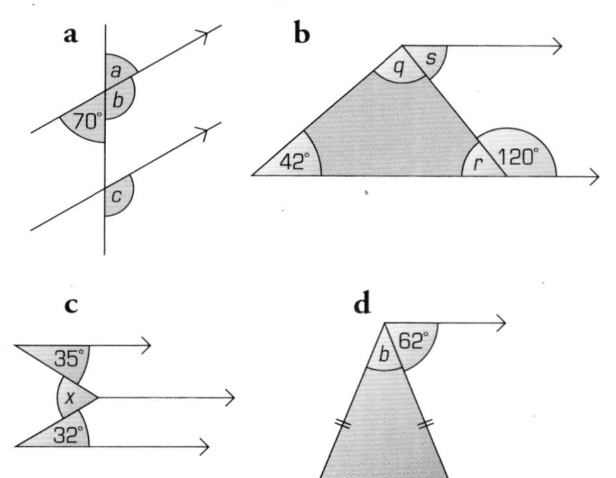

2 Use algebra to help you find these unknown angles.

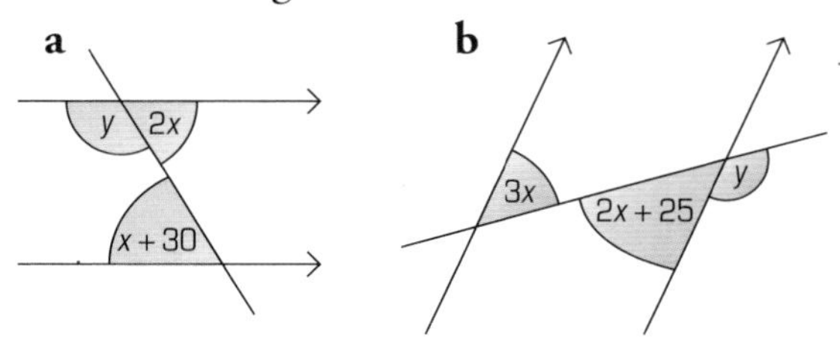

3 **a** Find a, b and c.

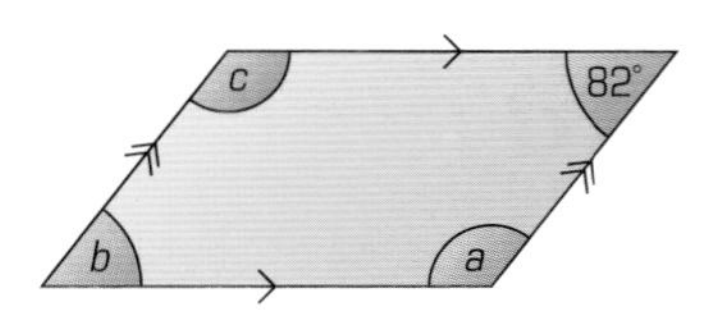

b Name this shape and find ∠PQR.

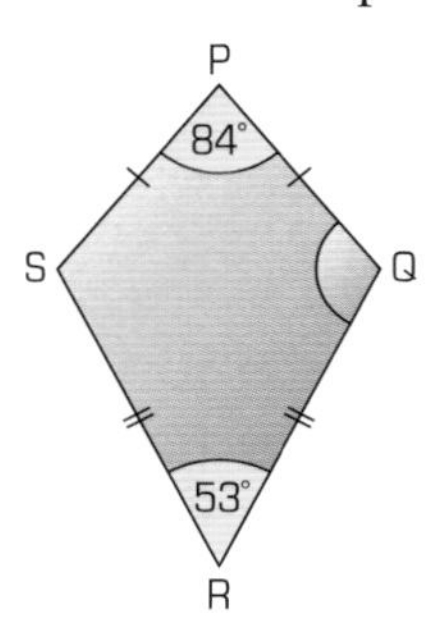

4 The shape below has three identical blue tiles and three identical red tiles. The sides of the tiles are all the same length. Opposite sides of the tiles are parallel.

a Name the shape of the tiles.

b One angle is 65°. Calculate angle x.

c Calculate angle y.

Show all your working.

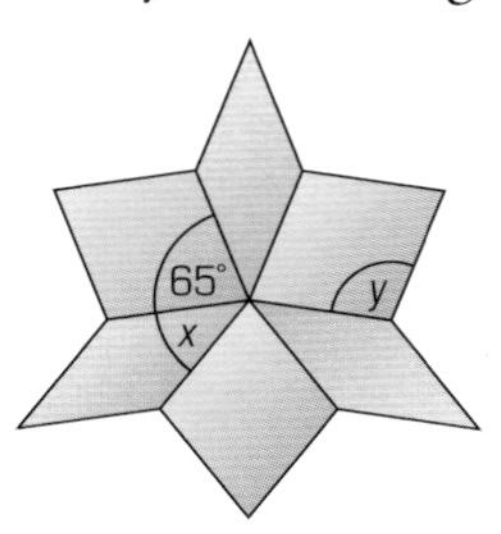

5 Two isosceles triangles have the same base PR, so that PQ = QR and PS = SR.

a Show that $a = 15°$.

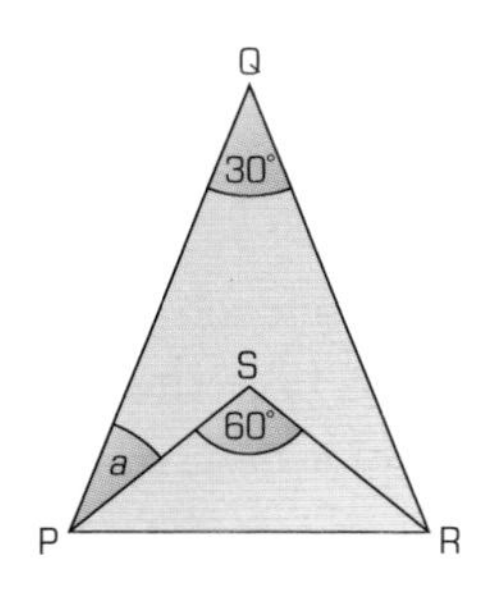

b Other pairs of isosceles triangles can be drawn from the same base PR.
∠PSR is twice the size of ∠PQR.
Call these $2y$ and y.
Prove that angle x is always half angle y.

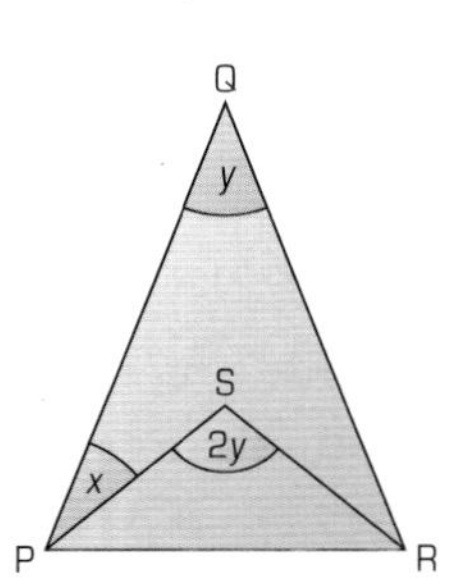

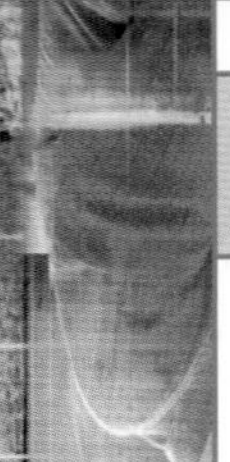

S1.4 Pythagoras' theorem

This spread will show you how to:

- Understand and apply Pythagoras' theorem.

KEYWORDS
Hypotenuse
Pythagoras' theorem
Square root

The longest side of a right-angled triangle is called the **hypotenuse**.
The hypotenuse is always opposite the 90° angle.

Take a right-angled triangle ...

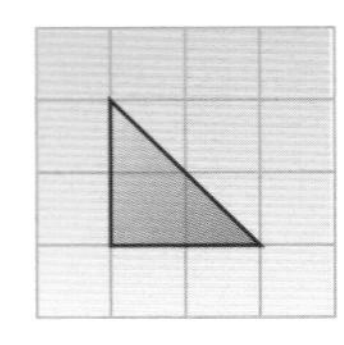

Construct squares on all three sides ...

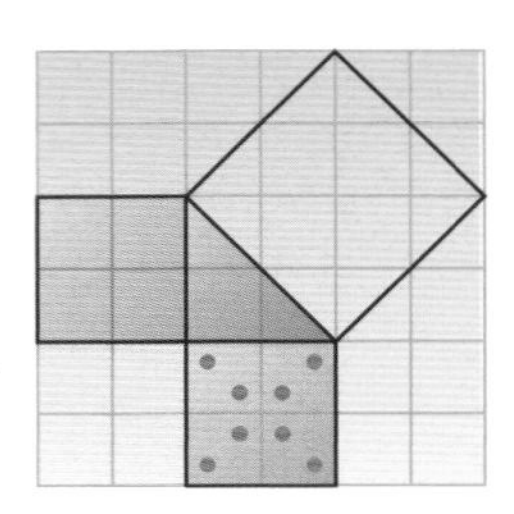

The two smaller squares fit exactly inside the larger one.

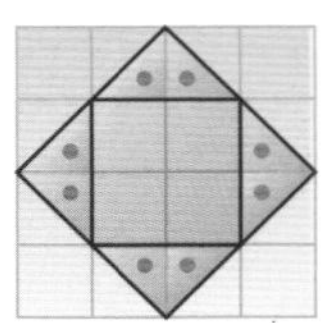

- For a right-angled triangle, **Pythagoras' theorem** says that:
  $a^2 = b^2 + c^2$
  (a, b and c are the lengths of the sides).
  Note that a is the hypotenuse.

It does not matter which sides you label b and c as long as a is the hypotenuse.

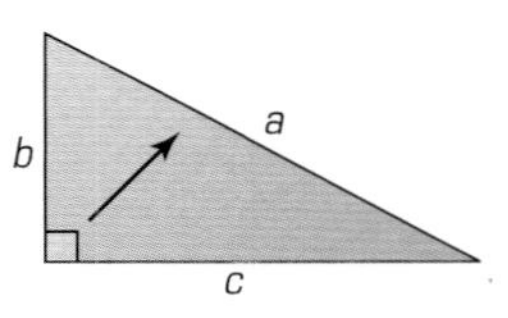

You can use Pythagoras' theorem to find lengths in right-angled triangles.
You need to remember to take the **square root**.

example

a Find the length XY in this right-angled triangle.

X
3 cm
Y 4 cm Z

b Find the length PQ in this triangle. Give your answer to 2 decimal places.

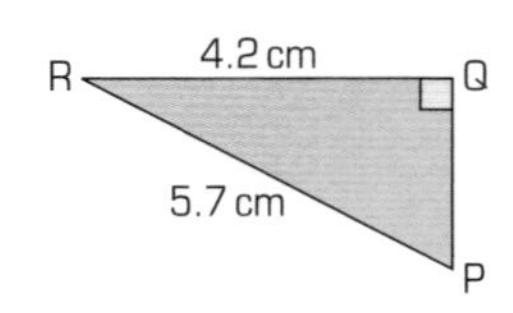

a Label the sides a, b and c:

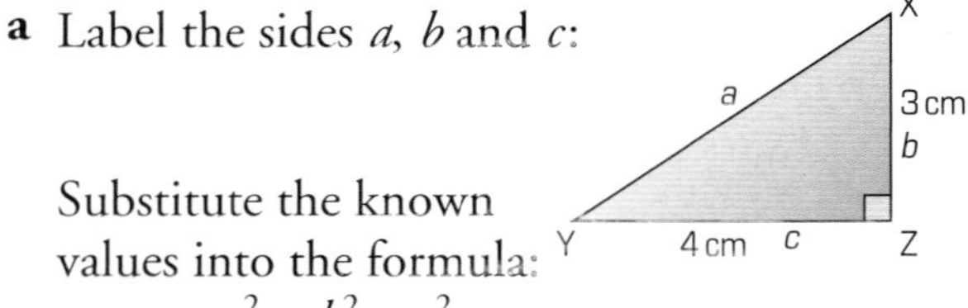

Substitute the known values into the formula:

$a^2 = b^2 + c^2$
$a^2 = 3^2 + 4^2$
$a^2 = 9 + 16$
$a^2 = 25$
$a = \sqrt{25} = 5$

$\Rightarrow$ The length XY is 5 cm.

b Label the sides:

R 4.2 cm c Q
b
5.7 cm a
P

Substitute the values: $5.7^2 = b^2 + 4.2^2$
Rearrange: $b^2 = 5.7^2 - 4.2^2$
$b^2 = 14.85$
$b = \sqrt{14.85}$
$= 3.85$ (to 2 dp)

$\Rightarrow$ PQ = 3.85 cm

Exercise S1.4

1 Calculate the unknown sides in these right-angled triangles.

a

4 cm
6 cm
x

b

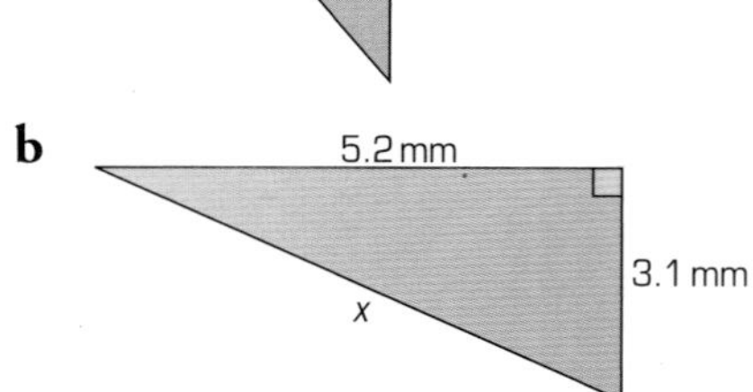

c

20 m
35 m
x

d

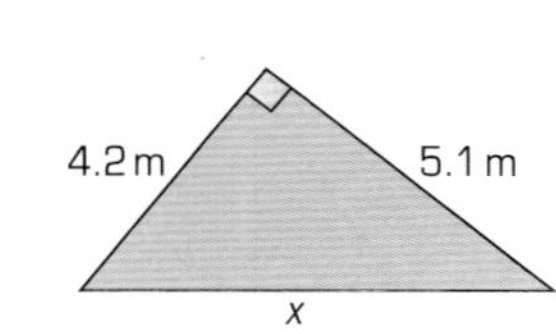

2 Calculate the unknown sides in these right-angled triangles.

a

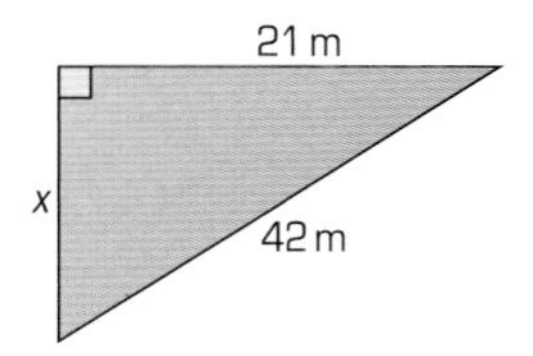

b

3.8 m
4.2 m
x

c

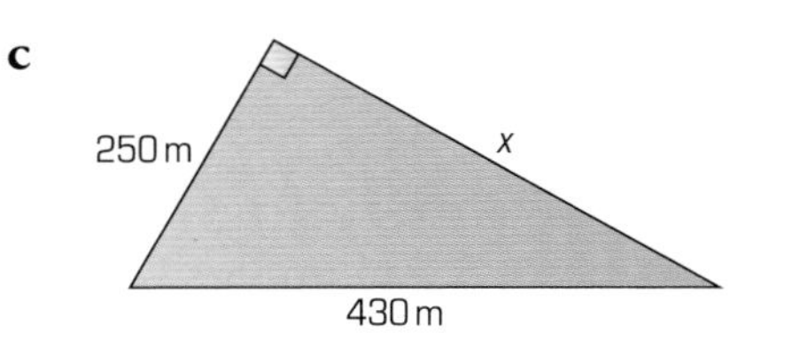

3 Calculate the unknown sides in these right-angled triangles.

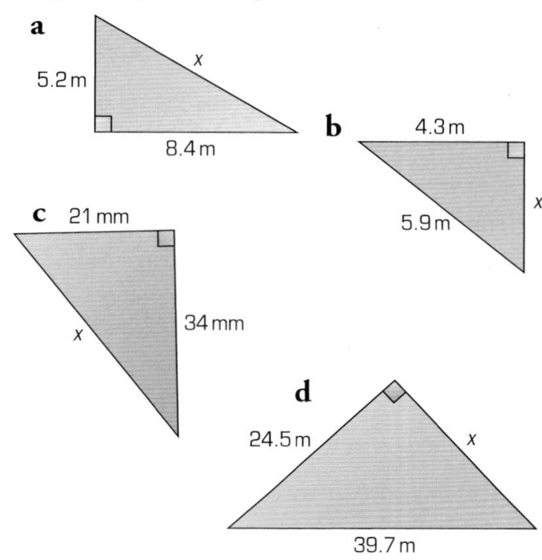

4 Here is a shape made from two triangles:

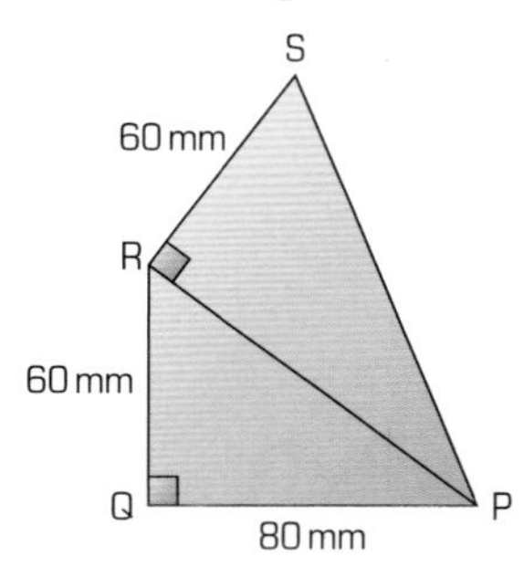

a Explain why PR is 100 mm.

b Calculate the length of PS. Show your working.

5 Which of these triangles are right-angled? Use Pythagoras' theorem to check.

a

6 m
10 m
8 m

b

7 m
4 m
10 m

c

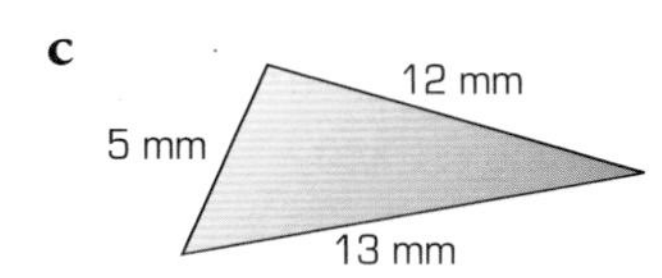

Pythagoras problems

This spread will show you how to:
- Understand and apply Pythagoras' theorem.

KEYWORDS
Pythagoras' theorem
Hypotenuse

Remember:
In Pythagoras' theorem, a is the hypotenuse.

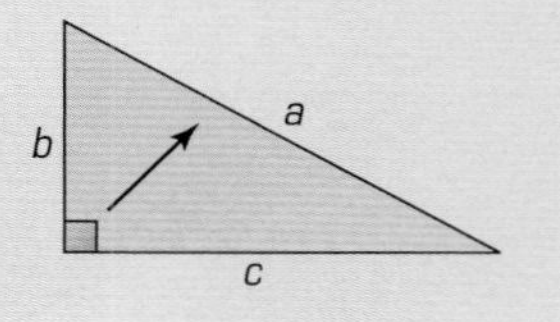

You can use Pythagoras' theorem to decide whether a triangle is right-angled.

example

A triangle has sides 70 m, 240 m and 250 m.
Prove that the triangle is right-angled.

The hypotenuse is the longest side, so $a = 250$.
$a^2 = 250^2 = 62\,500$ $\quad b^2 + c^2 = 70^2 + 240^2 = 62\,500$
So $a^2 = b^2 + c^2$ $\quad \Rightarrow$ The triangle is right-angled.

Sometimes you can solve a problem by splitting a shape into right-angled triangles.

example

Find the vertical height h of the isosceles triangle.
Give your answer to 1 decimal place.

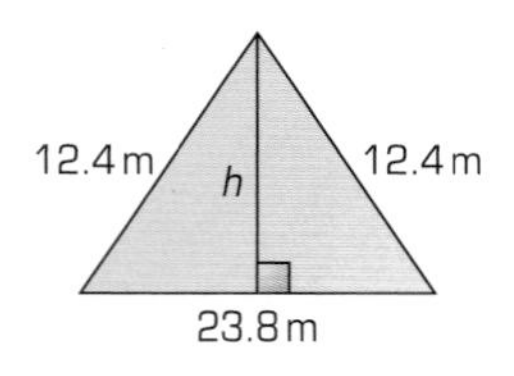

Split into two right-angled triangles:

12.4 m, h, 11.9 m — h, 12.4 m, 11.9 m

$a^2 = b^2 + c^2$
$12.4^2 = h^2 + 11.9^2$
$h^2 = 12.4^2 - 11.9^2 = 12.15$
$h = \sqrt{12.15} = 3.485\,685\,0 = 3.5$ m (to 1 dp)

Pythagoras' theorem can be useful in solving real-life problems.

example

a A man rows 6 km west, then 5 km south. How far is he from his original position? Give your answer to the nearest km.

b A 6 m ladder leans against a wall. Its foot is 1.8 m away from the wall. How far up the wall does the ladder reach?

a Sketch a diagram:

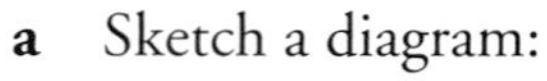

6 km b, 5 km c, a

$a^2 = 6^2 + 5^2$ (Pythagoras' theorem)
$= 36 + 25 = 61$
$a = \sqrt{61} = 7.810\,249\,6$
The man is 8 km from his original position (to the nearest km).

b Sketch a diagram:

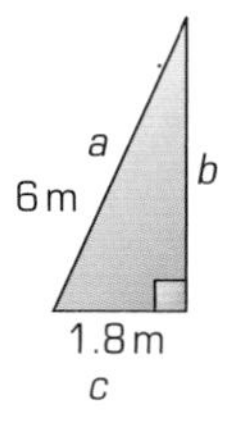

$6^2 = b^2 + 1.8^2$ (Pythagoras' theorem)
$36 = b^2 + 3.24$
$b^2 = 36 - 3.24 = 32.76$

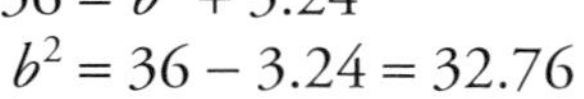

$b = \sqrt{32.76} = 5.723\,635\,2$
The ladder reaches 5.7 m up the wall (to 1 dp).

Exercise S1.5

1 Which of these triangles have right angles?

Triangle	Side 1	Side 2	Side 3
a	4 cm	5 cm	6 cm
b	10 mm	24 mm	26 mm
c	30 cm	40 cm	50 cm
d	2.4 m	1 m	2.6 m

2 Calculate the unknown sides in these triangles.

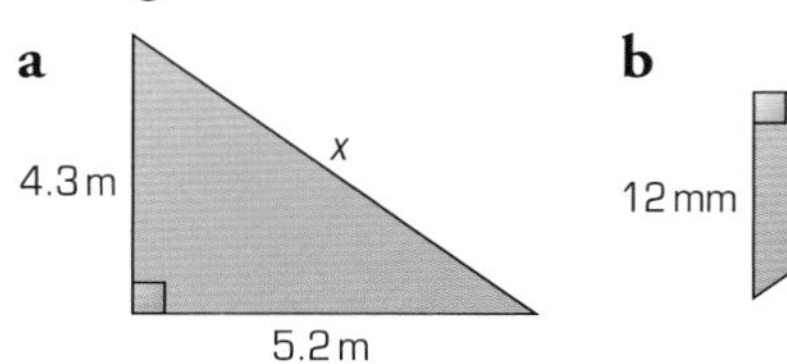

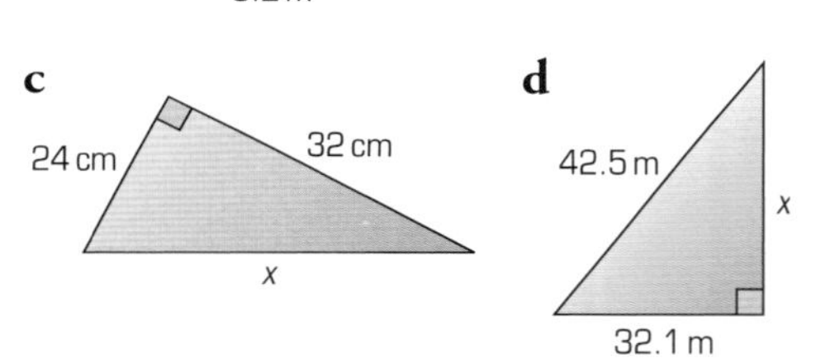

3 Calculate the unknown lengths in these shapes.

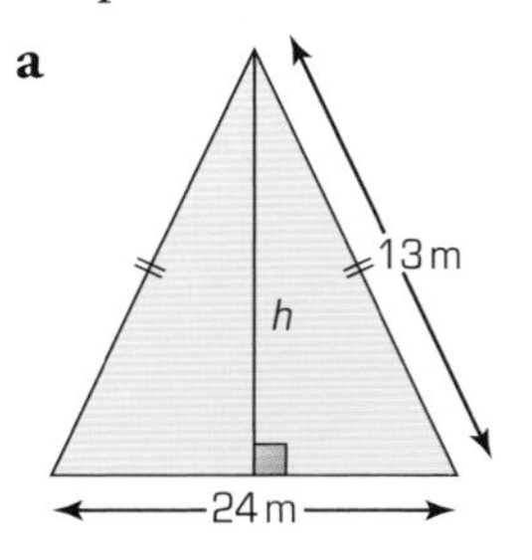

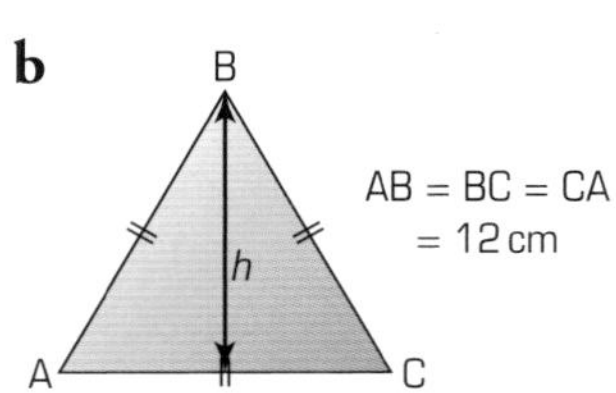

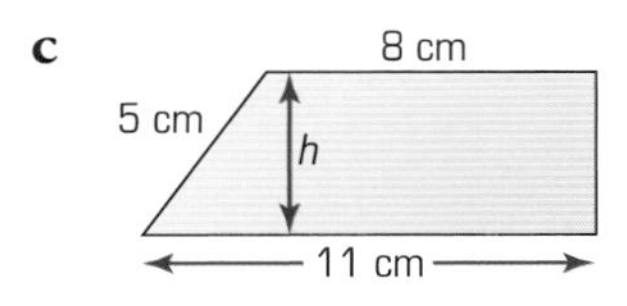

4 Find the area of an equilateral triangle of side 15 mm.

$A = \frac{1}{2}bh$

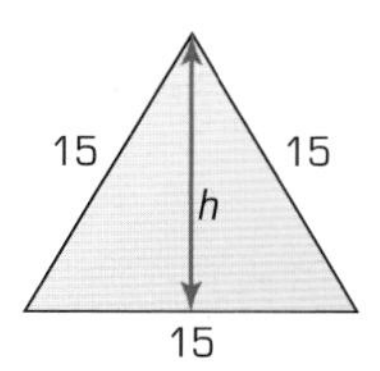

5 A man travels 5.2 km north and then 4.3 km east.
How far is he from his original position?

6 A 7.5 m ladder leans against a wall.
It rests on the wall at a point 6 m above the ground.
How far away from the wall is the foot of the ladder?

7 Ramps help people get into buildings.
A ramp that is 10 m long must not have a height greater than 0.83 m.
Here are plans for a ramp.

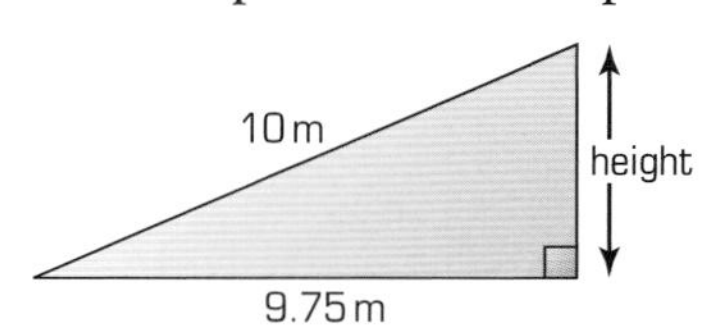

Is this ramp too high?

8 **a** Find the perimeter of triangle PSR.

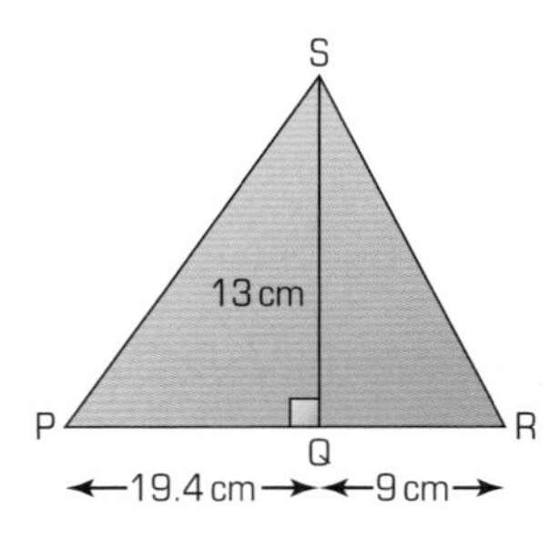

b Is △PSR a right-angled triangle?
Explain your answer.

S1.6 Circle properties

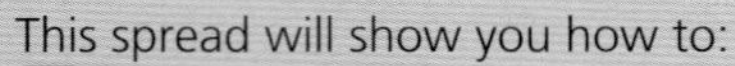

This spread will show you how to:

- Know the definition of a circle and the names of its parts.
- Explain why inscribed regular polygons can be constructed by equal divisions of a circle.
- Know that the tangent at any point on a circle is perpendicular to the radius at that point.
- Explain why the perpendicular from the centre to a chord bisects the chord.

KEYWORDS

Chord, Bisect, Perpendicular, Sector, Inscribe, Tangent, Radius, Centre, Segment

Remember:
You can use a circle to inscribe a regular polygon (for example, an octagon):

- Draw a circle.

- An octagon has 8 sides, so divide the centre into eight equal angles ($360° \div 8 = 45°$).

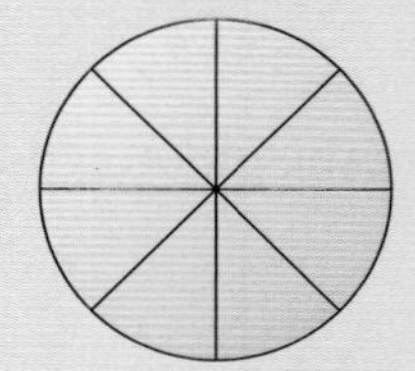

- Join the points where the radii meet the circumference.

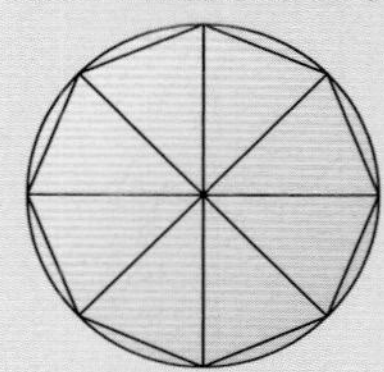

The resulting shape is a regular octagon.

You should remember these names of parts of a circle:

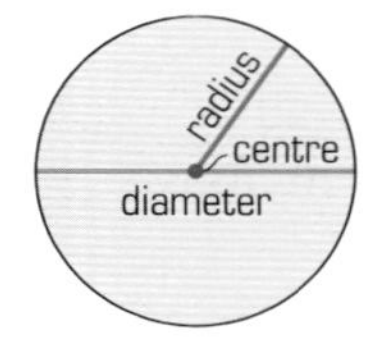

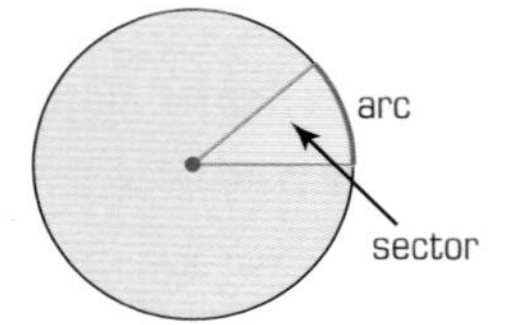

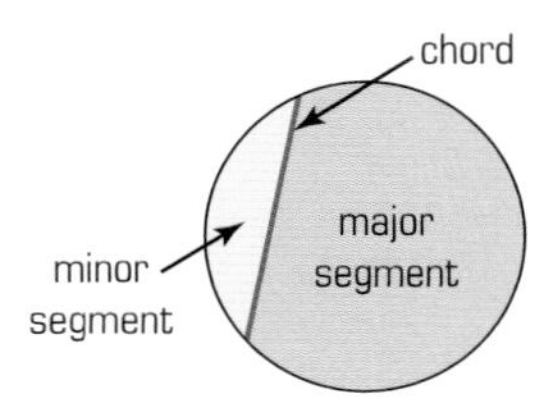

- A **tangent** is a line that touches a circle at a single point.

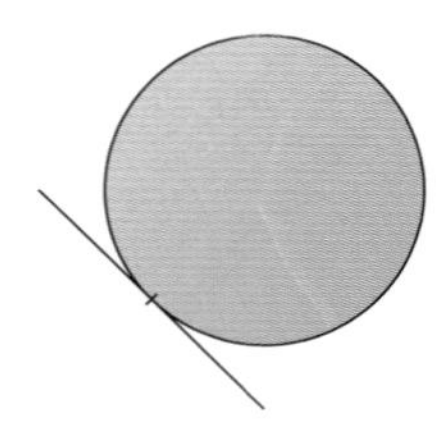

Tangents have an interesting property ...

- The radius OP is perpendicular to the tangent at P.

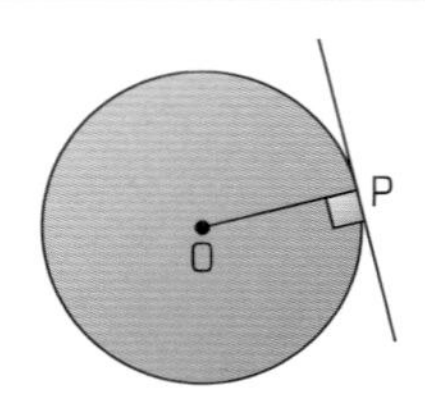

- The perpendicular from the centre to a chord bisects the chord.

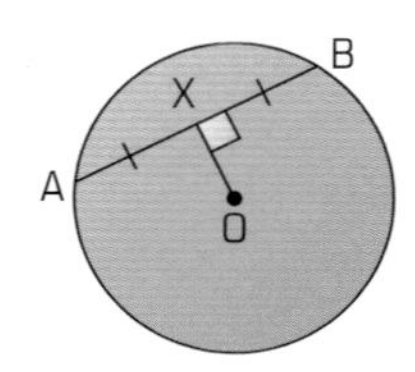

You can use Pythagoras' theorem to prove this result (see question 10 in the exercise).

Exercise S1.6

1 **a** Draw an arc of radius 5 cm.
On this arc draw a sector with angle 45°.

b Draw an arc of radius 8.2 cm.
On this arc draw a sector with angle 123°.

2 Draw a circle with diameter 4.5 cm.
On your circle label:

a a segment **b** a radius
c a chord **d** a tangent.

3 Give the definition of:

a a circle **b** a tangent
c a segment **d** a semicircle
e an arc
(you may use diagrams to help if you wish).

4 Draw a circle of radius 6.5 cm and construct a regular hexagon.
Add the diagonals.

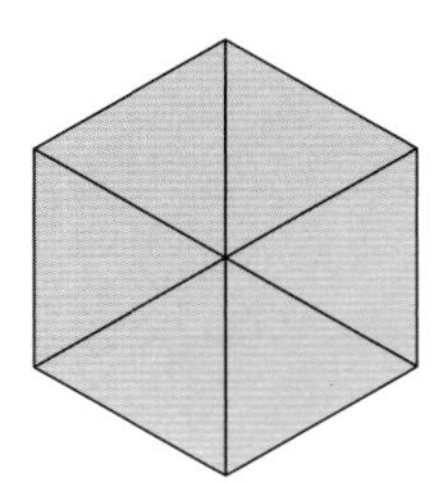

Explain why a regular hexagon is made up of six equilateral triangles.

5 Construct a regular nonagon by dividing the centre of the circle into equal parts.

6 **a** Construct any triangle and the perpendicualr bisectors of its sides.

b Construct the **circumcircle** – centre where the perpendicular bisectors cross (the **circumcentre**), passing through all the vertices.
Repeat for different triangles.

'Circumcircle' is short for circumscribed circle – it is drawn around the outside of a shape

7 In this circle the length of the chord, PQ, is 10 cm. The perpendicular distance AO, from the chord, to the centre of the circle is 3.5 cm. Find the radius of the circle.

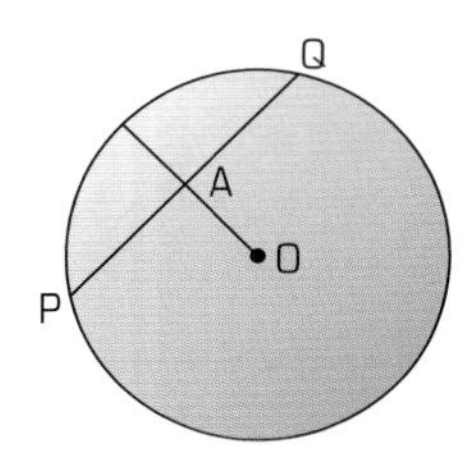

8 The radius of a circle is 5.6 cm and the length of a chord is 8 cm.
Find the perpendicular distance from the chord to the centre of the circle.

9 In the circles below, AB = 8 cm and PQ = 10 cm.
Find the radius of the circles.

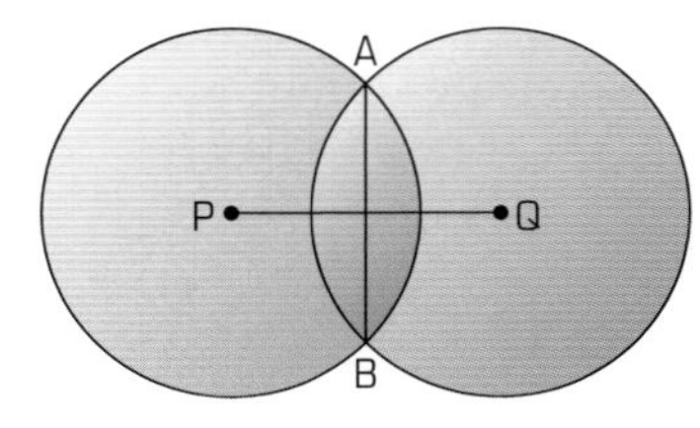

P, Q are the centres.
The circles have equal radii.

10 Show that the perpendicular from the centre of a circle to a chord bisects the chord.
(Show that AX = XB.)

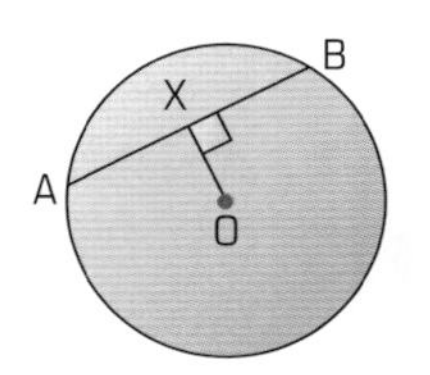

Hint: Draw in the lines OA and OB.
Use Pythagoras' theorem in △AOX and △BOX.

S1.7 Constructing triangles

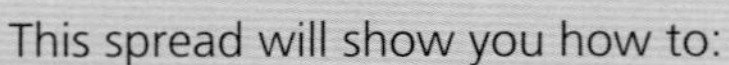

This spread will show you how to:

- Use straight edge and compasses to construct a triangle
- Know that triangles given SSS, SAS, ASA or RHS are unique but that triangles given SSA or AAA are not.

KEYWORDS

Construct
SSS
SAS
ASA
RHS

You should be able to construct a triangle given:

1 All three sides (SSS)

example

Construct △PQR where PQ = 8.7 cm, QR = 5.9 cm and RP = 6.3 cm.

First draw a sketch. Draw the baseline PQ. Draw arcs at 6.3 cm and 5.9 cm. Join the lines to form △PQR.

2 Two sides and the included angle (SAS)

example

Construct △ABC where AB = 8.5 cm, BC = 7.5 cm and ∠B = 53°.

First draw a sketch. Draw the baseline AB. Draw BC at a 53° angle. Join AC to form △ABC.

3 Two angles and the included side (ASA)

example

Construct △LMN where LM = 5.6 cm, ∠L = 95° and ∠M = 42°.

First draw a sketch. Draw the baseline LM. Draw 95° at L. Draw 42° at M to form △LMN.

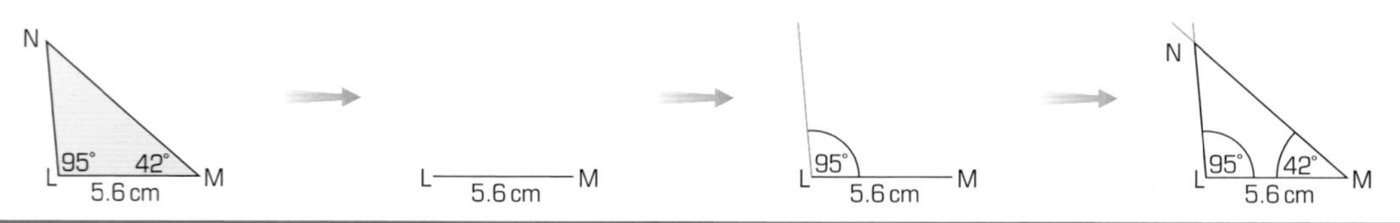

4 A right-angle, hypotenuse and side (RHS)

example

Construct △EFG where EF = 10 cm, ∠G = 90° and EG = 6 cm.

First draw a sketch. Draw GE, with a 90° angle at G. Draw a 10 cm arc from E. Join EF to form △EFG.

Exercise S1.7

1 Construct the following triangles.
Measure the angle or side marked x on your construction.

a

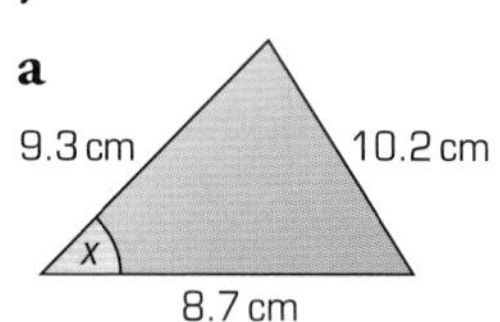

b

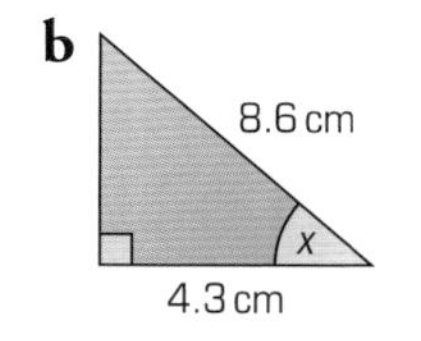

c

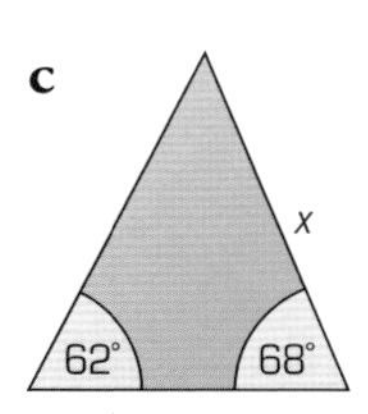

d

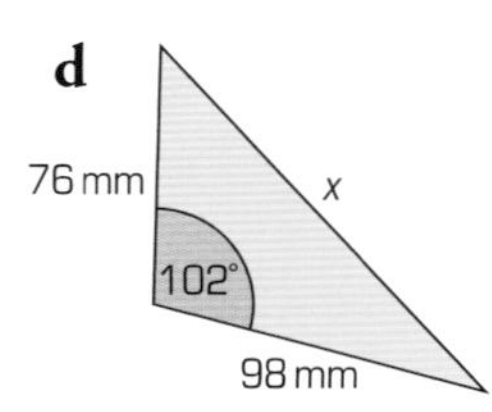

2 Construct the following quadrilaterals.
Measure x and y.

a

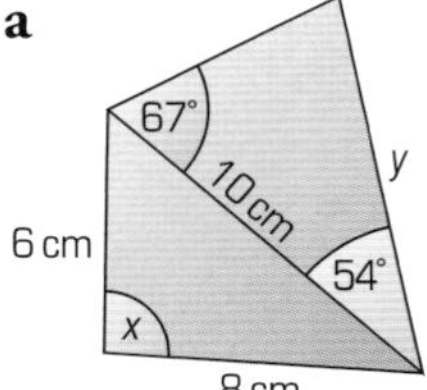

b

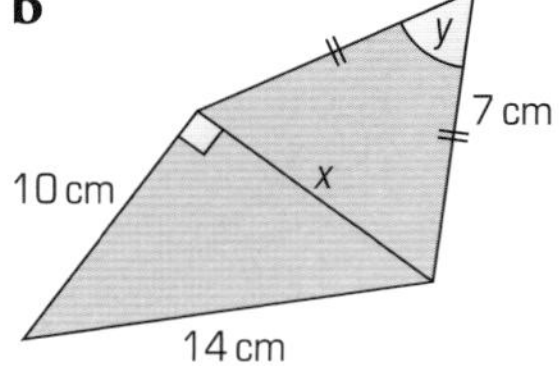

c

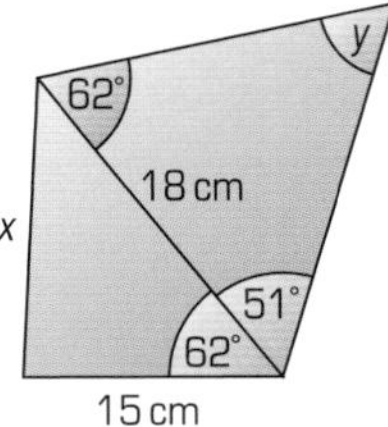

d

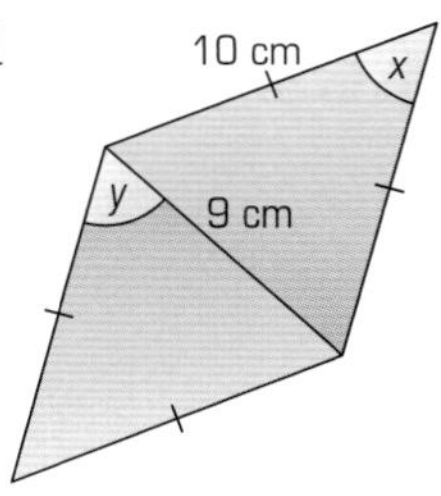

3 ABC is a triangle.
Is it possible to construct a unique triangle with only the following information?
If so, construct it.

a AB = 6 cm, BC = 12 cm, AC = 8 cm
b AB = 5.2 cm, BC = 4.8 cm, AC = 12.4 cm
c $\angle A = 50°$, $\angle B = 70°$, $\angle C = 60°$
d $\angle A = 120°$, $\angle B = 20°$, $\angle C = 40°$
e $\angle A = 50°$, $\angle B = 55°$, AB = 6.5 cm
f $\angle A = 60°$, $\angle B = 48°$, BC = 73 mm

4 Use the measurements provided to construct each triangle.
In each case, say whether there is a way of constructing a different triangle using the measurements given.

a

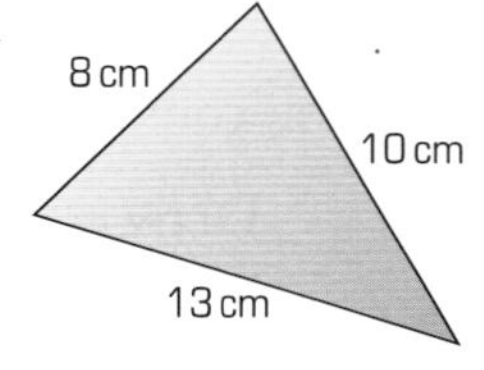

b

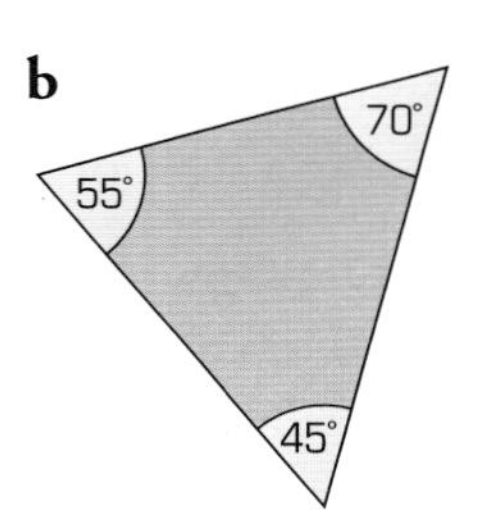

c

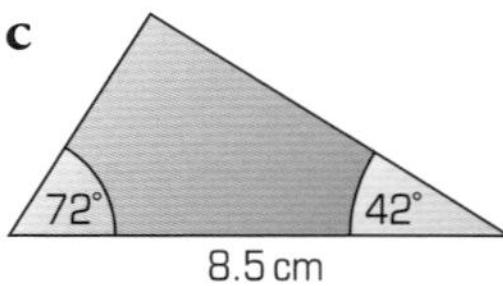

d

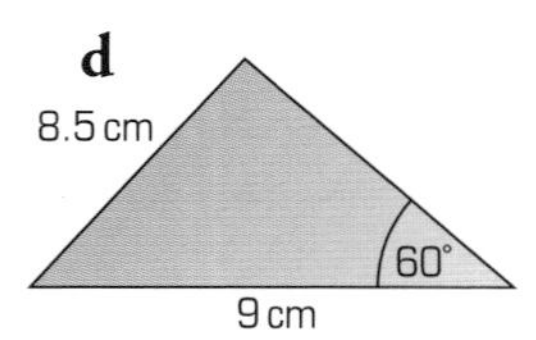

e

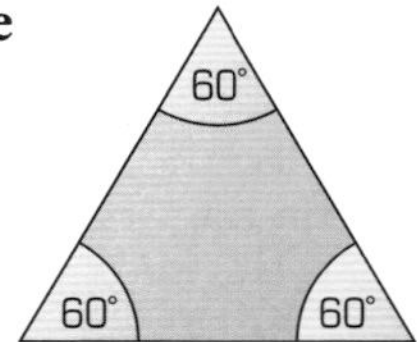

f

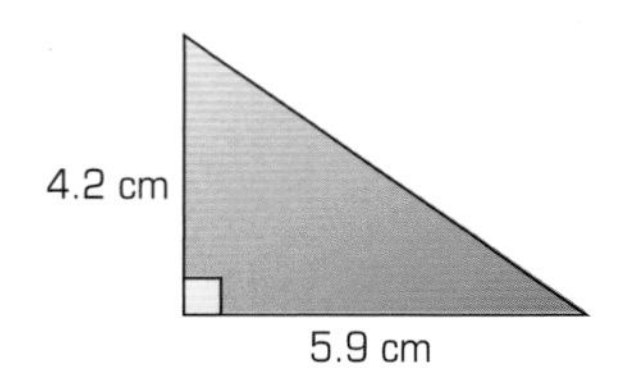

S1.8 Drawing loci

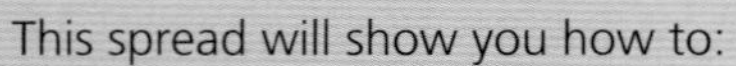

This spread will show you how to:

- Find the locus of a point that moves according to a simple rule.

KEYWORDS

Locus, Equidistant, Radius, Angle bisector, Perpendicular bisector

A **locus** is a set of points that satisfies given conditions.

- ▶ **The locus of a set of points that are a given distance *r* from a fixed point is a circle radius *r*.**

Note: The plural of locus is **loci** (like the plural of radius, radii).

example

Weed killer is sprayed within 5 cm of a plant. Construct the locus of the region that is sprayed.

The boundary of the region forms a circle.
Construct a circle with radius 5 cm.

- ▶ **The locus of a set of points that are equidistant from two points A and B is the perpendicular bisector of the line AB.**

example

A lighthouse is 10 km east of an island. A boat's course is equidistant from the lighthouse (L) and the island (I).
Using a scale of 1 cm to 1 km, construct accurately the locus of the boat's course.

Draw points I and L 10 cm apart (scale 1 cm : 1 km). → Draw equal arcs from I and L. → Join the intersection of the arcs.

The black line is the locus of the boat's path.

- ▶ **The locus of a set of points that are equidistant from two adjacent sides is the angle bisector of the two sides.**

example

A beetle crawls across a paving slab ABCD so that it is always equidistant from the edges AB and AD. Accurately construct its path.

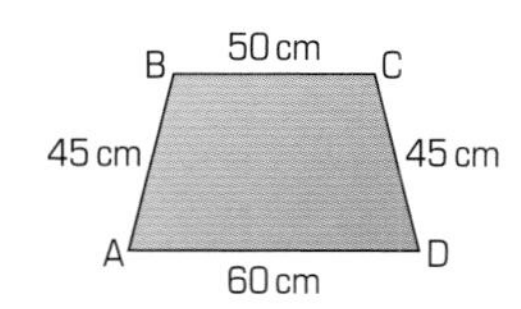

Draw an arc from A.

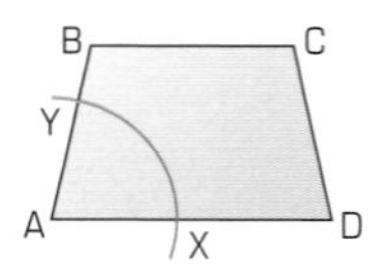

Draw equal arcs from X and Y.

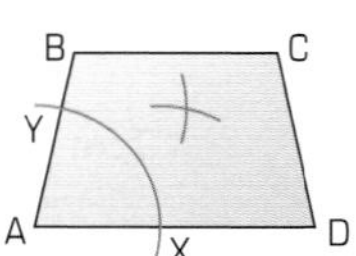

Join the intersection of the arcs to A.

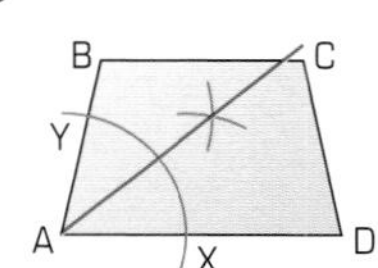

The red line is the locus of the beetle's path.

Exercise S1.8

1 Netta News delivers within a 7 km radius. Zetta News delivers within a 4 km radius. The shops are 5 km apart.
Draw a scale diagram to show the region which both shops serve.

2 Here is a scale drawing of a field LMNO.

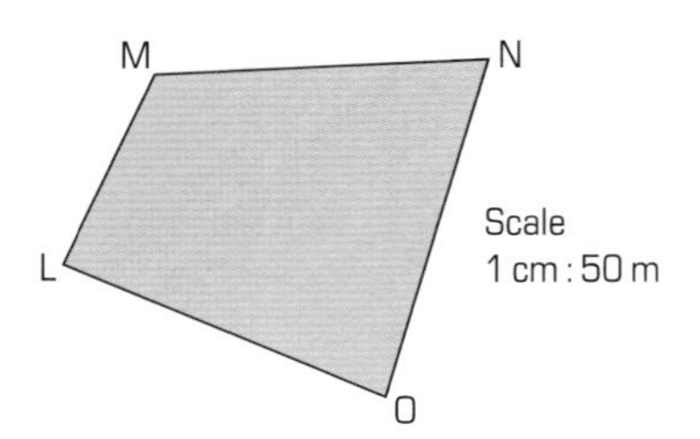

Trace the field and show the path that Ted, the shepherd, walks if he is always equidistant from O and N. Bill the cowman walks so that he is always equidistant from the walls ON and OL. Show his path on the same diagram.

3 A spider is dangling motionless on a single strand of web. Describe the path a fly makes when he is buzzing 6 cm away from the spider. Don't forget the fly moves in 3-D space.

4 A black stick is stuck to a piece of card, the card is held upright and the stick is spun as fast as possible. Which of these shapes would you see?

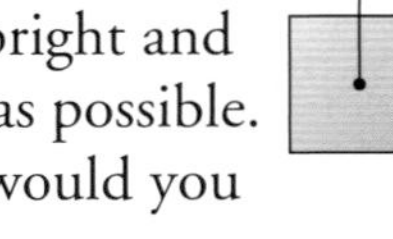

a

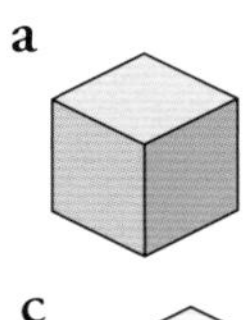

b

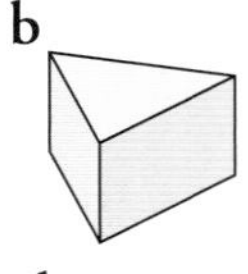

c

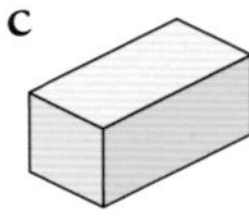

d

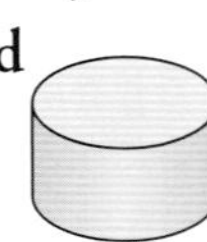

e

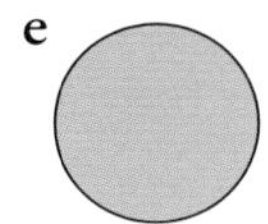

5 Two walls are put round a garden.

a The walls cast a shadow 1 m into the garden during the day. Shade the region where the shadow is cast.

b A superstitious cat will only walk across the garden if it is equidistant from both walls.
Copy the diagram and mark the path that the cat takes.

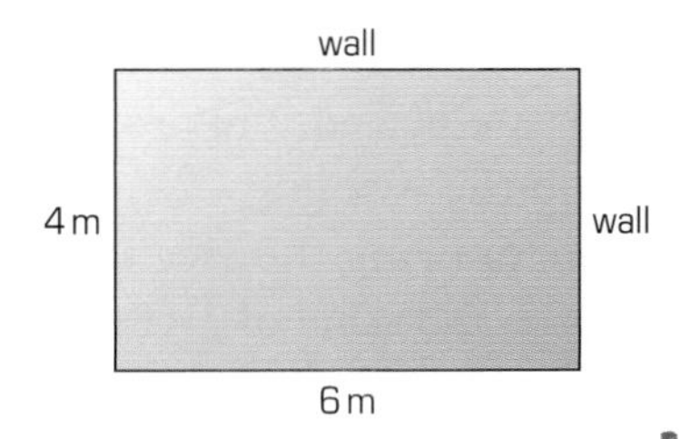

6 A gardener is planning to erect a fence in her garden. Here are her two plans. Both fences will cast a shadow 50 cm wide.
Copy the plans and mark where the shadow will fall.
Which fence will give more shade?

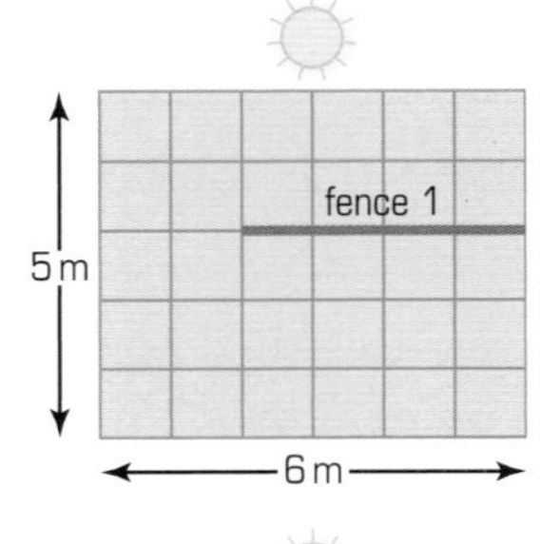

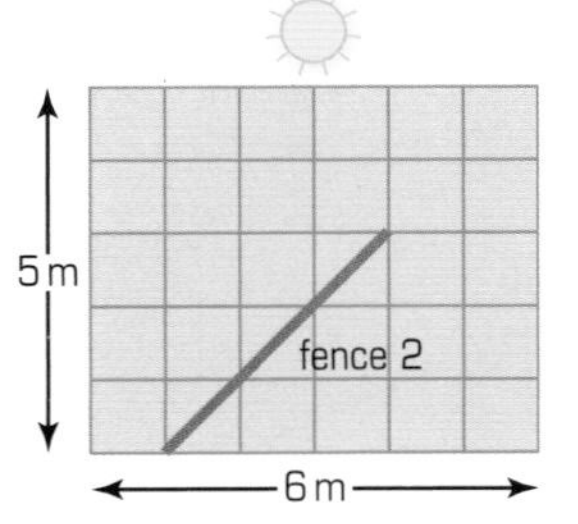

7 Give some examples where the locus is a sphere.

S1.9 More complex loci

This spread will show you how to:

- Extend to more complex rules involving loci and simple constructions.

KEYWORDS

Locus, Region, Construction, Radius, Scale drawing

Sometimes you may need to combine two or more constructions to form a complex locus.

example

The diagram shows a sketch plan of a playground.
A and B are swings and X is a sandpit.
Wood shavings are placed 2 m around the sandpit and 4 m around the swings.
Show on a scale drawing the area that is covered by wood shavings. Use a scale of 1 cm : 2 m.

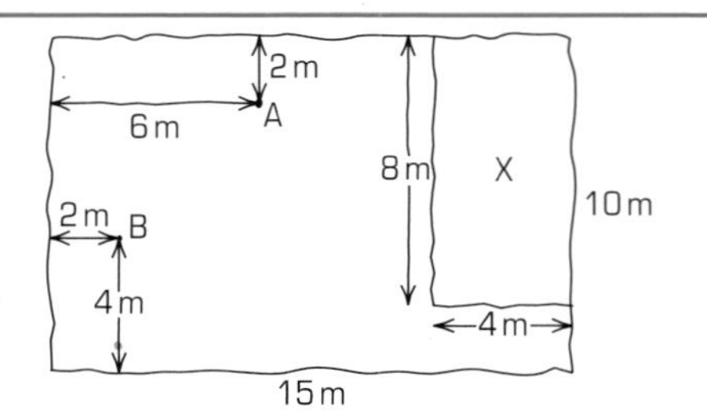

Swings: circles radius 4 m (= 2 cm on drawing).

Sandpit: straight lines 2 m away from edge (= 1 cm on drawing) ...

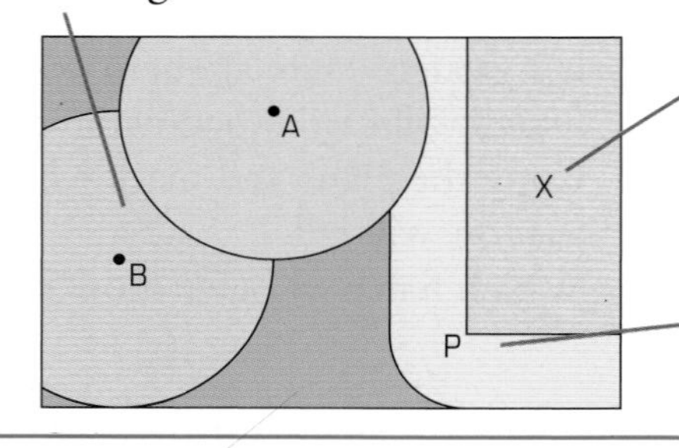

This diagram is not drawn to full size.

... and a quarter circle, radius 1 cm centre P.

The previous example shows that you need to take care with corners when drawing loci.

example

Buster the bulldog is tied to a shed as shown in the sketch plan.
The dog is tied at the point A by a 2 m rope.

Show the region that Buster can reach on a scale drawing.
Use a scale of 1 cm : 1 m.

5 m
3 m
1 m
A

At B, the rope is caught on the edge of the shed, so reduce the radius to 1 cm and draw a quarter-circle, centre B.

5 m
3 m
B A

Note: This diagram is not drawn to full size.

Buster is tied to a fixed point, so draw a semicircle, radius 2 cm centre A.

Exercise S1.9

1 The plan shows the position of three towns labelled A, B and C. The scale of the plan is 1 cm to 10 km.

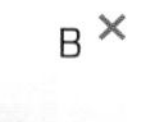

The towns need a new TV aerial. The aerial must be closer to A than B and less than 55 km from C. On a copy of the plan, show the region where the new TV aerial can be erected. Leave in your construction lines.

2 Here is the plan of a new semicircular swimming pool. There is to be a paved area all around the pool. The edge of the paved area will always be 3 m from the pool. Using a scale of 1 cm to 3 m, draw the pool and show the paved area.

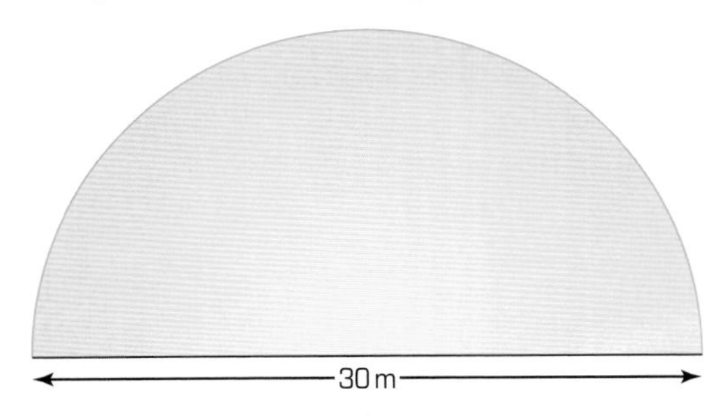

3 Two trees are planted 10 m apart in a field. By drawing a scale diagram (1 cm to 1 m), show two points that are both 8 m away from the trees.
Is it possible to plant another tree that is 4 m from both trees?
Explain your answer.
Now draw the locus of all points that are the same distance from both trees.

4 Draw x- and y-axes from $^{-}5$ to 5.
Plot the points A (3, 2) and B (3, $^{-}4$).
Draw the line that shows the locus of the points which are the same distance from A and B.

5 **a** Copy the diagram.
Use a straight edge and compasses to draw the locus of all points that are the same distance from P as from Q.
Leave in your construction lines.

b Shade in the region where points are closer to P than Q.

6 **a** Copy the diagram and, using a scale of 1 cm to 1 m, show the region where points are less than 6 m from G.

b Construct the locus of points which are equidistant from G and H.

c Mark on your diagram the points that are 6 m from G and 6 m from H.

Summary

You should know how to ...

1 Understand and apply Pythagoras' theorem.

2 Know, from experience of constructing them, that triangles given SSS, SAS, ASA or RHS are unique but triangles given SSA or AAA are not.

3 Generate fuller solutions to problems.

Check out

1 Find the unknown sides (to 2 dp).

a

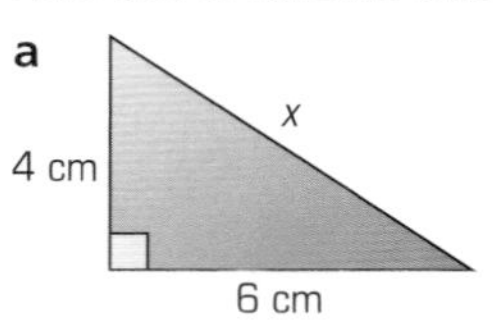

b

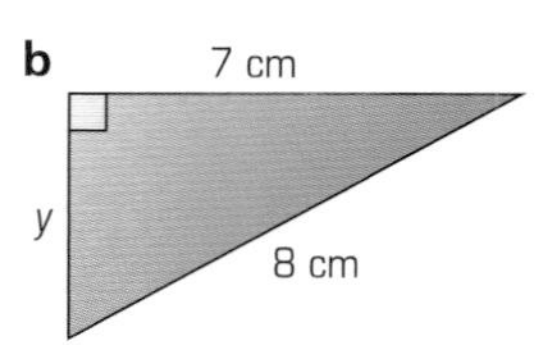

c

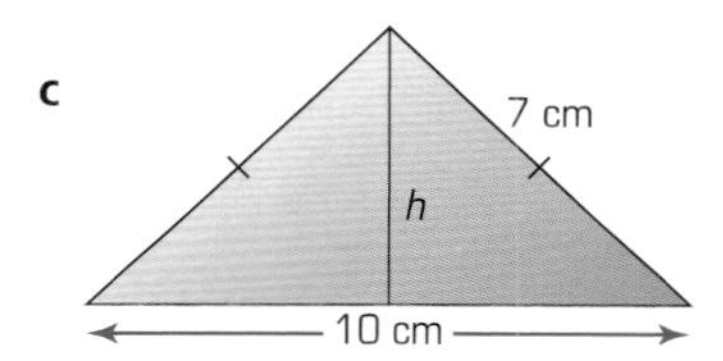

2 Construct the following triangles.
For each triangle, state whether it is unique or not.

a $\triangle ABC$, AB = 4 cm, BC = 6 cm, CA = 10 cm
b $\triangle PQR$, $\angle P = 42°$, $\angle Q = 58°$, $\angle R = 80°$
c $\triangle XYZ$, $\angle X = 90°$, YZ = 10.4 cm, XY = 5.9 cm

3 Find the area of an isosceles triangle with two equal sides of 10 cm and the third side 8 cm.
Give your answer to 2 dp.

1 Handling data

This unit will show you how to:

- Suggest a problem to explore using statistical methods.
- Identify possible sources of bias and plan how to minimise it.
- Identify what extra information may be required to pursue a further line of enquiry.
- Find the median and quartiles for data sets.
- Estimate the mean, median and interquartile range of a large set of grouped data.
- Select, construct and modify suitable graphical representation to progress an enquiry.
- Look for cause and effect and try to explain anomalies.
- Identify key features present in the data.
- Analyse data to find patterns and exceptions.
- Compare two or more distributions and make inferences.
- Examine critically the results of a statistical enquiry, and justify choice of statistical representation in written presentations.
- Generate fuller solutions to problems.
- Recognise limitations on accuracy of data.

When you collect data in an experiment, try to keep the units consistent.

Before you start

You should know how to ...

1 Calculate the mean, median and range of a small set of data.

2 Draw a scatter diagram.

Check in

1 Find the mean and range of these data.
24 30 27 25 25 28

2 The table shows the percentages achieved by 10 students in the calculator and non-calculator mathematics exams.

Calculator %	54	68	47	79	82	57	56	63	65	72
Non-calculator %	46	54	43	76	84	49	50	61	59	64

Draw a scatter diagram to show these results.

D1.1 The handling data cycle

This spread will show you how to:

- Suggest a problem to explore using statistical methods, frame questions and raise conjectures.
- Identify possible sources of bias and plan how to minimise it.

KEYWORDS

Bias, Census, Raw data, Sample, Conjecture, Interpret

Lucy is investigating how long it takes boys and girls to 'spot the difference' between two pictures.

There are four main stages in a statistical enquiry.

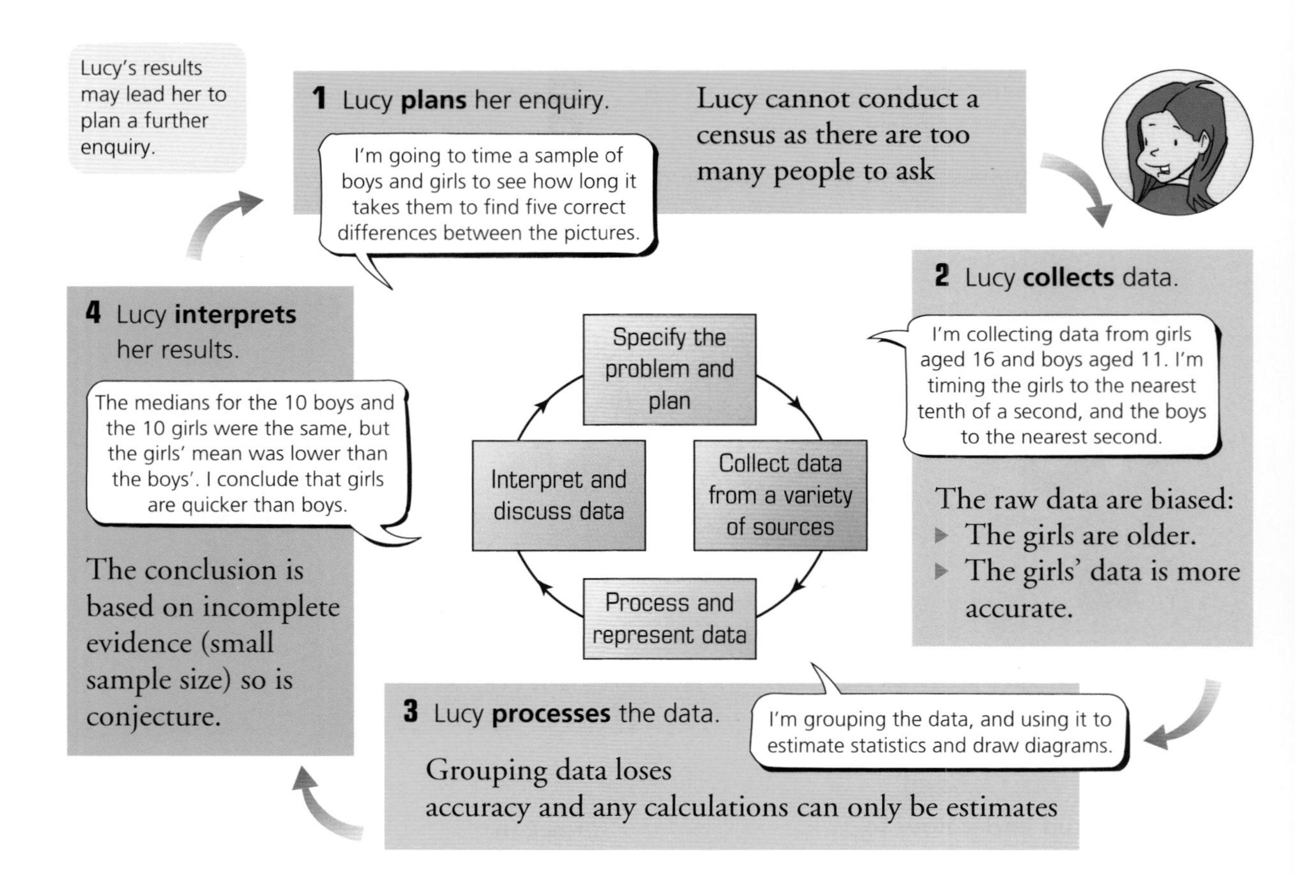

The results of any enquiry are based on the data collected.
If the data is biased, the results may suggest an incorrect conclusion.

- **Bias** is the distortion of a result due to factors that have not been accounted for.
- You need to plan each stage of the handling data cycle to minimise bias.

Exercise D1.1

1 For each of these situations, discuss:

a how the data are biased

b how you could change the situation to minimise the bias.

i Zach is conducting a survey into how students travel to school.
He goes to the school entrance two minutes before school begins and asks the first 15 students that pass.

ii Yasmin is carrying out a survey of school canteen purchases.
She writes down the names of all the girls in her class, puts the names in a hat and chooses 10.
She asks the first 10 boys that go to the canteen at lunchtime.

iii Sandy is carrying out a survey into the typical number of text messages sent per day.
She asks a group of friends, all of whom use the same network on a 'pay as you go' tariff.

2 Dan plays seven games of cricket.
In six games he scores between 50 and 60 runs.
In one game he scores no runs.

a Discuss whether the mean, the median or the mode would best describe Dan's average number of runs.

b The range of the number of runs is 54.
Explain how this gives a biased picture of Dan's cricket scores.

3 The graph shows the percentages of a group of boys and girls that own pets.

Critically analyse each of these statements:

a 'More boys than girls own dogs.'

b 'The same number of boys and girls own rabbits.'

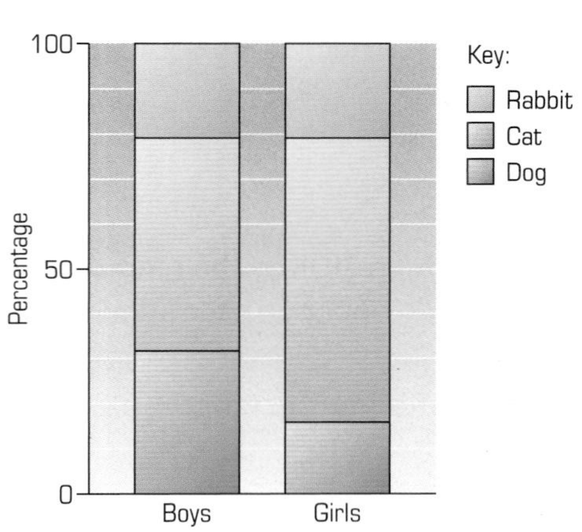

D1.2 Calculating statistics

This spread will show you how to:

- Find the median and quartiles for data sets.

KEYWORDS

Mean
Median
Interquartile range
Lower quartile
Upper quartile
Range

Lucy timed how long it took each of 15 friends to spot the difference between two pictures.
These are her results, in order (times given in seconds):

1.4 3.9 4.2 4.3 4.4 4.7 4.7
4.8 4.9 5.0 5.1 5.2 5.9 7.3 15.2

Lucy works out the mean and range:

Remember: To work out the mean you add up the values and divide by the number of values.

Sum of values = 81
81 ÷ 15 = 5.4 The mean time is 5.4 seconds.

Largest value – smallest value = 15.2 – 1.4 = 13.8 The range is 13.8 seconds.

- **To describe a data set you need a measure of average and a measure of spread.**

The mean and range of Lucy's data are quite large – they are distorted by the extreme value recorded (15.2).

To describe data sets that have a small number of extreme values, you use the **median** and the **quartiles**.

The median time is 4.8 seconds.

Remember: To work out the median you put all the values in order and find the value in the middle.

1.4 3.9 4.2 4.3 4.4 4.7 4.7 4.8 4.9 5.0 5.1 5.2 5.9 7.3 15.2

- **The lower quartile is the value that is $\frac{1}{4}$ of the way along a set of data.**
  The lower quartile is 4.3.

- **The upper quartile is the value that is $\frac{3}{4}$ of the way along a set of data.**
  The upper quartile is 5.2.

- **The interquartile range (IQR) is the range of the middle 50% of the data.**
  Interquartile range = upper quartile – lower quartile

For Lucy's data, the interquartile range = 5.2 – 4.3 = 0.9

This means that the range of the central 50% of times is 0.9 seconds.

Exercise D1.2

1 For each set of data, find:

i the median
ii the lower quartile
iii the upper quartile
iv the interquartile range.

> **Hint**: In parts **c** and **d** you will need to calculate halfway values between two values.

a 6 8 6 7 4 9 8 7 9 4 5 7 5 6 7
b 59 47 38 67 37 46 26 38 56 24 21
c 61 56 59 49 62 48 57 53 52 63 56 61 60
d 4 6 7 9 11 13 15 16 17 19

2 Dan plays seven games of cricket.
The numbers of runs scored in each game are:

54 52 50 53 0 55 51

a Find the mean and the median number of runs scored.
b Find the range and the interquartile range of the number of runs scored.
c Which average and which measure of spread should you use to best describe Dan's score?
Explain your answer.

In his next 12 matches Dan makes the following numbers of runs:

26 32 10 8 46 22 16 39 42 3 13 36

d Find the overall mean and median number of runs scored.
e Find the overall range and interquartile range of the number of runs scored.
f Comment on which average and measure of spread you would now use to best describe Dan's overall score. Explain your answer.

3 Sandy collects data on the number of phone calls made by a sample of girls and boys.
She summarises her results in a stem-and-leaf diagram.

Find, for both the girls and boys:

a the median
b the lower quartile
c the upper quartile
d the interquartile range.

Girls		Boys
8 5 0	8	2 5 9
7 6 6 4 2 1	7	1 3 3 9 9
9 8 3 3 2	6	2 4 4 4 5 9
6 5 4 0	5	1 2 2 3 8
7 3 2	4	0 5 9
2 0	3	1 4 8

Key:
2 | 4 | 0
42 – girls
40 – boys

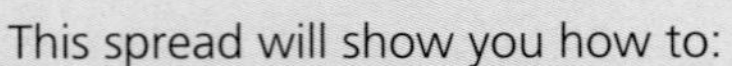Representing data

This spread will show you how to:

- Estimate the mean, median and interquartile range of a large set of grouped data.

KEYWORDS
Cumulative frequency
Interquartile range
Estimate of the mean/median
Lower quartile
Upper quartile

Lucy collects more data for her 'spot the difference' survey. She times 140 people and groups the data.

Time (seconds)	0–2	2–4	4–6	6–8	8–10	10–12	12–14	14–16
Frequency	8	16	21	35	25	19	11	5

Lucy calculates the running total, or **cumulative frequency**.

Cumulative frequency	8	24	45	80	105	124	135	140

To work out the cumulative frequency, just add up the frequencies:
$8 + 16 = 24$
$24 + 21 = 45$ ...

To estimate the median and interquartile range Lucy draws a **cumulative frequency graph**.

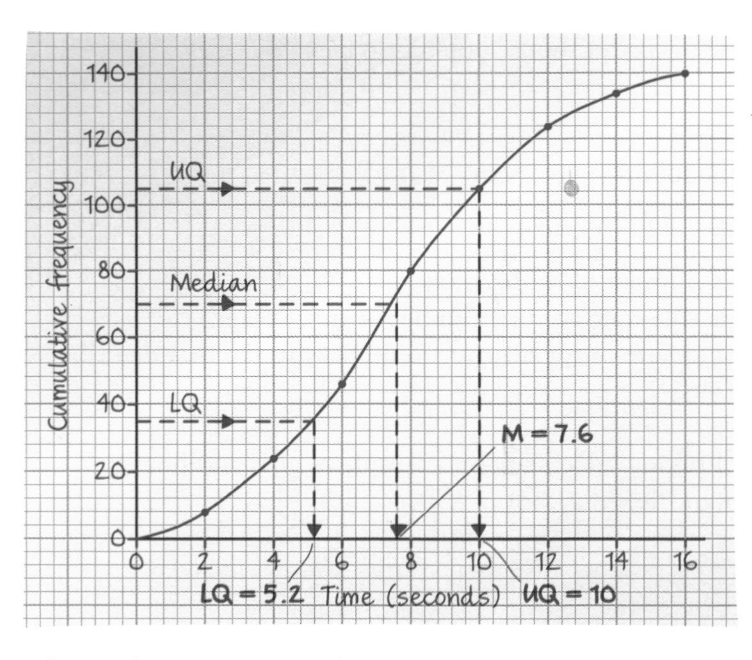

To draw a cumulative frequency graph, plot cumulative frequency against **upper class boundary** (the highest value of each interval): (2, 8) (4, 24) (6, 45) ...

- To find the ...
  - **median**, go halfway up the vertical axis. Read off the corresponding value on the horizontal axis.
  - **lower quartile (LQ)**, go a quarter of the way up the vertical axis.
  - **upper quartile (UQ)**, go three-quarters of the way up the vertical axis.

The median is 7.6 seconds.
The interquartile range is $10 - 5.2 = 4.8$ seconds.

- To find the mean when data are grouped, you use the middle value of each class to represent the class.

- To find the range of grouped data, subtract the lowest value of the first group from the highest value of the last group.

Exercise D1.3

1 40 students took part in a survey of how much time they spent using a computer on one evening.

Time, t, spent using computer (minutes)	$0 \leqslant t \leqslant 30$	$30 < t \leqslant 60$	$60 < t \leqslant 90$	$90 < t \leqslant 120$
Frequency	4	11	19	6

a Estimate the range and mean time spent using the computer on that evening.

The data were used to draw a cumulative frequency graph.

b Use the diagram to estimate:

i the median

ii the interquartile range

of times spent using the computer on that evening.

c Which average, the mean or the median, gives a better representative value for these data?
Give a reason for your answer.

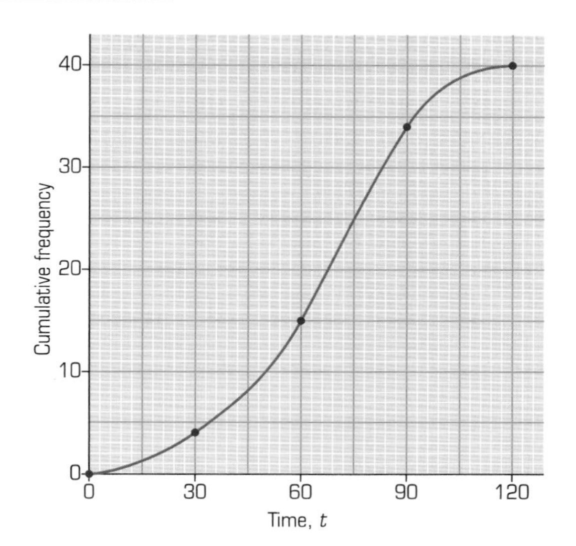

2 100 students took part in a survey of time spent watching television one evening.

Time, t, watching TV (minutes)	$0 \leqslant t \leqslant 60$	$60 < t \leqslant 120$	$120 < t \leqslant 180$	$180 < t \leqslant 240$	$240 < t \leqslant 300$
Frequency	13	17	28	35	7

a Estimate the range and mean time spent watching television that evening.
The data were used to draw a cumulative frequency graph.

b Use the diagram to estimate:

i the median

ii the interquartile range

of times spent watching television that evening.

c Which average, the mean or the median, gives a better representative value?
Give a reason for your answer.

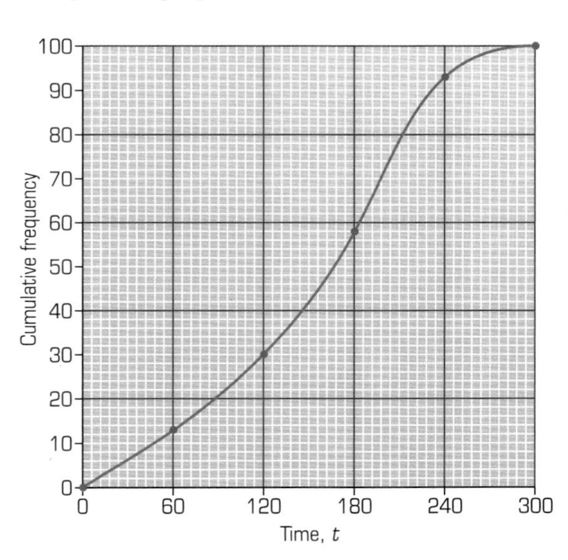

D1.4 Comparing distributions

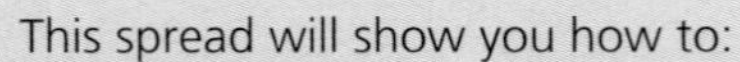

This spread will show you how to:

- Compare two distributions and make inferences.
- Identify key features present in the data.

KEYWORDS
Cumulative frequency
Median

Lucy has drawn a cumulative frequency graph for her 'spot the difference' survey.
She uses her graph to estimate:

- how many people took more than nine seconds to spot the difference
- the percentage of people that took less than three seconds.

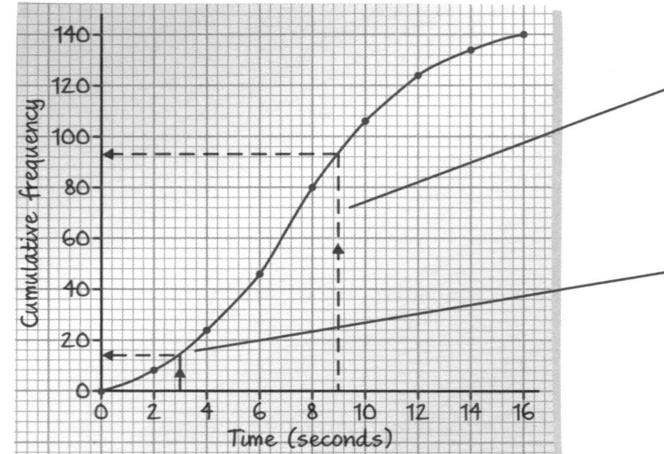

Lucy draws a line at 9 seconds up to the curve and then across. The reading is 94.
140 – 94 = 46
46 people took longer than 9 seconds.

Lucy draws a line at 3 seconds up to the curve and then across. The reading is 14.
$\frac{14}{140} \times 100 = 10\%$
10% of people took shorter than 3 seconds.

Luke is carrying out his own 'spot the difference' survey.
He plots his results on the same graph as Lucy's so they can compare distributions.

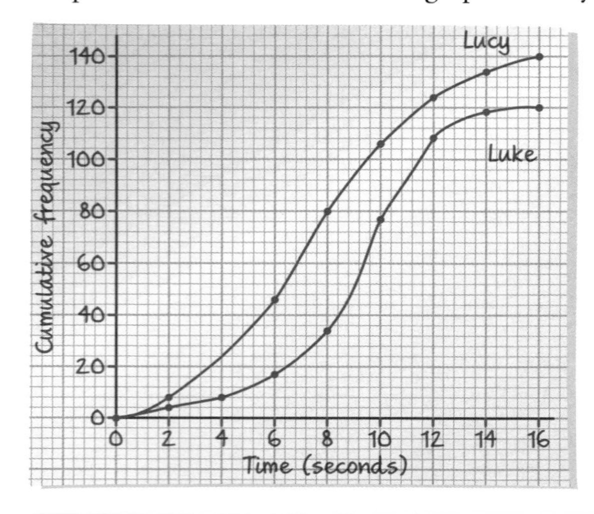

There are 120 people in Luke's survey, 20 fewer than in Lucy's.

Luke's median is 9.4 seconds.
Lucy's median is 7.6 seconds.
On average, the people Luke surveyed took longer.

Luke's IQR is 3.4 seconds.
Lucy's IQR is 4.8 seconds.
Luke's times are more consistent than Lucy's.

- **When you compare distributions, comment on an average and a measure of spread.**
- **Write your comments in terms of the variables measured.**

In Lucy's and Luke's surveys, the variable is time.

Exercise D1.4

1 Use the cumulative frequency diagram in Exercise D1.3 question 1 to estimate how many students used the computer for:

a 45 minutes or less

b 75 minutes or more.

2 Use the cumulative frequency diagram in Exercise D1.3 question 2 to estimate the percentage of students that watched television for:

a 30 minutes or less

b $2\frac{1}{2}$ hours or more.

3 A leaflet is written in two different languages. The cumulative frequency graph shows the number of words per sentence.

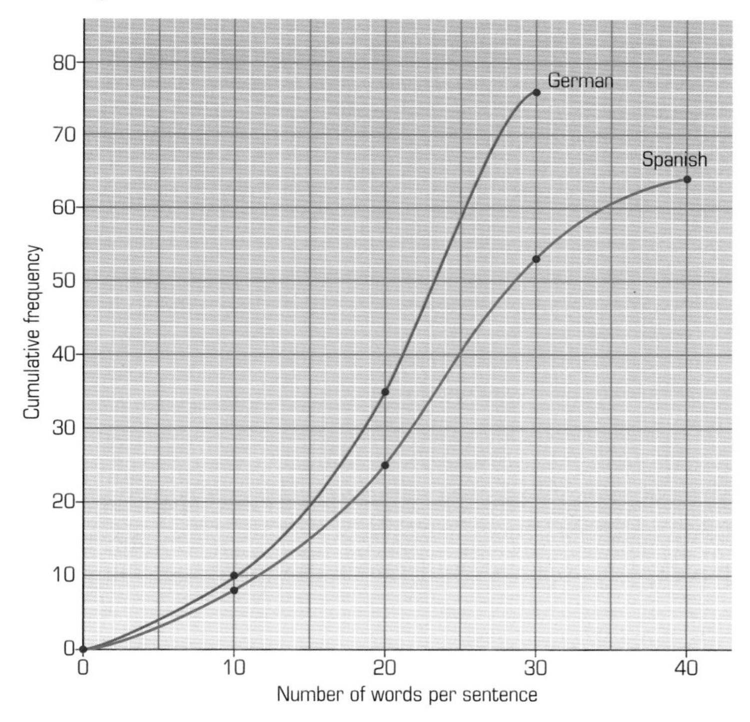

a Copy and complete the table, using the graph to estimate the values.

	Number of sentences	Median	Upper quartile	Lower quartile	IQR
German					
Spanish					

b Comment on differences and similarities between the two languages for this leaflet.

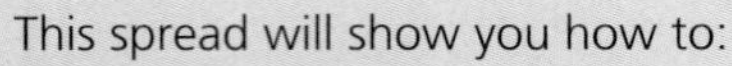

Interpretation of data

This spread will show you how to:

- Select, construct and modify scatter graphs and lines of best fit by eye.
- Identify key features present in the data.
- Analyse data to find patterns and exceptions.
- Look for cause and effect and try to explain anomalies.

KEYWORDS

Scatter diagram
Correlation
Line of best fit
Predict
Anomaly

Lucy gives 15 friends two different sets of pictures and times how long it takes them to spot the differences for each set.

First set	1.4	3.9	4.2	4.3	4.4	4.7	4.7	4.8	4.9	5.0	5.1	5.2	5.9	7.3	15.2
Second set	1.9	3.0	3.9	4.1	4.0	4.6	4.4	4.5	8.4	4.9	5.3	4.9	6.0	7.5	6.4

Lucy draws a scatter diagram to summarise the data.
The graph shows positive correlation.

If there is correlation you can draw a line of best fit.

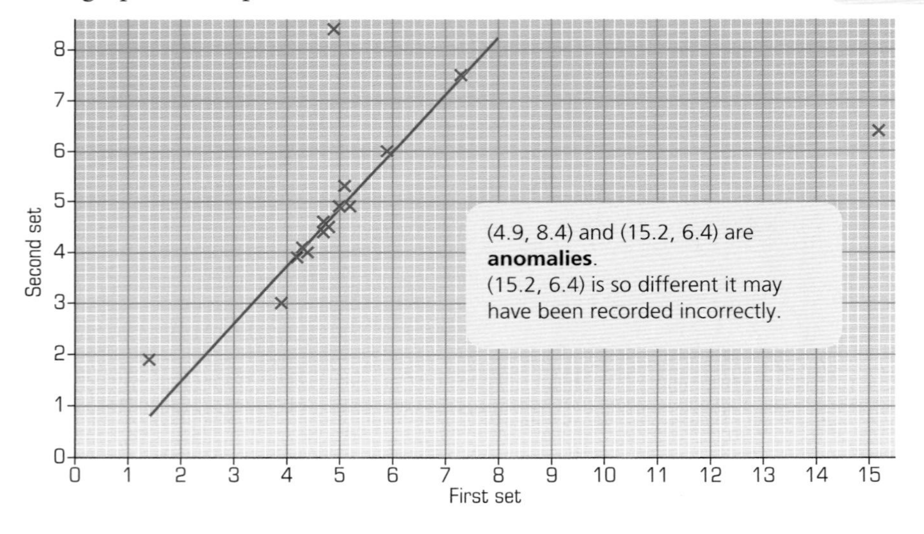

To draw a line of best fit:

- find the mean of each data set.
  First set mean = 5.4
  Second set mean = 4.92
- Draw a straight line that passes through (5.4, 4.92), and has roughly the same number of data values on either side.

Lucy times two more friends on the first set of pictures.
Martha takes 4.6 seconds and Nina takes 8.3 seconds.

Lucy can use her graph to **predict** the time Martha might take on the second set of pictures.
Lucy estimates that Martha would take 4.4 seconds.

Lucy cannot predict Nina's time on the second set, because 8.3 seconds is outside of the plotted points (the value at 15.2 is an anomaly).

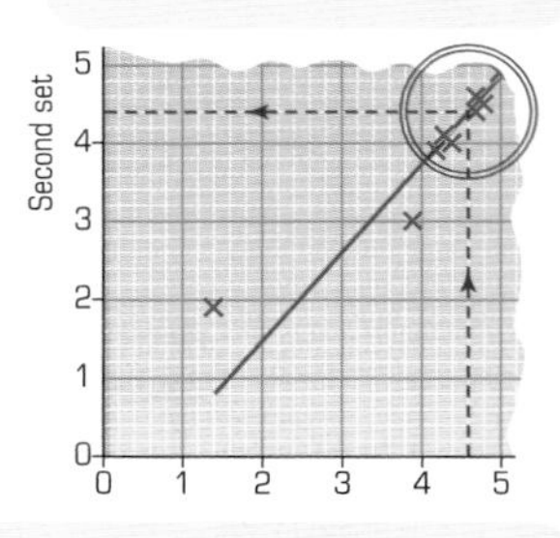

- **You can use a line of best fit to predict values, as long as they lie within the range of data.**

In general, the second set times are quicker. The second set of pictures may have been easier, or Lucy's friends may have improved with practice.

Exercise D1.5

1 Angie drew a scatter graph of average weekly expenditure on durable household goods and average weekly income for different regions of Great Britain.

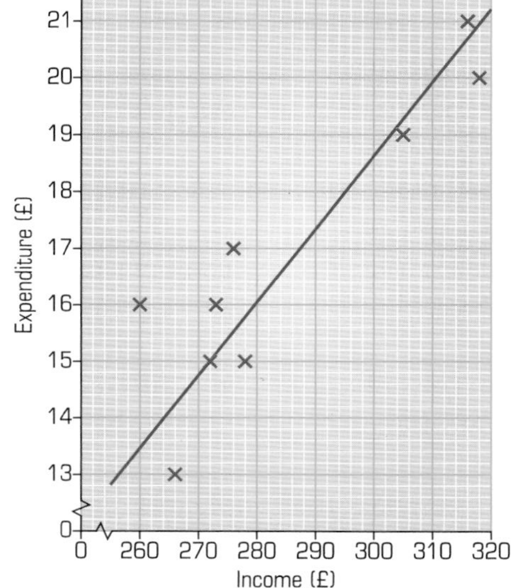

a Describe the relationship between average income and expenditure.

b Use the graph to estimate the average income of a region with average expenditure of £18.

c Explain why you would not use this graph to estimate the average income of a region with average expenditure of £28.

d One region spent less than expected on durable household goods. Identify the point and explain why you chose this point.

2 The scatter graph shows the length of pregnancy and birth weight of all the babies born one day at a hospital.

a Estimate the weight of a baby born after a pregnancy of

i 272 days

ii 284 days.

b Give an interpretation of the line of best fit.

c Explain why it would not be sensible to extend the line of best fit.

3 Adrian noted the weight and bleep test scores of a group of boys. (A bleep test is a multi-stage fitness test.)

Weight (kg)	36	38	42	33	26
Bleep test score	7.6	8.0	7.2	8.2	8.4

Weight (kg)	44	50	45	59	34
Bleep test score	7.4	7.0	7.1	6.4	8.5

a Draw a scatter diagram.

b Describe the correlation shown by the graph.

c Calculate the mean point and use this to draw a line of best fit.

d Give an interpretation of the scatter graph and the line of best fit.

Misleading statistics

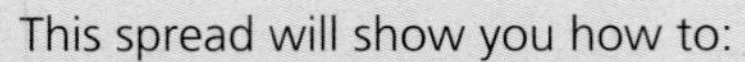

This spread will show you how to:

- Examine critically the results of a statistical enquiry.
- Justify choice of statistical representation.
- Recognise limitations of any assumptions and their effect on conclusions drawn.
- Identify what extra information may be required to pursue a further line of enquiry.

KEYWORDS
Raw data
Misleading

Lucy	**Luke**
Median 7.6 s	Median 9.4 s
IQR 4.8 s	IQR 3.4 s

Here are the results of the 'spot the difference' survey carried out by Lucy and Luke.

The results suggest that the people in Lucy's survey were quicker than those in Luke's survey.

But you do not know:

- whether Lucy and Luke used the same pictures, or pictures of the same level of difficulty
- if anyone was allowed a practice go on a different set of pictures
- the ages of the people, so the surveys may not be comparing similar abilities
- how accurate Lucy and Luke were at timing
- if all the data was recorded correctly.

> When you examine the results of a statistical enquiry, you need to know how the raw data were collected.

Graphs can summarise raw data.
They should be chosen to highlight key points and to make comparisons easier.

> You can give a misleading impression by:
> - highlighting a selected part of a graph
> - only reporting part of the results of an enquiry.

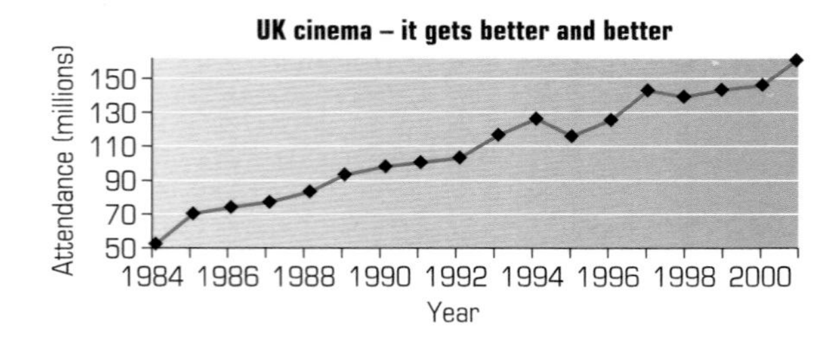

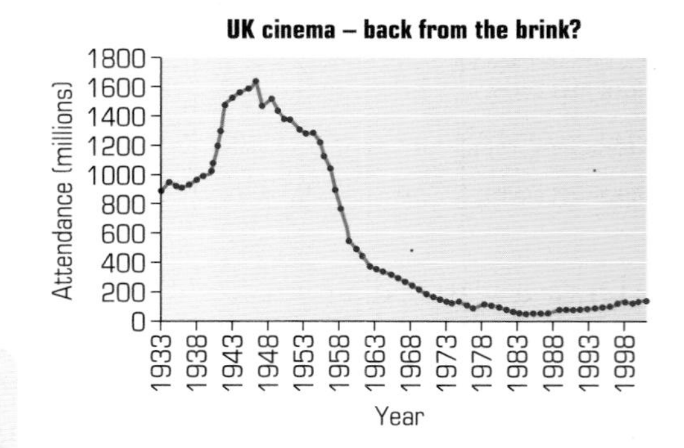

> When you compare distributions, you should draw the same type of graph using the same scale, and if possible the same axes.

Exercise D1.6

1 A local council issued this report.

50% of households in Queenstown are recycling paper, glass and plastic

a Can you tell from the report:
  i how many households there are
  ii the percentage of households that are recycling paper, glass and plastic?

b If not, what extra information would you need?

Rosalind and Simon decided to do a survey to find the actual percentage of households that recycle.

c Rosalind stood in a supermarket car park next to the paper recycling bin and counted how many of the shoppers brought paper for recycling.
Give three different reasons why this may not give very good data.

d Simon asked 10 of his friends how many of their parents recycled paper, glass and plastic.
Give two different reasons why this method may not give very reliable data.

2 This headline was printed in a newspaper. Explain why this is not shocking news.

33% of schoolchildren achieve less than average marks in an exam

3 These two bar charts represent exactly the same information.

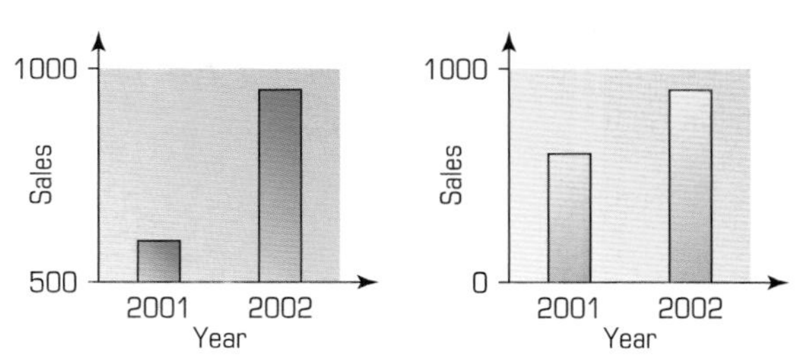

Comment critically on the bar charts.

4 Warren FC has a seating capacity for 10 000 fans at its ground.
David drew this graph to show the average home attendance for five seasons.

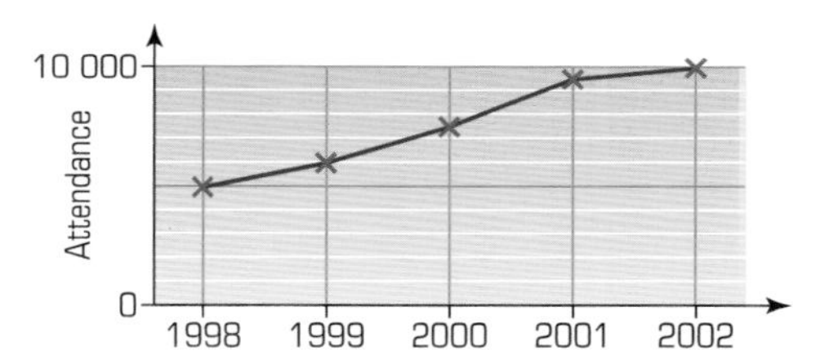

He predicts that in 2003 and 2004 the average attendance will increase.
Explain why David must be wrong.

5 The pie charts show the proportions who voted for different political parties in the local elections of two towns, A and B.

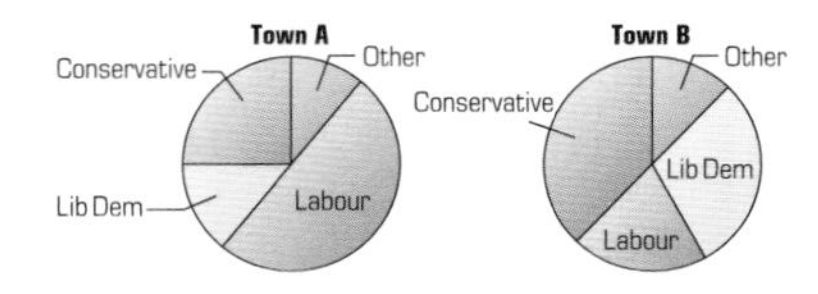

a Robbie says: 'The pie charts show that more people voted Conservative in Town B than in Town A.'
Why may Robbie be wrong?

b Nina says: 'Half the people eligible to vote in Town A voted Labour!'
Identify an assumption that Nina is making, and suggest why it may not be true.

D1 Summary

You should know how to ...

1 Identify possible sources of bias and plan how to minimise it.

2 Examine critically the results of a statistical enquiry, and justify choice of statistical representation in written presentations.

3 Recognise limitations on accuracy of data and measurements.

4 Generate fuller solution to problems.

Check out

1 Thao was timing how long it took for students to recite the alphabet and to count backwards from 20 to 0.

Plan Thao's survey so that any bias is minimised.

2 These are the results from Thao's pilot survey.

Alphabet, seconds	5.7	4.9	6.5	6.3	7.0
Countdown, seconds	6.9	5.4	6.5	6.9	7.5

Alphabet, seconds	5.6	5.2	6.7	12.1	6.8
Countdown, seconds	7.0	6.8	7.9	7.9	8.1

What measures and diagrams should Thao use to represent these data?
Justify your answers.

3 Describe why the results of Thao's survey may not be accurate.

4 Explain how Thao could generate fuller solutions in her investigation.

2 Perimeter, area and volume

This unit will show you how to:

- Use units of measurement to calculate, estimate, measure and solve problems.
- Convert between area measures and between volume measures.
- Know and use the formulae for the circumference and area of a circle.
- Know and use the formulae for length of arcs and area of sectors of circles.
- Find points that divide a line in a given ratio, using the properties of similar triangles.
- Given the coordinates of points A and B, calculate the length of AB.
- Understand and use measures of speed (and other compound measures) to solve problems.
- Solve problems involving rates of change.
- Calculate lengths, areas and volumes in right prisms, including cylinders.
- Generate fuller solutions to problems.
- Recognise limitations on accuracy of data and measurements.

Wheels have many uses.

Before you start

You should know how to ...

1. Find the area of a triangle, a parallelogram and a trapezium.
2. Convert between metric units of length.
3. Round measurements to a number of decimal places.
4. Find the coordinates of the midpoint of a line segment.
5. Use Pythagoras' theorem in simple cases.

Check in

1. Find the area of:

   a 

   b 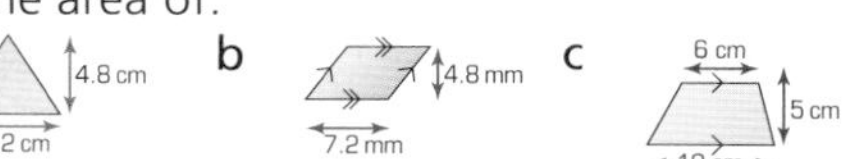

   c 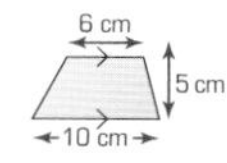

2. Convert to the units given in brackets

   a 4 cm (mm) b 5.2 m (cm) c 2.4 km (m)
3. Round to the number of decimal places given:

   a 4.285 (2 dp) b 1.999 (1 dp)

   c 0.0432 (2 dp)
4. Find the midpoint of the line joining A(2, 4) to B(3, 8).
5. Use Pythagoras' theorem to find a and b (to 1 dp):

   a 

   b  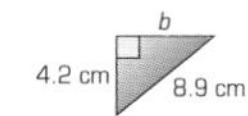

S2.1 Metric measures

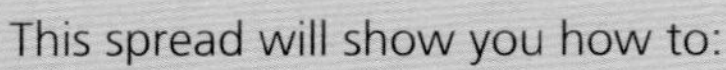
This spread will show you how to:

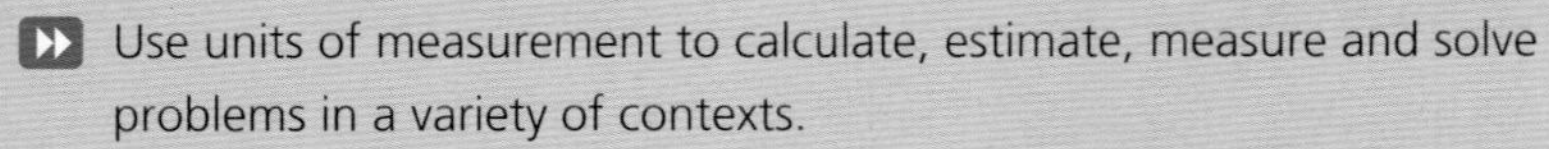
- Use units of measurement to calculate, estimate, measure and solve problems in a variety of contexts.

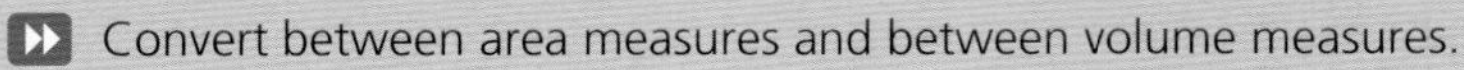
- Convert between area measures and between volume measures.
- Recognise that measurements given to the nearest whole unit may be inaccurate by up to one half of the unit in either direction.

KEYWORDS

Area, Volume, Capacity, Hectare, Conversion

Remember:
The unit ha stands for 'hectare'.
1 ha = 10 000 m^2

The metric units of area are mm^2, cm^2, m^2, ha and km^2.
You should be able to derive these area conversions:

$1\ cm^2 = 10\ mm \times 10\ mm \quad 1\ m^2 = 100\ cm \times 100\ cm$

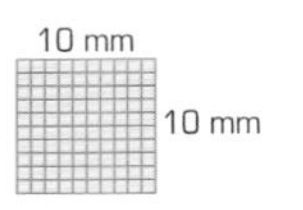

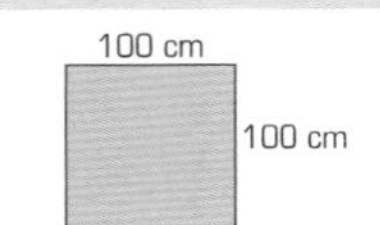

► $1\ cm^2 = 100\ mm^2 \qquad 1\ m^2 = 10\ 000\ cm^2$

The metric units of volume are mm^3, cm^3, m^3 and km^3.
You should be able to derive these volume conversions:

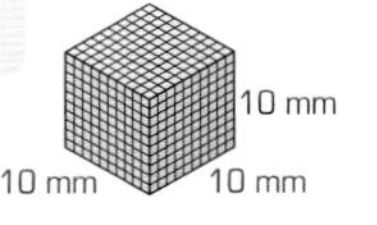

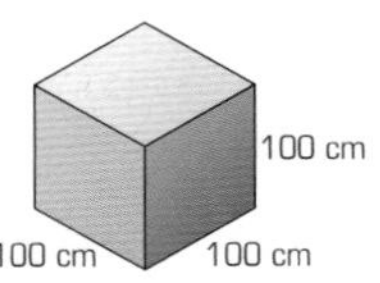

$1\ cm^3 = 10\ mm \times 10\ mm \times 10\ mm \quad 1\ m^3 = 100\ cm \times 100\ cm \times 100\ cm$

► $1\ cm^3 = 1000\ mm^3 \qquad 1\ m^3 = 1\ 000\ 000\ cm^3$

Remember:
Capacity is the amount of liquid a 3-D shape can hold.
It is related to volume.
► 1 ml = 1 cm^3
► 1 l = 1000 cm^3

Measurements are only as accurate as the units used.

These three pencils measure 12 cm to the nearest cm.
However they are not exactly the same length.

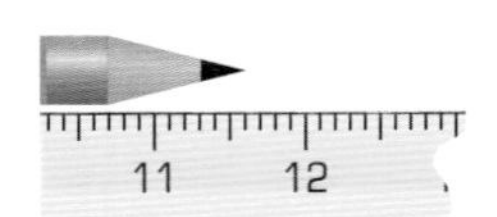

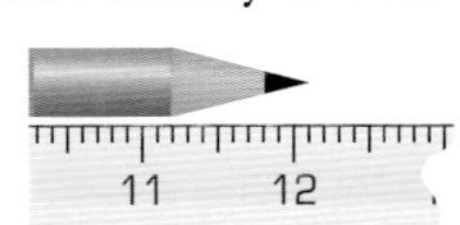

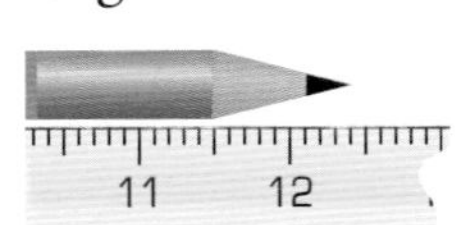

A measurement of 12 cm to the nearest cm could be anything from 11.5 cm to 12.5 cm.

► Measurements given to the nearest whole unit may be up to one half of the unit longer or shorter.

Even measurements given to 1 decimal place can be inaccurate.

example

Curtis buys a bag of apples. The bag weighs 3.4 kg to the nearest 0.1 kg.

a What is the most that the apples could weigh?
b What is the least that the apples could weigh?

a The most the apples could weigh is 3.45 kg.

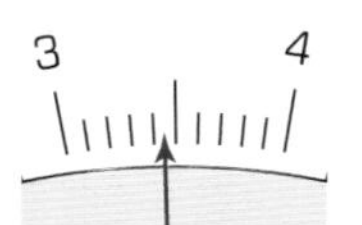

b The least the apples could weigh is 3.35 kg.

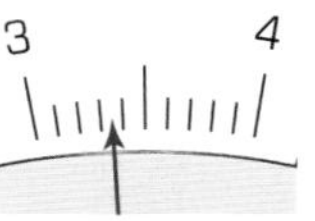

Exercise S2.1

1 Calculate the area in cm^2 of each of these shapes.

a

4.7 cm

15.3 cm

b

3 cm

6.5 cm

c

42 cm

43 cm

89 cm

2 Change these areas to mm^2.

a $8\ cm^2$ **b** $56\ cm^2$ **c** $2.4\ cm^2$
d $13.89\ cm^2$ **e** $345\ cm^2$

3 Change these areas to cm^2.

a $0.3\ m^2$ **b** $45\ m^2$ **c** $7\ m^2$
d $34.9\ m^2$ **e** $0.789\ m^2$

4 Change these volumes to **i** m^3 **ii** litres.

a $5567\ cm^3$ **b** $3600\ cm^3$
c $345\ 000\ cm^3$

5 Calculate the area in m^2 of each of these shapes.

a

4.2 m

125 cm

b

2.5 m

1.7 m

98 cm

c

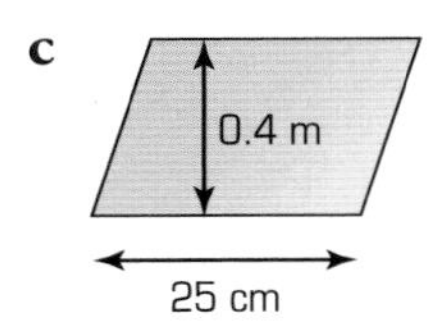

6 Each shape in this question has an area of $40\ cm^2$. They are all sketches.

a Calculate the height of the parallelogram.

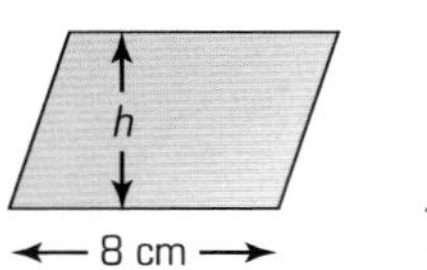

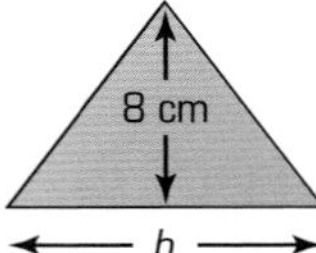

b Calculate the length of the base of the triangle.

c What might be the values of *h*, *a* and *b* in this trapezium?
Write four different answers.

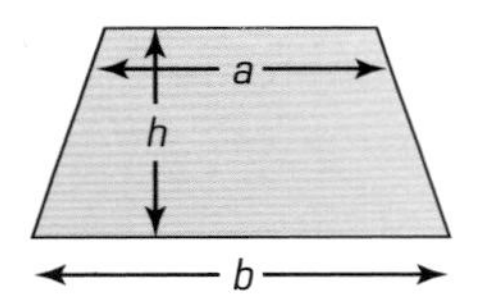

7 A table top is in the shape of a trapezium. Calculate the area of the table top.

Hint: You will need Pythagoras' theorem to work out the height of the trapezium.

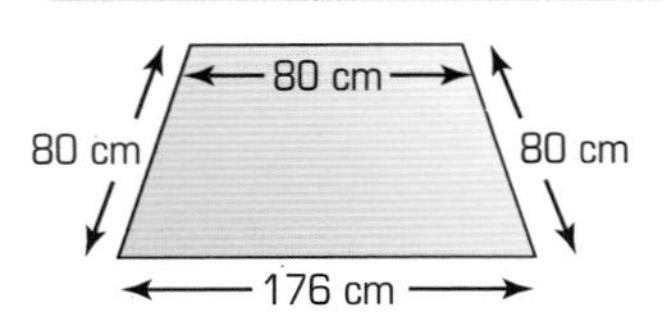

8 The length of a pencil is 35 cm to the nearest cm.

a What is the shortest length that the pencil could be?

b What is the longest length it could be?

9 The number of matches in a box is 120 to the nearest 10.

a What is the smallest number of matches that there could be?

b What is the largest number there could be?

Circumference and area of a circle

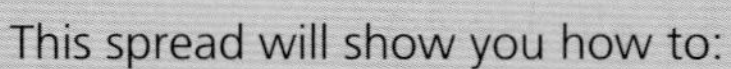

This spread will show you how to:

- Know and use the formulae for the circumference and area of a circle.

KEYWORDS

Circumference
Diameter
Pi (π)
Approximate
Radius

This circle has diameter *d*, radius *r* and circumference *C*.

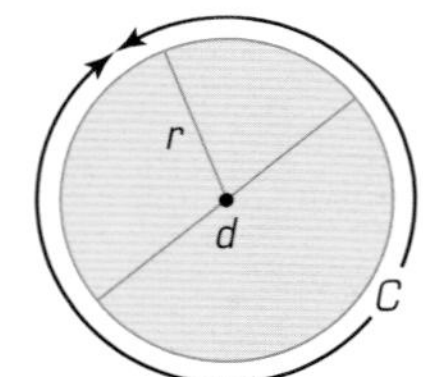

Remember:

- $C = \pi d$ or $C = 2\pi r$

π (pronounced 'pie') cannot be expressed exactly because it is irrational.

- As a decimal, π approximates to 3.14 (to 2 dp).
- As a fraction, π approximates to $\frac{22}{7}$.

You can find a more accurate value for π on a scientific calculator:

3.141592654

EXP

example

a A circle has a circumference of 20 cm. What is its diameter? Give your answer to 1 dp. (Use π =3.14)

b A wheel has a diameter of 2 m. How many complete rotations does it make when it travels 1 km? (Use π =3.14)

a $C = \pi d$

$20 = 3.14 \times d$

$d = 20 \div 3.14 = 6.369\ 42 \ldots$

$d = 6.4$ cm (to 1 dp)

b $C = \pi d = 3.14 \times 2\text{ m} = 6.28\text{ m}$

$1000 \div 6.28 = 159.2$

C fits into 1 km 159.2 times.

The wheel makes 159 complete rotations.

Hint: 1 km = 1000 m

Remember:

- $A = \pi r^2$ where *A* is the area of a circle.

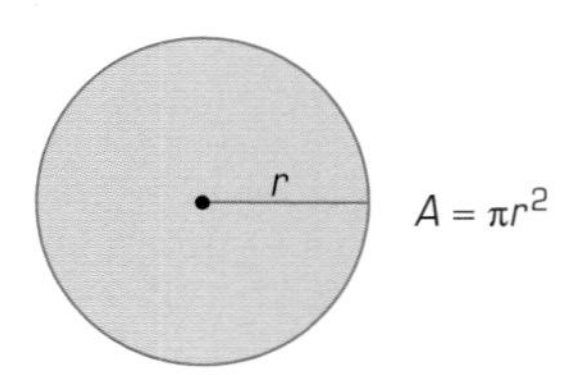

example

A circle has a radius of 14 cm.
What is its area?
Give your answer to the nearest square centimetre.

$A = \pi \times 14 \times 14$

$= 615.752 \ldots$

Area = 616 cm^2 to nearest cm^2.

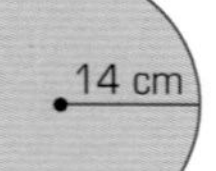

Exercise S2.2

1 Write down the diameter of each circle.

a

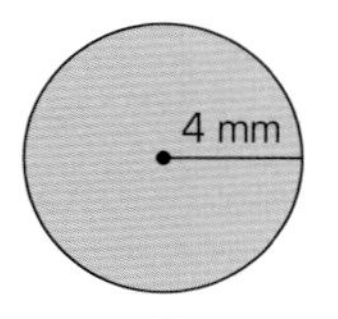

b

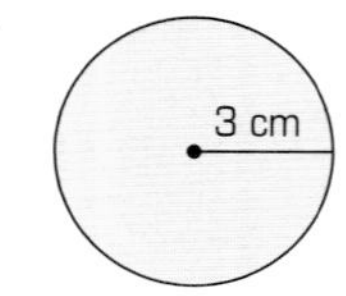

c

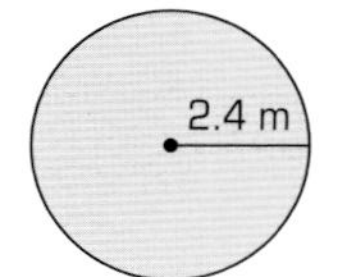

d

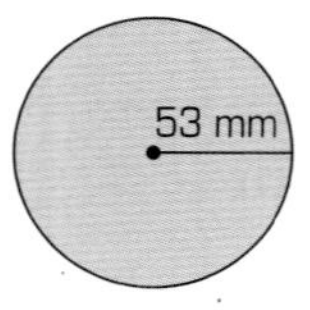

2 Calculate the circumference of each circle using $\pi = 3.14$.

a

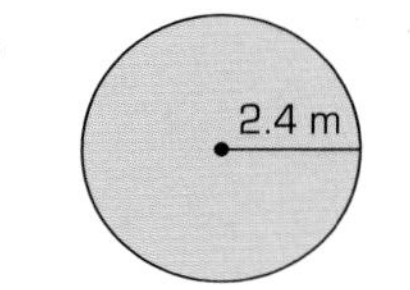

b

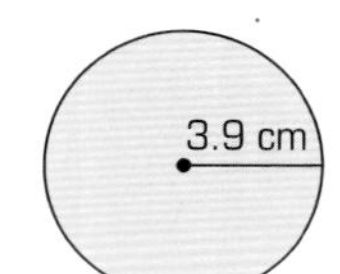

c

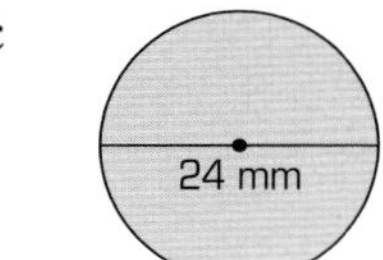

d

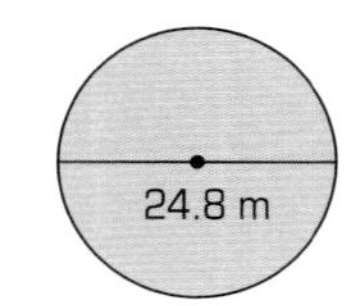

3 **a** Estimate the circumference of each circle.

i

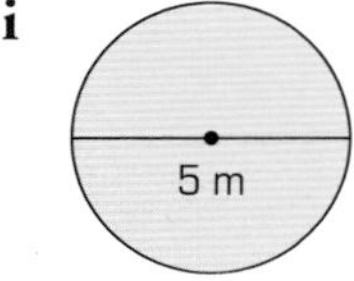

ii

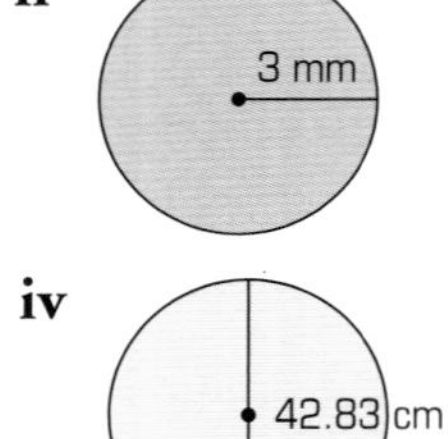

iii

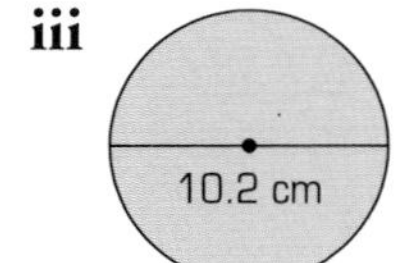

iv

b Now calculate the circumferences using the π button on your calculator.
Give your answers to 2 decimal places.

4 Find the diameters (to 1 dp) of circles with each of these circumferences.
(Use the π button on your calculator.)

a 23 cm **b** 567 mm
c 23.57 m **d** 84.345 km

5 The circumference of a jam jar lid is 12 cm. Find the radius of the lid.

6 The Zing Vaa restaurant has a round table with a diameter of 1.8 m. Each diner needs 45 cm of space. How many people can sit around the table?

7 Find the areas of these circles using $\pi = \frac{22}{7}$.

a diameter 14 mm
b diameter 70 m
c radius 35 cm
d radius 7.7 mm.

8 **a** Estimate the areas of these circles.

i **ii** **iii** **iv**

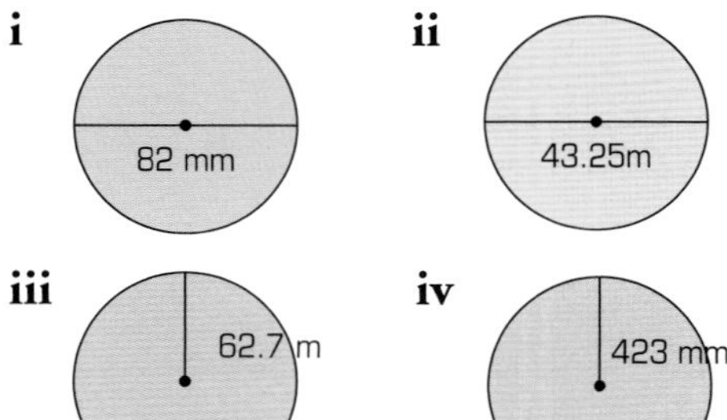

b Now calculate the areas using the π button on your calculator.

9 The largest fairground wheel in Brighton has an area of 500 m^2.
Find the radius of this wheel.

10 The inside lane of a mini-running track is 200 m long, 50 m on each straight and 50 m on each semicircular end.
What area is in the centre of the track?

11 Calculate the area of each shaded region.

a

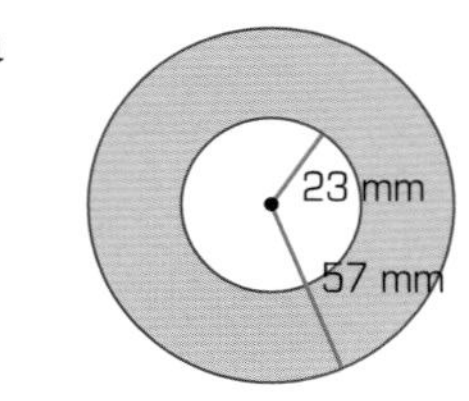

b

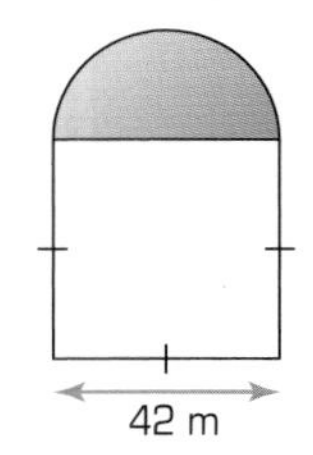

S2.3 Arcs and sectors

This spread will show you how to:

- Know and use the formulae for length of arcs and area of sectors of circles.

KEYWORDS

Arc, Sector, Circumference, Radii

You can use your knowledge of circles to find the area of a **sector**.

example

Find the areas of these sectors. Give your answers to 1 dp.

a

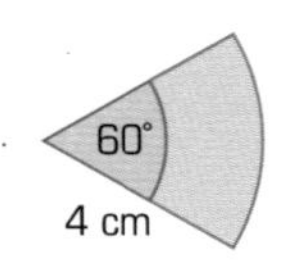

b

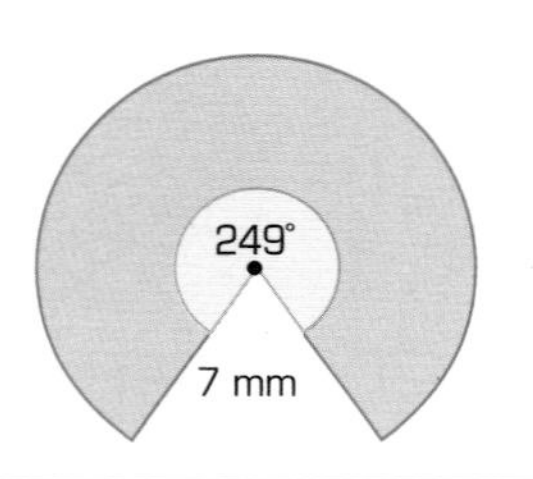

a Fraction of the whole circle is $\frac{60}{360}$.

Area of whole circle $= \pi r^2$

So area of sector $= \frac{60}{360} \times \pi r^2$

$= \frac{1}{6} \times \pi \times 4^2$

$= \pi \times \frac{16}{6}$

$= 8.4\text{ cm}^2$ (1 dp)

b Area of sector $= \frac{249}{360} \times \pi r^2$

$= \frac{249 \times \pi \times 7^2}{360}$

$= 106.5\text{ mm}^2$ (1 dp)

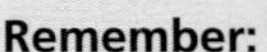

Remember:
A sector is a region of a circle bounded by an arc and two radii.

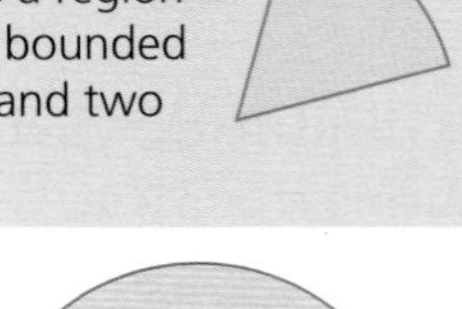

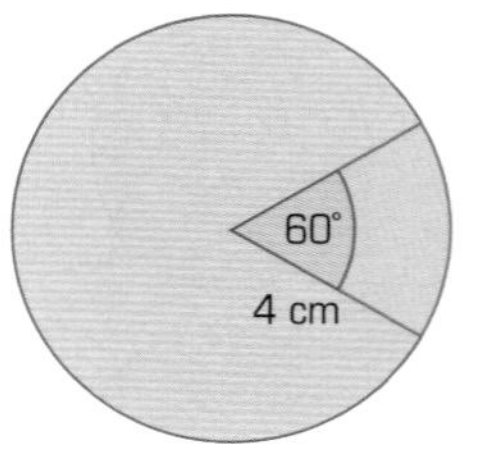

► Area of a sector, $A = \frac{\theta}{360} \times \pi r^2$
where θ is the angle between the two radii bounding the sector, and r is the radius.

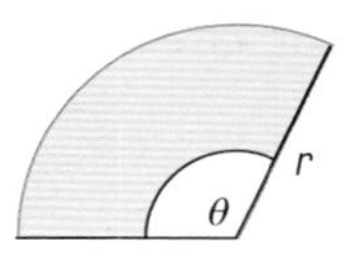

Similarly, the length of an arc is just a fraction of the circumference.

► Length of an arc, $l = \frac{\theta}{360} \times 2\pi r = \frac{\theta}{360} \times \pi d$

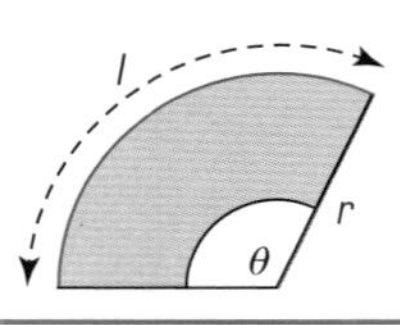

example

a Find the length of the arc. Give your answer to 2 dp.

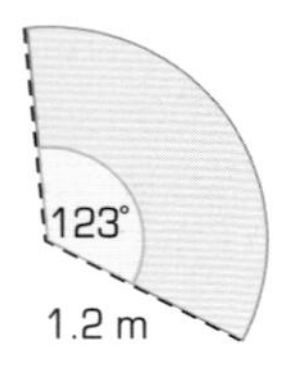

b Find the perimeter of the sector.
Give your answer to the nearest mm.

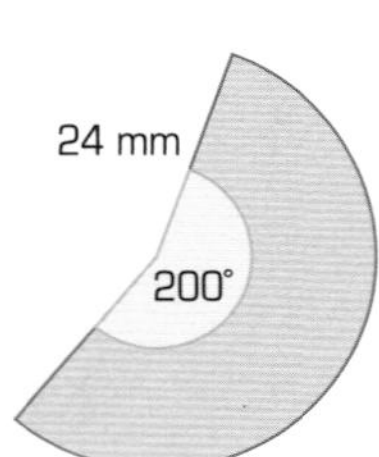

a Arc length, $l = \frac{\theta}{360} \times 2\pi r$

$= \frac{123}{360} \times 2 \times \pi \times 1.2$

$= 2.576\ 105\ \ldots$

$l = 2.58$ m (2 dp)

b Arc length, $l = \frac{\theta}{360} \times 2\pi r$

$= \frac{200}{360} \times 2 \times \pi \times 24$

$= 83.776$

Add lengths: $24 + 24 + 83.776$

$= 131.776$

Perimeter = 132 mm (nearest mm)

Exercise S2.3

1 Find the areas of these sectors.
Use $\pi = 3.14$.

a
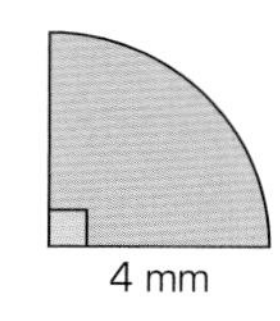

b
27 mm

c
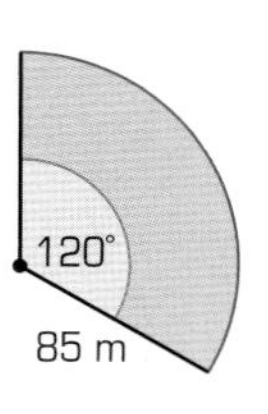

d
4.7 cm
260°

2 Find the lengths of these arcs.

a
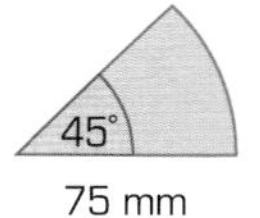

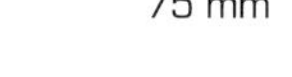

b
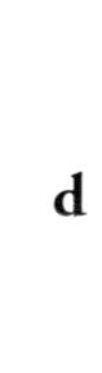
110°
8.2 cm

c
5 m

d
27 cm

3 Find the perimeters of these sectors.

a
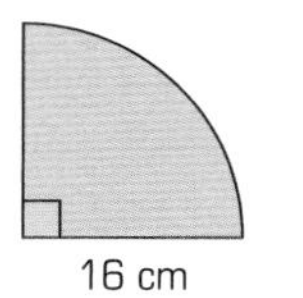

b
23 cm

c
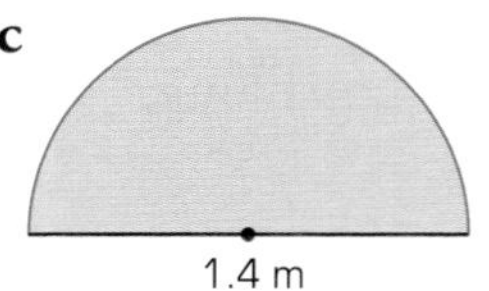

d
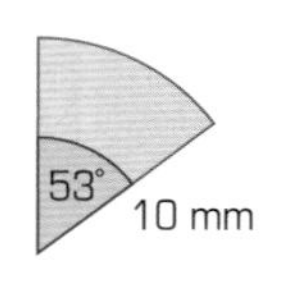

4 The area of this fan when fully open as shown is 32.5 cm^2. What is the radius of the sector it forms?

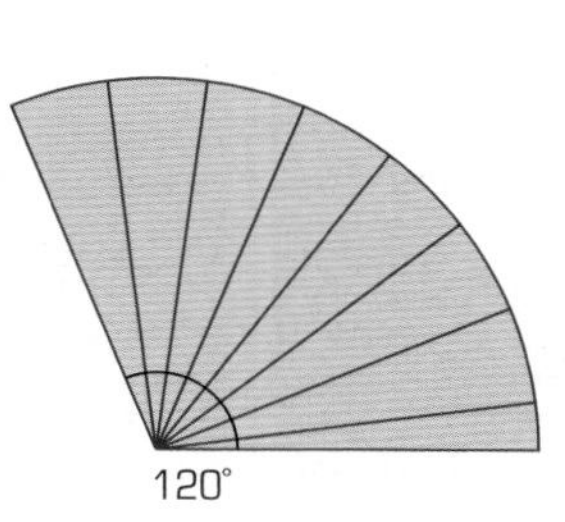

5 The area of this semicircular pond is 52 m^2. What is its radius?

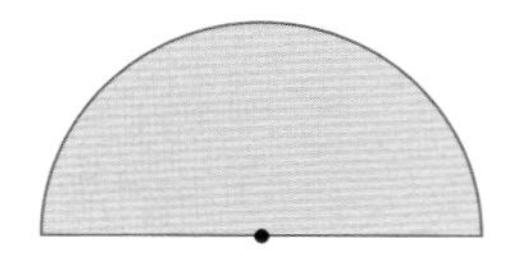

6 The minute hand of this clock is 10 cm long.

a How far does the tip of the minute hand travel in 1 hour?

The hour hand travels 4.2 cm in 1 hour.

b How long is the hour hand?

7 Find:

a the area

b the perimeter

of this shape.

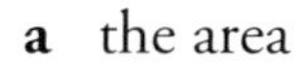
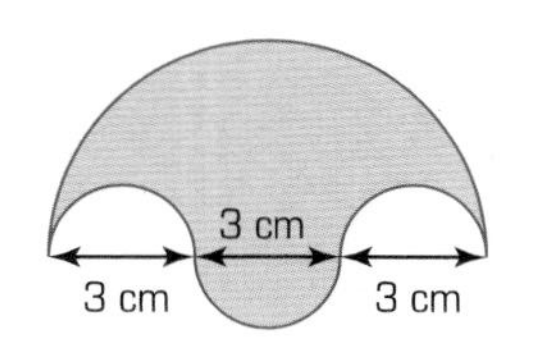

8 Find the area of the shaded region.

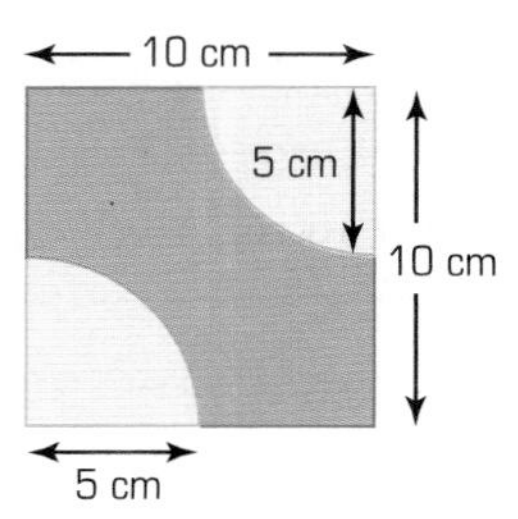

9 Sammy wants to make a poster. It is to be a semicircle. He needs an area of 500 cm^2. What diameter should he use?

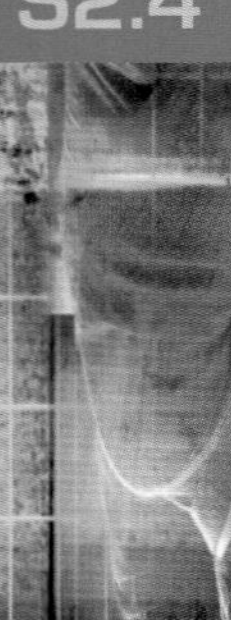

S2.4 Line segments

This spread will show you how to:

- Find points that divide a line in a given ratio, using the properties of similar triangles.
- Given the coordinates of points A and B, calculate the length of AB.

KEYWORDS

Line segment, Midpoint, Similar, Ratio, Pythagoras' theorem, Hypotenuse

Remember:
The midpoint, M, of the line segment joining A(x_1, y_1) to B(x_2, y_2) is given by:
M$\left(\frac{x_1 + x_2}{2}, \frac{y_1 + y_2}{2}\right)$

Remember: A line segment is just a finite part of a line.

For example, the midpoint of A(3, 2) and B(⁻1, 4) is: $\left(\frac{3 + {}^-1}{2}, \frac{2 + 4}{2}\right) = (1, 3)$

You can use similar triangles to divide a line in a given ratio.

example

A and B have coordinates (⁻2, 2) and (8, 7) respectively.
Find the point P that divides the line AB in the ratio 2 : 3.

First sketch AB on a grid.

Then draw triangles as shown.

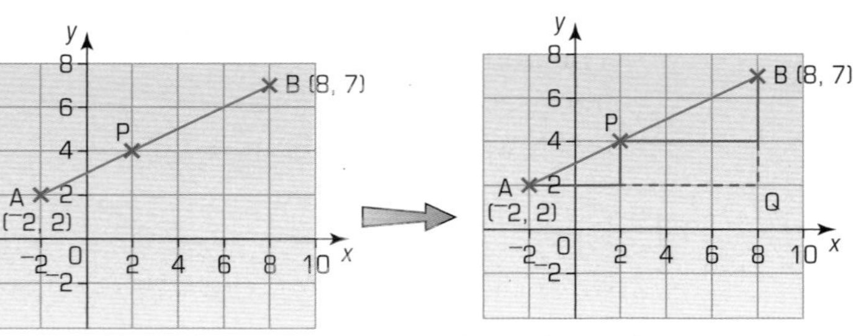

Mark P somewhere along the line.

The angles are the same, so the triangles are similar.

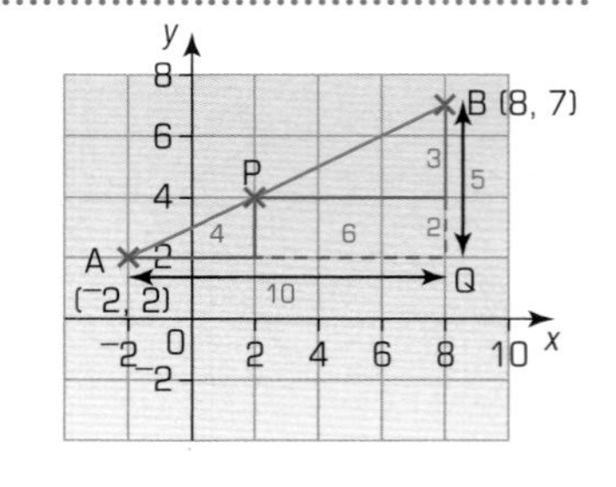

Divide AQ in the ratio 2 : 3
2 + 3 = 5 10 ÷ 5 = 2 2 × 2 = 4, and 3 × 2 = 6 ⟹ 4 : 6
Divide BQ in the ratio 2 : 3
5 ÷ 5 = 1 2 × 1 = 2 and 3 × 1 = 3 ⟹ 2 : 3

So P has coordinates (⁻2 + 4, 2 + 2) = (2, 4)

You can use Pythagoras' theorem to find the length of a line on a grid.

example

A and B have coordinates (1, 1) and (5, 4) respectively.
Find the length of the line segment AB.

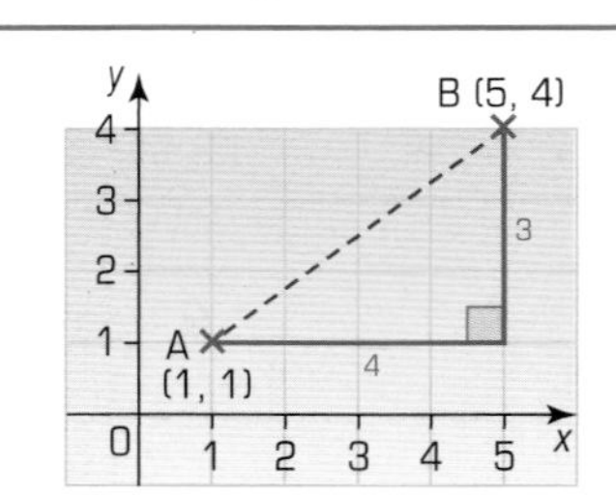

AB is the hypotenuse.
$AB^2 = 4^2 + 3^2 = 25$
$AB = \sqrt{25} = 5$
AB is 5 units long.

► The distance d between the points A(x_1, y_1) and B(x_2, y_2) is given by the formula:
$d^2 = (x_2 - x_1)^2 + (y_2 - y_1)^2$

Exercise S2.4

1 Find the midpoints of the lines joining these pairs of points.

a A(2, 4), B(5, 4)
b C(3, 1), D($^-$3, 1)
c E(2, 0), F(2, 10)
d G($^-$1, 4), H($^-$1, $^-$5)
e I(2, 4), J(3, 5)
f K($^-$2, 4), L($^-$8, 6)

2 A line AB joins the points A(4, 2) and B(x, 8). The midpoint is (5, 5). Find the value of x.

3 M is the midpoint of the line AB. Find x and y.

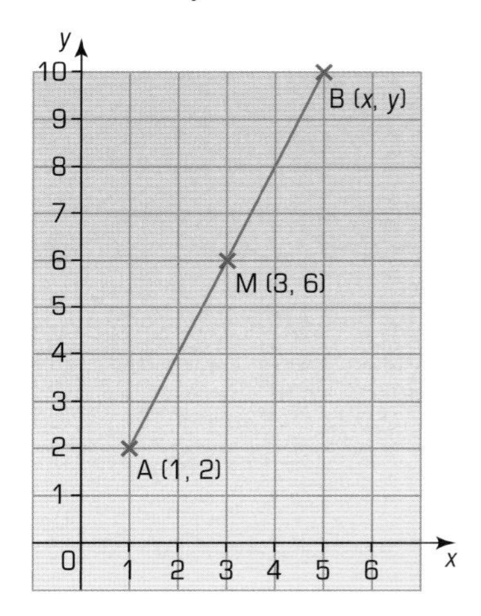

4 M is the midpoint of the line AB. Find x and y.

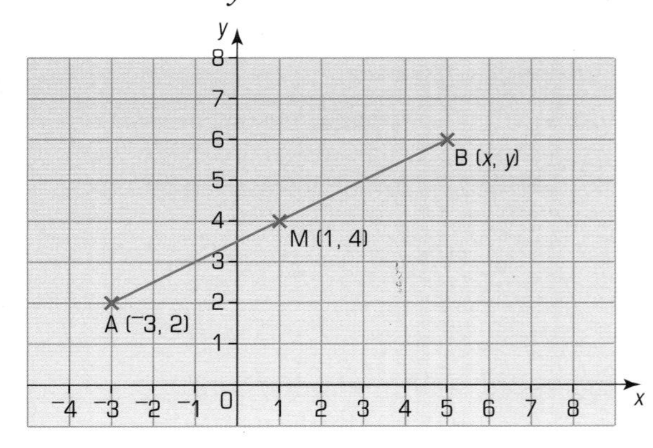

5 A is the point (1, 2), B is the point (9, 2) and C is the point (1, 10).

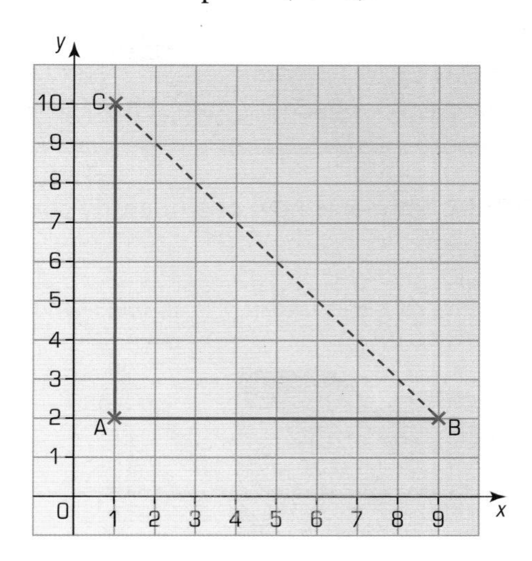

Find: **a** the midpoint of AC
b the midpoint of AB
c the midpoint of CB.

Draw these midpoints on a copy of the grid.
What do you notice about your results?

d Find the lengths of the sides AB, BC and AC.

6 Find the lengths of the lines joining these points.

Hint: Drawing a diagram will help.

a A(2, 4) B(3, 5)
b C(5, 6) D(3, 9)
c E(5, $^-$1) F($^-$2, 3)
d G($^-$2, $^-$3) H($^-$2, $^-$4)
e I(2, 5) J(2, 10)
f K(3, 9) L(5, 9)

7 The coordinates of A are (1, 4). The y-coordinate of B is 28. Line AB is 26 units long. Find the coordinates of the midpoint of AB.

Hint: Use Pythagoras' theorem first to find the x-coordinate of B. Draw a diagram.

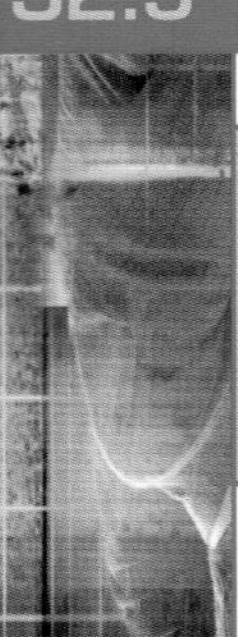

S2.5 Compound measures

This spread will show you how to:

- Understand and use measures of speed (and other compound measures) to solve problems.
- Solve problems involving constant or average rates of change.

KEYWORDS

Rate, Speed, Pressure, Average rate, Constant, Density, Direct proportion

- The **rate** compares how one quantity changes with another.

Note: Per means 'for every'.

The **speed** of this car is measured in miles (or kilometres) per hour.

Its **fuel economy** is measured in miles per gallon, or litres per 100 km.

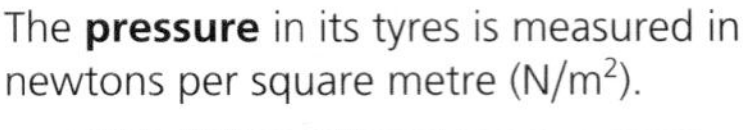

The **pressure** in its tyres is measured in newtons per square metre (N/m^2).

- If a rate is constant, then the two quantities are in **direct proportion.** You can connect them by a simple formula:
  - speed $= \dfrac{\text{distance travelled}}{\text{time taken}}$
  - density $= \dfrac{\text{mass of object}}{\text{volume of object}}$
  - pressure $= \dfrac{\text{force on surface}}{\text{surface area}}$

Note: Density is often measured in kg/m^3.
You can use density to compare the weight of objects that have the same volume. Concrete is denser than cotton wool.

example

a A mouse travels 5 m in 10 seconds at a constant speed.
Find the speed of the mouse.

b A uniform lump of metal weighs 912 g and has a volume of 40 cm^3.
Find its density.

a speed $= \frac{\text{distance travelled}}{\text{time taken}}$ so speed of mouse $= \frac{5}{10} = 0.5$ m/s

b density $= \frac{\text{mass of object}}{\text{volume of object}}$ so density of metal $= \frac{912}{40} = 22.8$ g/cm^3

Note: The units in the answer should match the units in the question.

If the rate varies, you can calculate an **average rate**.

example

Jana cycles 48 miles in four hours.
Find her average speed.

Average speed $= 48 \div 4$ miles/hour
$= 12$ miles/hour (or 12 mph)

Exercise S2.5

1 A train travels 280 miles in 5 hours. What is its average speed?

2 A car travels 200 km at an average speed of 50 km/h. How long does it take?

Remember: If speed = $\frac{\text{distance}}{\text{time}}$ then time = $\frac{\text{distance}}{\text{speed}}$.

3 A plane flies at 600 km per hour for 2 hours and 30 minutes. How far does it fly? (NB 2 hours and 30 minutes = 2.5 hours **not** 2.3 hours.)

Remember: If speed = $\frac{\text{distance}}{\text{time}}$ then distance = speed × time.

4 A runner runs 1500 m in 4.5 minutes. What is the runner's speed in metres per minute?

5 **a** A 2 cm cube of aluminium has mass 21.6 g. Calculate the density of aluminium in g/cm^3.

Hint: density = $\frac{\text{mass}}{\text{volume}}$

b Calculate the mass of a 3 cm cube of aluminium.

c The density of copper is 8.9 g/cm^3. Calculate the difference in mass between a 2 cm cube of copper and a 2 cm cube of aluminium.

6 The brick shown exerts a force of 30 N on the surface it stands on.

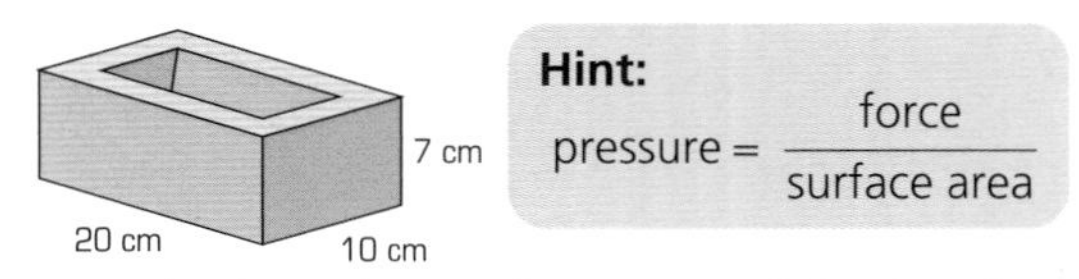

a Calculate the surface area of the brick, in m^2, that is in contact with the table when it stands **i** up on end **ii** as shown above.

b For which brick position is the pressure on the table greatest?

7 The diagram shows the distance between my home and two cities, Angereed (A) and Balderdash (B). It also shows the journey times.

A — 5 miles, 5 min — Home — 5 miles, 10 min — B

a What is the average speed of the journey from my home to Angereed in mph?

b What is the average speed of the journey from my home to Balderdash

c I drive from Angereed to home and then on to Balderdash. The journey time is 15 minutes. What is my average speed?

8 A satellite passes over both the North and South poles and it travels 700 km above the surface of the Earth. It takes 80 minutes to complete one orbit. Assume that the Earth is a sphere with diameter 12 800 km. Calculate the speed of the satellite, in kilometres per hour.

9 **a** A world-class sprinter can run 100 m in about 10 seconds. A bus takes 20 minutes to go 2 miles to the next village. Which average speed is faster?

b Ali walks 700 m to school. He timed himself with a stop watch. It took him 9 minutes 24.3 seconds. What was his average speed? Give your answer to a suitable degree of accuracy and explain why you have chosen it.

S2.6 Prisms

This spread will show you how to:

- Calculate lengths, areas and volumes in right prisms, including cylinders.

KEYWORDS

Prism, Cylinder, Surface area, Volume, Cross-section

- **Remember:** A prism is a 3-D shape with a uniform cross-section throughout its length.
- For any prism:
  - Volume = area of cross-section × length
  - Surface area = total area of all its faces

The cross-section of this pentagonal prism is uniform (the same) all the way along its length.

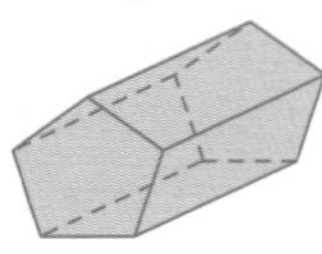

example

Find the volumes of these prisms.

a

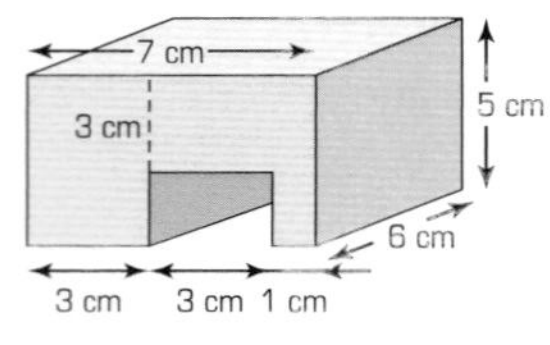

b

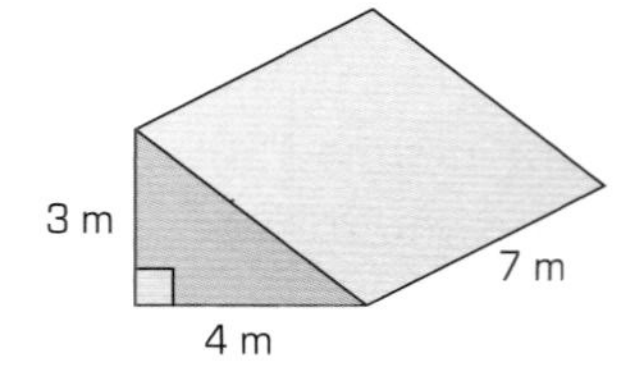

a Split the front into three rectangles.

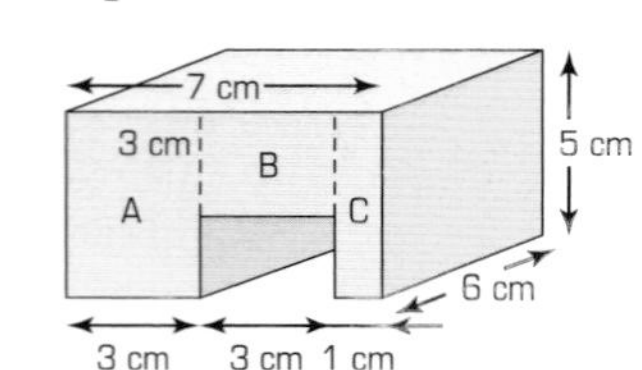

Area (A + B + C) $= (3 \times 5) + (3 \times 3) + (1 \times 5)$

$= 15 + 9 + 5 = 29$

Volume $= 29 \times 6 = 174\text{ cm}^3$

b The cross-section is a triangle.

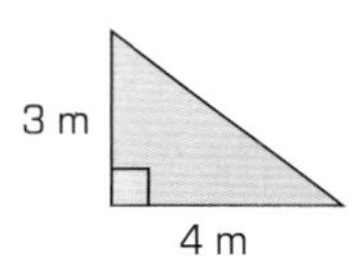

Area $= \frac{1}{2}$ base × height

$= \frac{1}{2} \times 3 \times 4 = 6$

Volume $= 6 \times 7 = 42\text{ m}^3$

A cylinder is a 3-D shape with a circular cross-section.

- Volume of a cylinder, $V =$ area of circular cross-section × height $= \pi r^2 h$

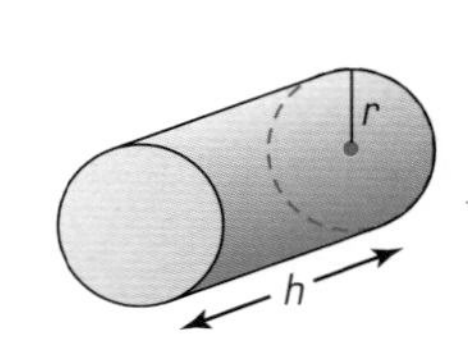

You can also derive a formula for the surface area of a cylinder.

- Total surface area of a cylinder, $A = 2\pi r^2 + 2\pi rh$

example

For this cylinder, find:

a the volume

b the surface area.

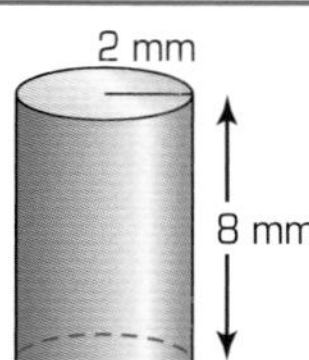

a $V = \pi r^2 h = \pi \times 2^2 \times 8$

$= 100.5\text{ mm}^3$ (1 dp)

b $A = 2\pi r^2 + 2\pi rh = 2 \times \pi \times 2^2 + 2 \times \pi \times 2 \times 8$

$= 125.7\text{ mm}^2$ (1 dp)

Exercise S2.6

1 Work out the volumes of these cuboids.

a

4 mm
3 mm
2 mm

b

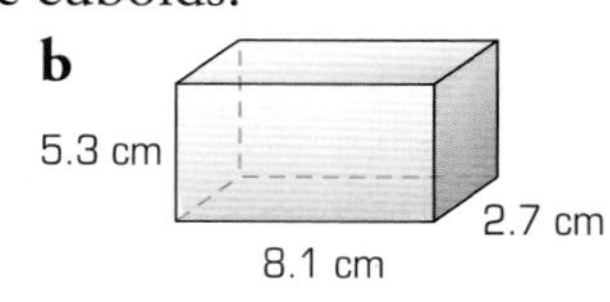

c

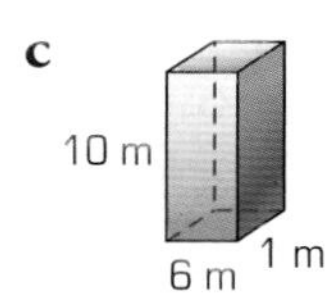

2 Calculate the surface areas of the cuboids in question 1.

3 Work out the volumes of these prisms.

a

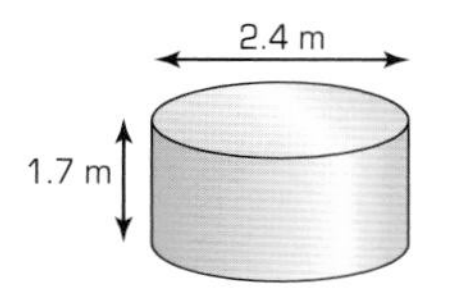

b

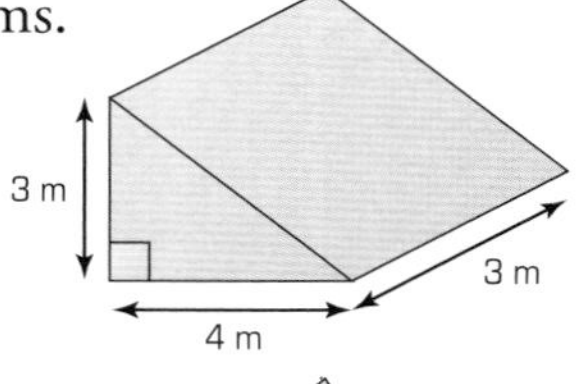

c

6 m
3 m
4 m
2 m
4 m

d

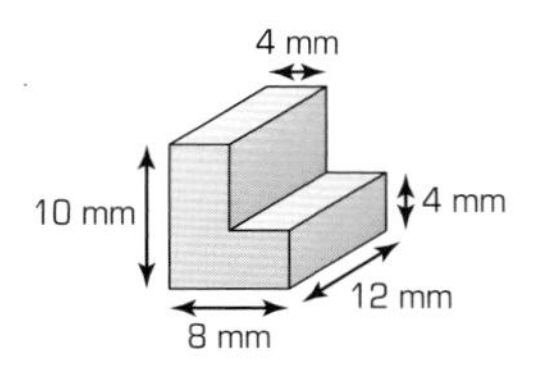

e

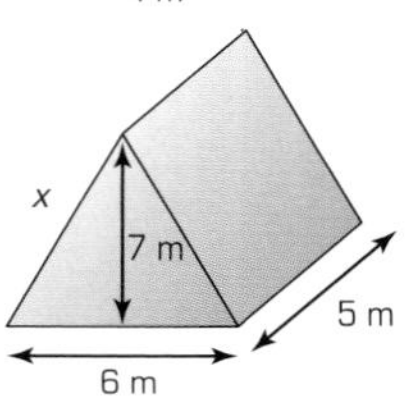

f

area = 42 mm²
12 mm

4 Find the surface areas of the prisms in question 3**a**–**e**.

In part **e** you need to find the length ***x*** first, using Pythagoras' theorem.

5 Find the unknown lengths marked in these diagrams.

a

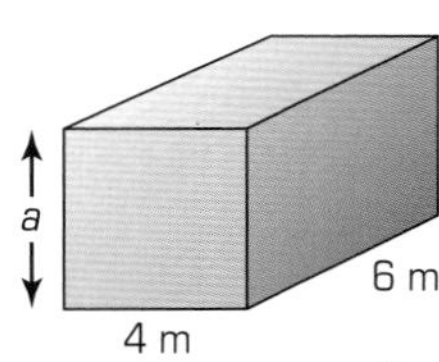

volume = 48 m^3

b

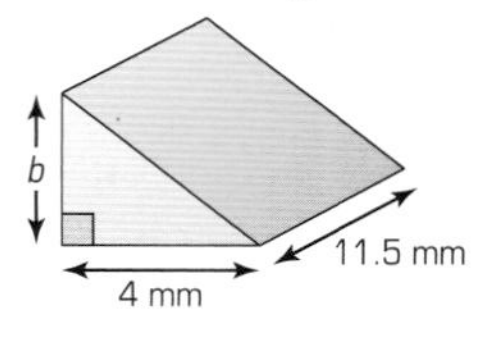

volume = 23 mm^3

c

10 cm
10 cm
c

volume = 5000 cm^3

6 Find the areas of the cross-sections of these shapes.

a

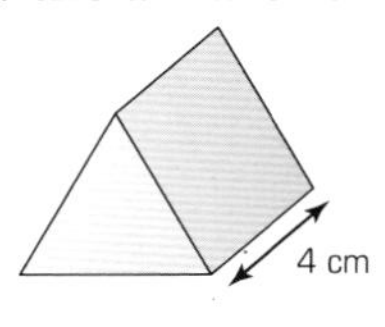

volume = 28.4 cm^3

b

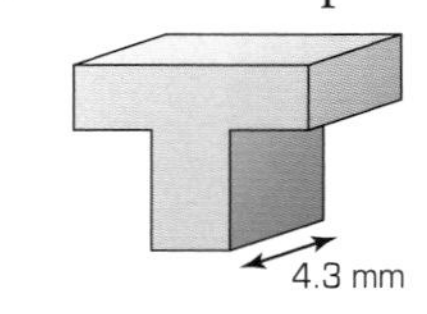

volume = 43 mm^3

c

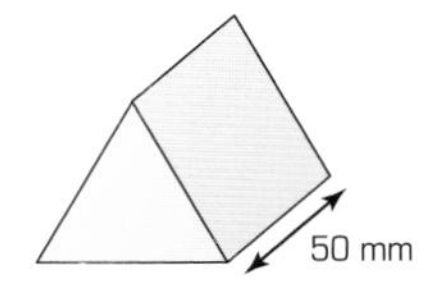

volume = 200 mm^3

7 A cuboid has a volume of 40 cm^3. Investigate its possible dimensions and cross-sections.
What dimensions give it its smallest surface area?
What is the largest surface area you can find for this cuboid?

8 Show that the total surface area of a cylinder is $2\pi r^2 + 2\pi rh$.

Summary

You should know how to …

1 Understand and use measures of speed (and other compound measures) to solve problems.

2 Generate fuller solutions to problems.

3 Recognise limitations on accuracy of data and measurements.

Check out

1 Which has the faster speed?

a A man who walks 10 km in 6 hours.

b An ant who walks 7 cm in 5 seconds.

2 Find the dimensions of a cube that has the same value of surface area as volume.

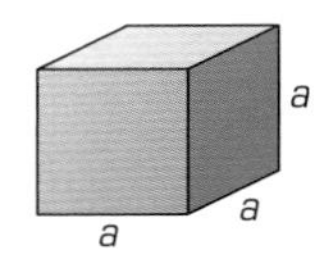

3 **a** A packet of biscuits weighs 200 g to the nearest 5 g.
What are the maximum and minimum possible weights?

b A rectangular tile measures 15 cm by 10 cm to the nearest cm.
What are the largest and smallest possible areas for the tile?

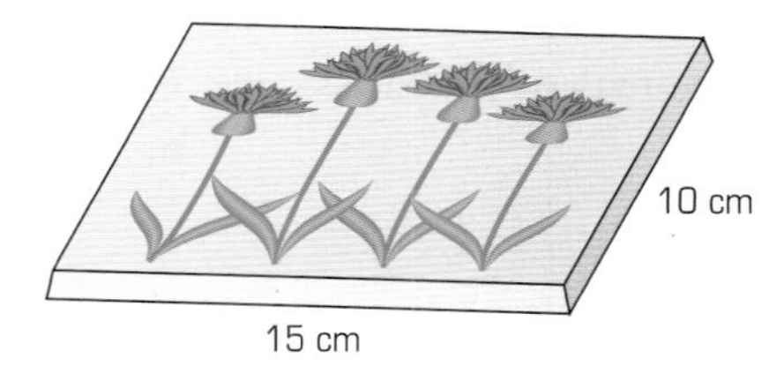

2 Number calculations

This unit will show you how to:

- Know and use the index laws for multiplication and division of positive integer powers.
- Begin to extend understanding of index notation to negative and fractional powers.
- Begin to write numbers in standard form.
- Round numbers to three decimal places and to a given number of significant figures.
- Understand upper and lower bounds.
- Extend mental methods of calculation, working with decimals, fractions, percentages, factors, powers and roots.
- Use standard column procedures to add and subtract integers and decimals of any size.
- Multiply and divide by decimals, dividing by transforming to division by an integer.
- Estimate calculations by rounding numbers to one significant figure.
- Multiply and divide fractions, interpreting division as a multiplicative inverse.
- Use algebraic methods to convert a recurring decimal to a fraction in simple cases.
- Recognise and use reciprocals.
- Use a calculator efficiently and appropriately.
- Generate fuller solutions to problems.
- Recognise limitations on accuracy of data and measurements.

The most accurate measurement is not always the best answer.

Before you start

You should know how to ...	Check in
1 Multiply and divide by any integer power of 10.	**1** Multiply each number by 10,100 and 1000: **a** 36.2 **b** 2150 **c** 0.0063
2 Round measurements to a given power of 10.	**2** Round each of these measurements. **a** 0.618 m to the nearest cm **b** 6724 km to the nearest 100 km.
3 Use standard procedures to add, subtract, multiply and divide decimals.	**3** Use a written method to work out these. **a** 23.628 + 1.483 **b** 82.97 – 2.345 **c** 62.4 × 0.94 **d** 59.22 ÷ 6.3

N2.1 Powers and roots

This spread will show you how to:

- Know and use the index laws for multiplication and division of positive integer powers.
- Extend understanding to negative and fractional powers.
- Extend mental methods, working with powers and roots.

KEYWORDS

Power | Index laws
Surd | Square root
Index, indices

You use a power, or **index**, to show how many times a number is multiplied by itself.

- Any number raised to the power zero = 1
  - $a^0 = 1$, for any number a.
- Powers can also be negative.
  - $a^{-n} = \frac{1}{a^n}$ for any numbers a and n.
- You may already know the three index laws.
  - $x^a \times x^b = x^{a+b}$
  - $x^a \div x^b = x^{a-b}$
  - $(x^a)^b = x^{ab}$

$5^3 = 5 \times 5 \times 5$

$5^2 = 5 \times 5$

$5^1 = 5$

$5^0 = 1$

$5^{-1} = \frac{1}{5}$

$5^{-2} = \frac{1}{25}$

(÷5 at each step)

As you reduce the power by 1, you divide by the base.

Powers can also be fractional.

$\sqrt{4} \times \sqrt{4} = 4^1 = 4$ by definition,

$4^{\frac{1}{2}} \times 4^{\frac{1}{2}} = 4^1 = 4$ from the laws of indices. So $4^{\frac{1}{2}} = \sqrt{4}$

- $a^{\frac{1}{2}} = \sqrt{a}$ where a is a number.

example

Evaluate:

a $2^3 \times 2^{-5}$ **b** $4^2 \div 4^3$ **c** $(3^2)^{-2}$

d $9^{\frac{1}{2}}$ **e** $3^2 \times 121^{\frac{1}{2}}$ **f** $144^{-\frac{1}{2}}$

a $2^3 \times 2^{-5} = 2^{3+-5} = 2^{-2} = \frac{1}{2^2} = \frac{1}{4}$

b $4^2 \div 4^3 = 4^{2-3} = 4^{-1} = \frac{1}{4}$

c $(3^2)^{-2} = 3^{2 \times -2} = 3^{-4} = \frac{1}{3^4} = \frac{1}{81}$

d $9^{\frac{1}{2}} = 3$

e $3^2 \times 121^{\frac{1}{2}} = 9 \times 11 = 99$

f $144^{-\frac{1}{2}} = \frac{1}{144^{\frac{1}{2}}} = \frac{1}{12}$

Many square roots cannot be written **exactly as decimals**.
For example, $\sqrt{2} = 1.414\,213\,562\ldots$

The dots indicate that the digits carry on forever randomly.

Square roots that cannot be written exactly as decimals are called **surds**.
For example: $\sqrt{2}$, $\sqrt{3}$, and $\sqrt{10}$ are surds.

Exercise N2.1

1 Calculate each of these, leaving your answer in index form where appropriate.

a $4^2 \times 4^3$ **b** $7^2 \times 7^6$

c $8^{10} \div 8^6$ **d** $4^7 \div 4^2$

e $6^4 + 6^7$ **f** $4^5 - 4^3$

g $5^6 \times 5^{-4}$ **h** $4^{-8} \div 4^5$

2 Demonstrate, for the values $a = 2$, $b = 3$ and $c = 4$, that these rules are valid.

$a^b \times a^c = a^{b+c}$

$a^b \div a^c = a^{b-c}$

$(a^b)^c = a^{bc}$

3 Calculate each of these, leaving your answer in index form where appropriate.

a $5^2 \times 5^3 \div 5^3$

b $8^2 \times 8^6 \div 8^6$

c $(7^6 \div 7^{-4}) \times 7^2$

d $(4^7 \div 4^2) \div 4^3$

e $10^3 \times 10^{-6} \div 10^{-4}$

f $3^{-2} \times 3^{-6} \div 3^{-5}$

g $9^2 \div (9^6 \times 9^6)$

h $5^2 \times (5^6 \div 5^6)$

4 Write each number as a power of 2.

a 16 **b** 64

c $\frac{1}{16}$ **d** 256

e $\frac{1}{64}$ **f** 1

g $\frac{1}{32}$ **h** 2

5 Investigate how you could enter fractional powers such as $3^{\frac{1}{2}}$ into your calculator.

Hint: You may have an x^y key.

6 Use your calculator to decide if the following is true:

$\sqrt[3]{64} = 64^{\frac{1}{3}}$

Try other examples to decide if the generalisation:

$\sqrt[b]{a} = a^{\frac{1}{b}}$

is true.

7 You can multiply square roots.
For example:

$\sqrt{6} \times \sqrt{6} = \sqrt{(6 \times 6)} = \sqrt{36} = 6$

$\sqrt{6} \times \sqrt{3} = \sqrt{18}$

$= \sqrt{(9 \times 2)}$

$= \sqrt{9} \times \sqrt{2}$

$= 3\sqrt{2}$

Multiply these square roots, leaving your answer in surd form as simply as possible.

a $\sqrt{8} \times \sqrt{5}$ **b** $\sqrt{8} \times \sqrt{4}$

c $2\sqrt{7} \times \sqrt{3}$ **d** $6\sqrt{4} \times 5\sqrt{10}$

8 **a** Show that the area of this triangle is $8\ m^2$.

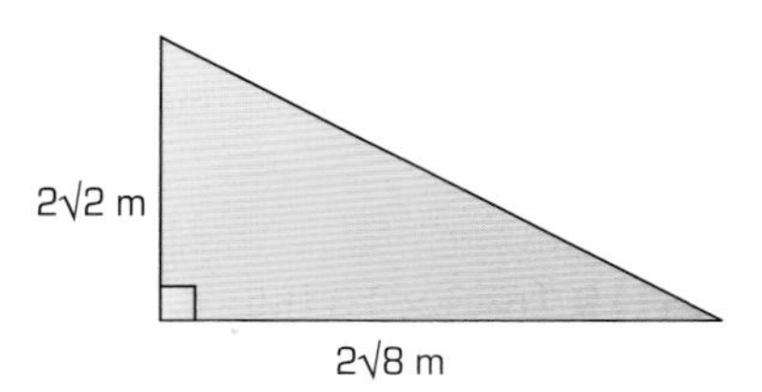

b Show that the length $2\sqrt{8}$ can also be expressed as $4\sqrt{2}$.

c Use Pythagoras' theorem to calculate the length of the hypotenuse, leaving your answer in surd form as simply as possible.

9 Harry says that he has drawn a square with an area of $40\ cm^2$.

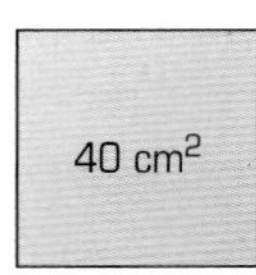

Write in surd form as simply as possible:

a the length of each side of the square

b the perimeter of the square.

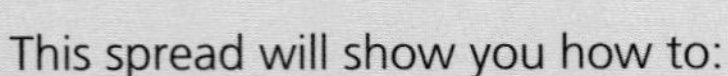

N2.2 Standard form

This spread will show you how to:

- Begin to write numbers in standard form.
- Enter numbers in standard form and interpret the display.

KEYWORDS

Standard form
Power
Index notation

This table shows some powers of 10 written in index notation.

← Numbers getting smaller   Numbers getting larger →

Power of 10	10^{-3}	10^{-2}	10^{-1}	10^{0}	10^{1}	10^{2}	10^{3}
Number	$\frac{1}{1000}$ or 0.001	$\frac{1}{100}$ or 0.01	$\frac{1}{10}$ or 0.1	1	10	100	1000

You can write **any** number using powers of 10.

$524 = 5.24 \times 100$
$= 5.24 \times 10^{2}$

$3980 = 3.98 \times 1000$
$= 3.98 \times 10^{3}$

$56.3 = 5.63 \times 10$
$= 5.63 \times 10^{1}$

$0.0467 = 4.67 \times 0.01$
$= 4.67 \times 10^{-2}$

▸ You can write any number in the **standard form** $A \times 10^{n}$.

$1 \leqslant A < 10$ and n is an integer.

example

Write these numbers in standard form.

a 91 643 **b** 0.65 **c** 76.5×10^{2}

a $91\,643 = 9.1643 \times 10\,000$
$= 9.1643 \times 10^{4}$

b $0.65 = 6.5 \times 0.1$
$= 6.5 \times 10^{-1}$

c 76.5×10^{2}
$= (7.65 \times 10) \times 10^{2}$
$= 7.65 \times 10^{3}$

You can apply this technique in reverse.

example

Convert these to ordinary numbers.

a 5.4×10^{3} **b** 2.81×10^{7} **c** 9.053×10^{-2}

a 5.4×10^{3}
$= 5.4 \times 1000$
$= 5400$

b 2.81×10^{7}
$= 2.81 \times 10\,000\,000$
$= 28\,100\,000$

c 9.053×10^{-2}
$= 9.053 \times 0.01$
$= 0.090\,53$

You may need to remind yourself how to multiply by powers of 10.

Standard form is useful in describing very large and very small quantities.

example

a Light travels at a speed of 3×10^{8} metres per second (m/s). Write this as an ordinary number.

b The diameter of a human hair is roughly 0.000 15 m. Write this in standard form.

a 3×10^{8} m/s $= 3 \times 100\,000\,000$ m/s $= 300\,000\,000$ m/s

b $0.000\,15$ m $= 1.5 \times 10^{-4}$ m

Check how to input 3×10^{8} on your calculator. For some calculators you press [3] [EXP] [8].

Exercise N2.2

1 Write each of these numbers in standard form.

a 456 **b** 24.56
c 894.5 **d** 0.0178
e 0.00 367 **f** 732.7
g 43.51 **h** 39 000

2 Write each of these standard form numbers in the ordinary way, for example, $45 \times 10^3 = 45 \times 1000 = 45\ 000$.

a 7.4×10^8 **b** 4.1×10^{-3}
c 1.82×10^{-5} **d** 8.7×10^{-2}
e 3.6×10^{-2} **f** 3.4×10^{-2}
g 5.96×10^{-1} **h** 9.112×10^{-6}

3 Write these numbers as ordinary numbers.

a 54×10^6 **b** 71×10^{-4}
c 8×10^{-3} **d** 47×10^5
e 0.36×10^7 **f** 1.4×10^2
g 5×10^{-1} **h** 91×10^0

4 Copy and complete each of these statements.

a $0.81 \times 10^? = 810$
b $? \times 10^{-2} = 5.46$
c $0.007 \div 10^3 = ?$
d $4.5 \times 10^7 = ?$
e $0.98 \div 10^? = 0.000\ 98$
f $? \div 10^{-3} = 5.6$

5 **Puzzle**
Arrange these numbers from largest to smallest to spell the name of a Greek mathematician.

E $0.0016 \div 0.001$
D 2.3×10^{-3}
I $300 \div 10^5$
U $0.0023 \div 10^{-2}$
L 37.2×10^{-4}
C 10.2×10^{-3}

6 Write the key sequence that you need to input each of these standard form numbers into your calculator and write the actual display your calculator shows.

a 5.9×10^3
b 5.6×10^{-7}

7 Write each of these numbers in standard form.

a The average diameter of a unicorn hair is 0.00 043 m.
b The height of the highest cloud above Aberdeen at noon yesterday was 24 132 m.
c The star 'Merrymaid' is 630 000 000 000 000 000 km from the Earth.

8 The mass of an object is 0.000 43 g and its volume is 0.8 m^3. Express its density using standard form to 3 significant figures.

Hint: To remind yourself about density, look at page 94.

9 Explain why each of these numbers is not in standard form. Then write them correctly using standard form notation.

a 562×10^{-3}
b 0.062×10^{-2}
c $10.2 \times 10^6 \times 10^{-3}$
d 0.4×10^8

10 Write these probabilities in standard form, to 2 significant figures.

a The probability of being selected at random from the UK population is $\frac{1}{57\,000\,000}$.
b The probability of being selected at random from a school with 936 pupils.

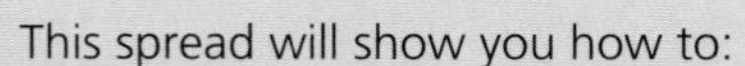

N2.3 Rounding

This spread will show you how to:

- Round numbers to three decimal places and to a given number of significant figures.
- Understand upper and lower bounds.

KEYWORDS
Significant figures
Upper bound
Lower bound
Discrete
Continuous
Accuracy

Significant figures are useful in rounding a number to a particular accuracy.

Number	1 sf	2 sf	3 sf
4.525	5	4.5	4.53
23.758	20	24	23.8
4679	5000	4700	4680
0.000 034 82	0.000 03	0.000 035	0.000 034 8

example

A number is written to 1 significant figure as 5.
Which two of these numbers could it possibly be?
4.2 4.6 5.9 0.5 50 5.37

4.6 and 5.37 both round to 5 (to 1 sf).

Measurements are often rounded, so you may not know what the true values are.
Rounded data can be: **discrete** or **continuous**

The population of Tibet is quoted as 5 000 000 to the nearest million.

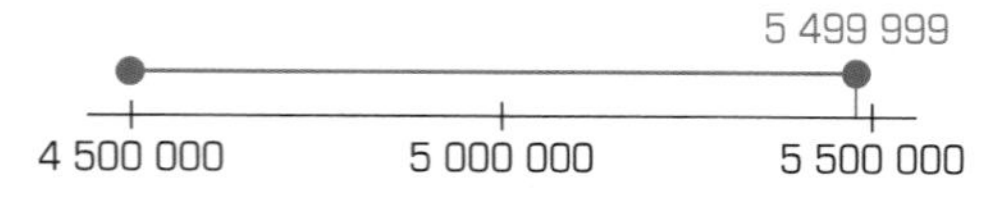

The population can be anywhere between 4 500 000 and 5 499 999 people.

You can write: $4\ 500\ 000 \leqslant p \leqslant 5\ 499\ 999$
or $4\ 500\ 000 \leqslant p < 5\ 500\ 000$
where p is the population.

The area of Tibet is quoted as 1 199 500 km^2 to the nearest square kilometre.

The land area may be anything between 1 199 450 km^2 and 1 199 550 km^2.

You can write:
$1\ 199\ 450 \leqslant A < 1\ 199\ 550$
where A is the area.

If A were actually equal to 1 199 550, it would round **up** not down.

- **The true value of a measurement can lie anywhere between an upper bound and a lower bound.**

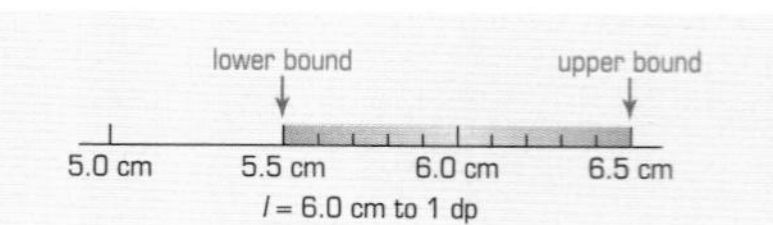

Exercise N2.3

1 Write each number correct to the number of decimal places indicated in brackets.

a 4.3637 (3 dp)
b 56.7382 (2 dp)
c 4.87 (1 dp)
d 34.862 345 (2 dp)
e 0.0048 (1 dp)
f 0.100 008 79 (3 dp)
g 1343.566 (2 dp)
h 4.999 999 (3 dp)

2 Write each number correct to:

a 2 significant figures
i 51 235 **ii** 1.386
b 3 significant figures
i 43.875 **ii** 33.999
c 1 significant figure
i 0.000 067 **ii** 2 640 000

3 **Puzzle**
A measurement is written to 1 sf as 8 cm. Which of the numbers in the cloud could have been the actual measurement?

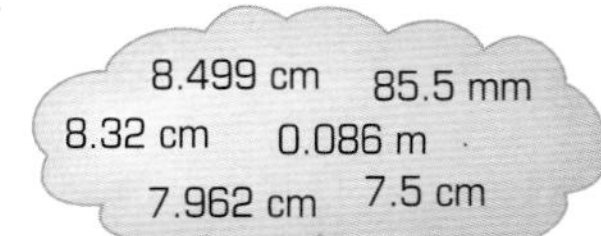

Hint: There is more than one answer.

4 The crowd numbers at Skill City Football Club are given to the nearest 50.
Find the largest and smallest each of the monthly crowds could be.

Sep	Oct	Nov	Dec	Jan	Feb	Mar	Apr	May
1300	1150	950	850	1450	1200	1350	1400	1250

5 Use a calculator to work out each of these calculations to the accuracy given in brackets.

a $\frac{4.63 \times 5.34}{4.56} + 5.72$ (to 3 sf)

b $\frac{46 + 37}{36 \times 58}$ (to 2 sf)

6 In parts **a** and **b**, state the upper and lower bounds.

a The range of shoe sizes (Z) in a class is given by: $3 \leqslant Z < 7$
b The heights (h) of children in a class are given by: $1.36 \text{ cm} \leqslant h < 1.6 \text{ cm}$
c What is the difference between discrete and continuous data?

7 **Puzzle**
Harry Gardener measures his sunflowers with a centimetre ruler and always gives their height to the nearest whole centimetre. He says that the difference between the tallest and shortest sunflower is 94 cm and that the height of the shortest is 1.43 m.

What is the greatest possible height of Harry's tallest sunflower?

8 The dimensions of a field are shown to the nearest metre.
Calculate the largest and smallest possible values for the area of the field.

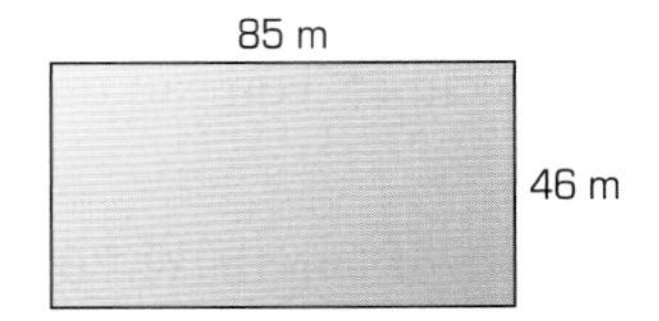

9 A building firm pays £13.46 per tonne (1000 kg) for stone.
Each tonne is calculated accurate to the nearest 10 kg.
How much money could the firm lose, in pounds, in an order of 10 tonnes?

10 A semicircular pond has a radius of 4.23 m and a depth of 0.92 m, both correct to the nearest centimetre.

Calculate the maximum and minimum amount of water, in litres, that is needed to fill the pond.

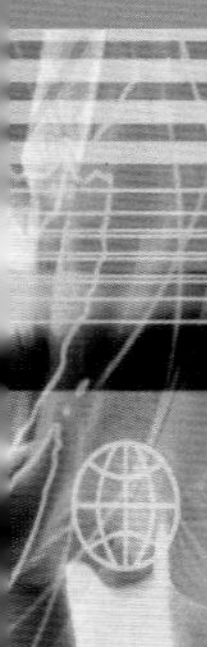

N2.4 Addition and subtraction

This spread will show you how to:

- Extend mental methods of calculation, working with decimals and fractions.
- Use standard column procedures to add and subtract decimals of any size.

KEYWORDS
Partitioning
Compensating
Common denominator

You should know these mental methods for adding and subtracting with decimals.

Partitioning

34.52 + 25.08

- Split into parts: = 34.52 + 25 + 0.08
- Do the easier part first: = 59.52 + 0.08
  = 59.6

Compensating

9.32 − 8.91

- Round up and compensate: = 9.32 − 9 + 0.09
- Do the easier part first: = 0.32 + 0.09
  = 0.41

These methods adapt well for problems relating to time.

example

Hernan arrives at the gym at 3.46 pm and leaves at 5.13 pm.
How long is Hernan at the gym?

- Split up the time line:

3.30, 3.46, 4.00, 4.30, 5.13, 5.30
14 mins | 1 hour | 13 mins

- Add up the parts: 14 min + 1 h + 13 min
  = 1 h 27 min

With harder decimals you can use the standard column method.

example

Two pipes are of length 34.765 m and 3.87 m.
Find the difference in length.

Use the zero as a place value holder.

$$\begin{array}{r} 34.765 \\ -\,03.870 \\ \hline 30.895 \end{array}$$

The difference in length is 30.895 m.

You can add and subtract a mixture of decimals and fractions using a written method.

example

Add the numbers $^{-}5.6$, $\frac{1}{8}$, $\frac{1}{3}$ and 9.3.

- Convert the decimals to fractions: $^{-}5\frac{3}{5} + \frac{1}{8} + \frac{1}{3} + 9\frac{3}{10}$
- Split into integers and fractions: $= {^{-}5} + 9 + \frac{^{-}3}{5} + \frac{1}{8} + \frac{1}{3} + \frac{3}{10}$
- Find the lowest common denominator: $= 4 + \frac{^{-}72 + 15 + 40 + 36}{120}$
  $= 4 + \frac{19}{120} = 4\frac{19}{120}$

There is usually more than one way to do a calculation:
- You could convert all the numbers to decimals.
- You could convert all the mixed numbers to fractions.

Exercise N2.4

1 Calculate each of these using a mental method.

a $13.7 + 3 + 0.03$ **b** $123 - 1\frac{2}{3}$
c $4.87 - 7 + 8.13$ **d** $10 - 4.862\,345$
e $^{-}87.345 - {}^{-}45$ **f** $0.1 + {}^{-}4 + 3.2$
g $13.12 - 7.9$ **h** $4.9 - 4.51 + 3.88$

2 What fraction of cake is left after slices of the following sizes have been removed?

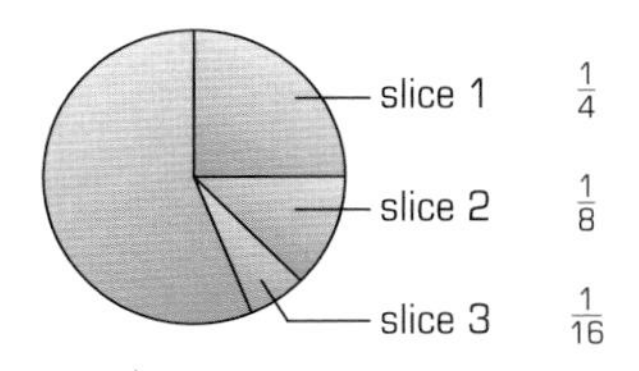

3 Work out the perimeter of this shape using a mental method.

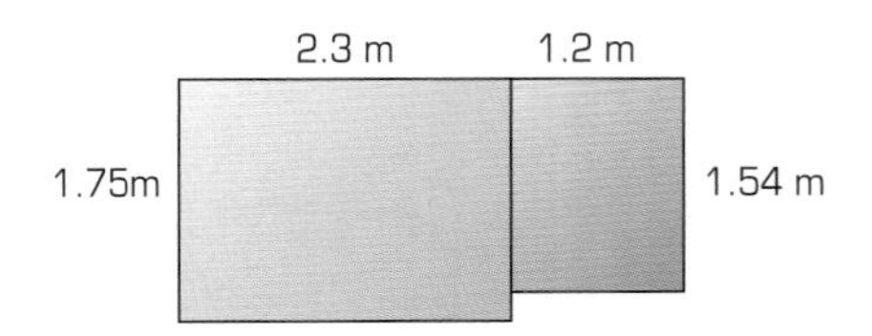

4 Look at these number cards:

a Work out the sum of all the cards.
b Work out the highest possible sum using only four cards.
c The card for 0.46 is removed. What card would you need to add for the sum of all the cards to be $^{-}5$?

5 **Puzzle**
Two fractions are represented by $\frac{a}{b}$ and $\frac{c}{d}$, where *a*, *b*, *c* and *d* are all different integers greater than 0 and less than or equal to 6.
If the sum of the two fractions is $\frac{9}{10}$ what are the values of *a*, *b*, *c* and *d*?

6 Use a written method to calculate each of these.

a $53.12 + 0.33 - 0.1113$

b $5123 + 2.9 - 7.35$

c $454.17 - 0.77 + 0.8813$

d $1.02 - 4.862345$

e $^{-}107.35 - {}^{-}45.7$

f $60.1 + {}^{-}4.678 + 132$

g $83.12 - 7.9999$

h $0.4489 - 4.5 + {}^{-}6.88$

7 **Puzzle**
A unit fraction has a numerator of 1, for example, $\frac{1}{5}$ and $\frac{1}{8}$ are unit fractions.
Which unit fraction is $\frac{1}{5}$ less than $\frac{9}{20}$?
Which three unit fractions have a sum of $\frac{33}{40}$?

8 **Investigation**
Here is a pattern using fraction subtraction:

$1 - \frac{1}{2} = \frac{1}{2}$

$1 - \frac{1}{2} - \frac{1}{3} = \frac{1}{2} - \frac{1}{3} = \frac{1}{6}$

$1 - \frac{1}{2} - \frac{1}{3} - \frac{1}{4} = ?$

Will this pattern ever reach zero?

9 **a** Use mental methods to calculate the length of wire needed to make this cuboid frame.

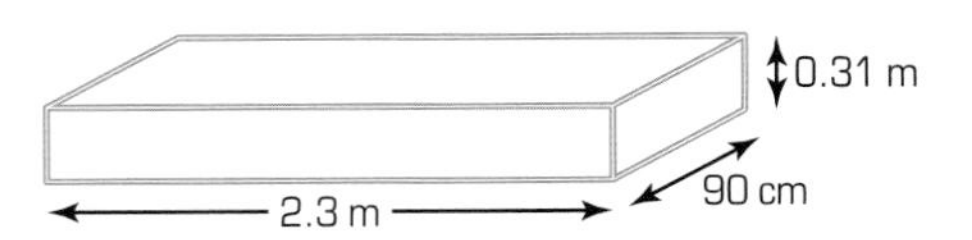

b Sushila has 13 m of wire. She makes a cuboid the same depth (90 cm) and height (0.31 m) as the one shown. How long is Sushila's cuboid?

Mental multiplication and division

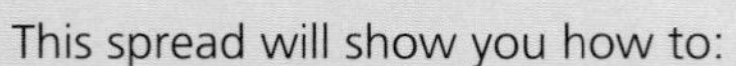

This spread will show you how to:

- Extend mental methods of calculation, working with decimals, fractions and percentages.
- Solve word problems mentally.

KEYWORDS
Partitioning
Factor

You should know these mental methods for multiplying and dividing with decimals.

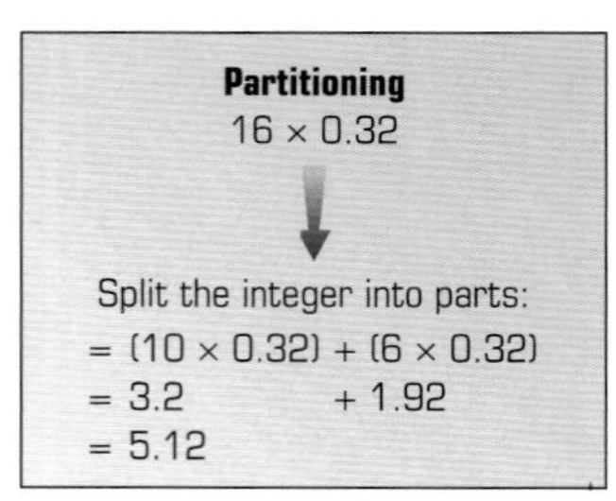

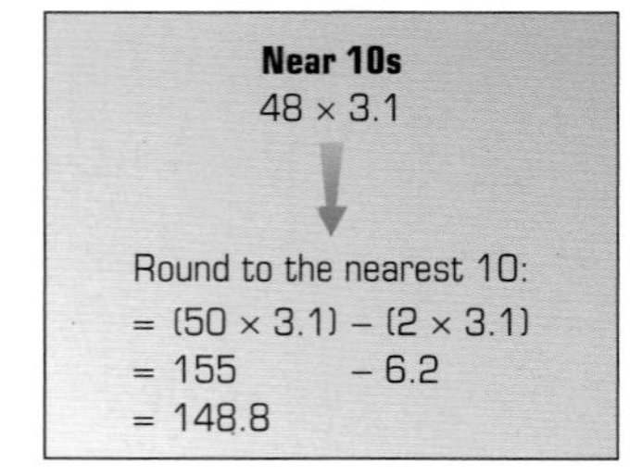

Factors
$700 \div 0.07$

Split the decimal into its factors:

$700 \div (7 \times 0.01)$
$= 700 \div 7 \div 0.01$
$= 100 \div 0.01$
$= 10\,000$

You can solve problems involving fractions and percentages using a variety of mental methods.

example

A length of string measures $1\frac{1}{4}$ m.
72% of it is cut off.
How much string is left?

Here are three different ways to solve this problem.

Convert to fractions:

$$72\% \text{ of } 1\tfrac{1}{4} = \frac{72}{100} \times \frac{5}{4} = \frac{9 \times 1}{10 \times 1} = \frac{9}{10}$$

$$\frac{5}{4} - \frac{9}{10} = \frac{25 - 18}{20} = \frac{7}{20}$$

So $\frac{7}{20}$ m of string remains.

Multiply by a single decimal:

Reduce by 72% $\Longrightarrow$ Find 28% of $1\frac{1}{4}$

$$0.28 \times 1.25 = (1 \times 0.28) + (0.25 \times 0.28) = 0.28 + 0.07 = 0.35$$

So 0.35 m of string remains.

Convert to decimals:

$$72\% \text{ of } 1\tfrac{1}{4} = 0.72 \times 1.25 = (1 \times 0.72) + (0.25 \times 0.72) = 0.72 + 0.18 = 0.9$$

$$1.25 - 0.9 = 0.35$$

So 0.35 m of string remains.

- You should consider a range of ways of solving a problem before tackling it.

Exercise N2.5

1 Calculate each of these using a mental method.

a 12×34 **b** 123×14
c $4.8 \div 0.3$ **d** $10.2 \div 4$
e $\frac{7}{20} \times 45$ **f** $^{-}0.1 \div 5$
g $3.2 \div 0.002$ **h** $19 \times {}^{-}3\frac{2}{3}$

2 Copy and complete the number snake.

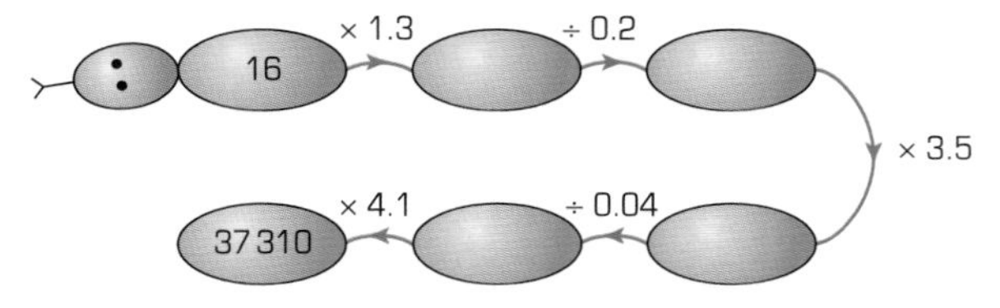

3 Find the value of m in each of these equations.

a $71.5 = 5m$ **b** $m \div 4.1 = 7$
c $\frac{4}{5} \times 3 = m$ **d** $6.46 \times 5.32 = m$
e $\frac{3}{8}$ of $m = \frac{3}{14}$ **f** $76.9 \div m = 153.8$

4 Callum works out 0.067×3.5 on his calculator and gets 0.023 45.
Explain how you know this cannot be right.

5 Look at these number cards:

a What is the lowest product you can make using only two cards?
b What is the mean value of the three cards?

6 Use your mental skills to complete this table:

	0.6	0.09	0.027	0.0018
35% of				
$\frac{2}{5}$ of				
0.34 times				
0.54 divided by				

7 Puzzle
Find the two numbers whose:

a sum is 0.8 and whose product is 0.15
b difference is 0.2 and product is 0.24
c product is $\frac{1}{2}$ and one of the fractions is the reciprocal of $\frac{8}{7}$
d quotient is 2 when the larger number is divided by the smaller and whose sum is 1.

8 Copy and complete the table to show the pay scale for each of the employees below:

	James	Leah	Jordan	Lisa	Lauren
Current pay per hour	£3.40	£3.80	£3.60	£2.70	£4.70
% increase	14	12	21	13	18
New pay per hour					

9 Work out the diameter of each circle, assuming $\pi = \frac{22}{7}$.
Give your answers to 2 dp.

Remember:
$A = \pi r^2$.

a

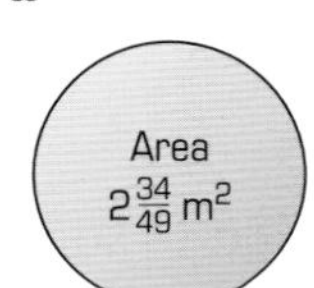

b

Area
10 m²

10 These two containers each hold 1 litre.
Use the information given to find the missing dimension in each shape.
Give your answers as mixed numbers.
Assume $\pi = \frac{22}{7}$.

a

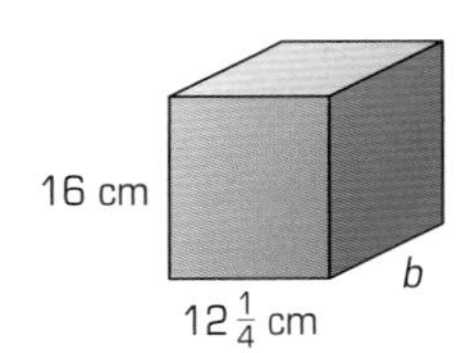

b

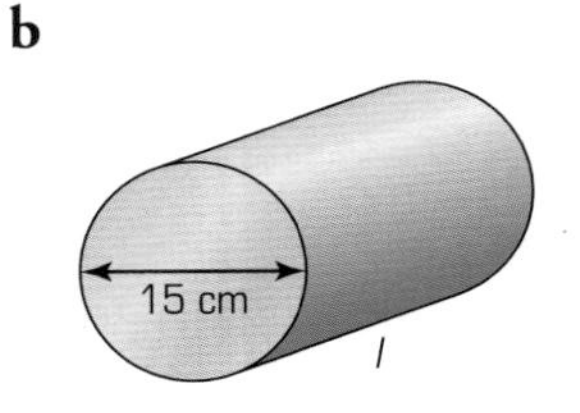

N2.6 Written multiplication

This spread will show you how to:

- Multiply by decimals.
- Estimate calculations by rounding numbers to one significant figure.

KEYWORDS
Estimate
Equivalent
Approximately

You may need to use a written method to multiply difficult decimals.

example

Calculate 0.0452×0.035, giving your answer to:

a 3 decimal places **b** 3 significant figures.

▶ Estimate first, by rounding both numbers to 1 sf: $0.05 \times 0.04 = 0.002$

▶ Find an equivalent calculation:
$0.0452 \times 0.035 = (452 \div 10\,000) \times (35 \div 1000)$
$= 452 \times 35 \div 10\,000\,000$

▶ Use the standard column method:

```
   4 5 2
 ×   3 5
 -------
 1 3 5 6 0
   2 2 6 0
 ---------
 1 5 8 2 0
```

▶ Convert back to a decimal: $15\,820 \div 10\,000\,000 = 0.00\,1582\,(0)$

Each digit has moved seven places to the right.

a To 3 dp the answer is 0.002.
b To 3 sf the answer is 0.001 58.

These answers compare well with the estimate.

You can use written multiplication to solve problems.

example

Gavin and Suzanne build a line of 5671 marbles for a charity event.
Each marble is 0.0091 m in diameter.
How long is the line of marbles?
Give your answer in metres correct to 3 significant figures.

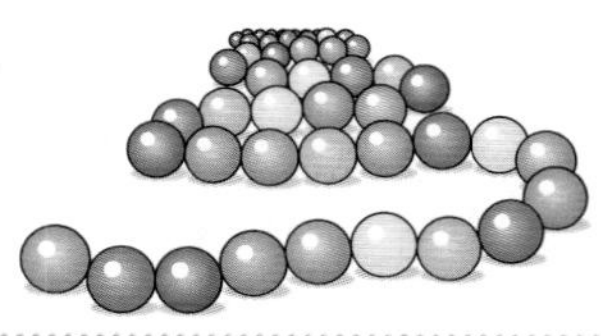

$5671 \times 0.0091 \cong 6000 \times 0.009 = 54$ The answer should be approximately 54 m.

$5671 \times 0.0091 = 5671 \times 91 \div 10\,000$

```
    5 6 7 1
  ×     9 1
  ---------
5 1 0 3 9 0
    5 6 7 1
-----------
5 1 6 0 6 1
```

$\Rightarrow 5671 \times 0.0091 = 516\,061 \div 10\,000 = 51.6061$

The line is 51.6 m long, to 3 sf.

Compare with the estimate of 54 m.

Exercise N2.6

1 Calculate each of these using a written method. Estimate first.

a 6.2×2.4
b 52.3×4.4
c 4.9×0.13
d 20.2×0.24
e 0.820×0.25
f 0.34×0.652
g $^{-}8.2 \times 0.072$
h $^{-}0.29 \times {}^{-}8.24$

2 Copy and complete these calculations.

a $? \div 13.8 = 3.4$
b $56 \times 0.00\,896 = ?$
c $7 + (? \div 53.6) = 3.4$
d $? + 6 = 4.3 \times 54.9$

3 Work out the area of each rectangle.

a

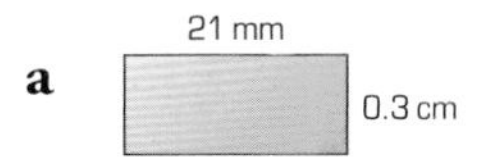

Give your answer in cm^2.

b

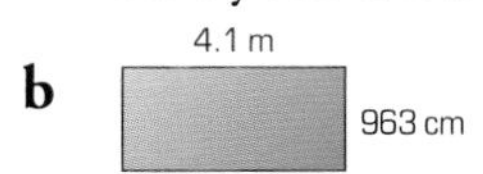

Give your answer in m^2.

4 Work out each of these using a written method.

a $2.1 \times 3.8 \times 0.86$
b $3.1 \times 0.002 \times 4.36$
c $^{-}2.12 \times 1.2 \times {}^{-}3.42$

Remember to estimate first.

5 If $a = {}^{-}0.54$, $b = {}^{-}0.078$ and $c = 0.63$, work out the value of:

a abc **b** a^2 **c** $a(b + c)$

6 Puzzle

Helen says that if she divides a mystery number by 34.4 and then multiplies the result by 5 she ends up with 43.5.
By forming and solving an equation, work out the mystery number.

7 a What is the cost of 6.5 m of tape at 7.6p per metre?

b If a snail takes, on average, 56.1 seconds to crawl 1 m, how long would it take it to crawl the length of a garden 21.6 m long?

c The area of a school desk is 1.82 m by 0.62 m. Calculate its area in square metres.

d Work out the cost of 563.2 kg of sugar at 18.6p per kilogram.

8 Puzzle

Using the digits 3, 4, 5, 6, 7 and 8, what is:

a the smallest
b the largest number you can make?

? ? . ? × 0 . ? ? ?

9 This trapezium has an area equal to that of a circle of radius 4.3 m.
Work out the height of the trapezium, using a written method. Let $\pi = 3.14$, and give your answer correct to 2 dp.

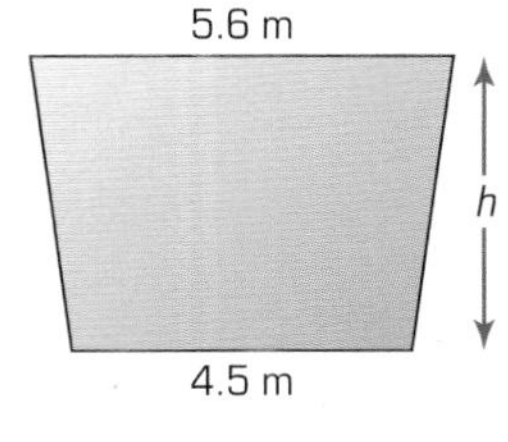

10 Work out each of these, giving your answers correct to:

i 3 decimal places
ii 3 significant figures.

Remember to estimate first.

a 701.5×5.34
b 0.0041×79.3
c 0.3434×0.43
d 64×0.45
e $^{-}0.004\,68 \times 0.000\,456\,7$
f $^{-}0.0789 \times {}^{-}0.3158$

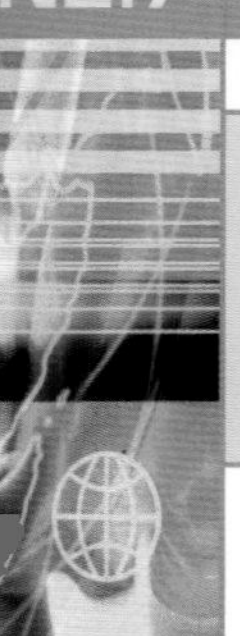

N2.7 Written division

This spread will show you how to:

- Divide by decimals by transforming to division by an integer.
- Estimate calculations by rounding numbers to one significant figure.
- Use efficient methods to divide fractions.

KEYWORDS

Divisor

Repeated subtraction

Multiplicative inverse

You can use a written method to divide difficult decimals. First change the divisor to an integer.

Remember:
The divisor is the number that you are dividing by.

example

Calculate $0.0452 \div 0.35$, giving your answer to:

a 3 decimal places (3 dp) **b** 2 significant figures (2 sf).

- First estimate: $0.0452 \div 0.35 \approx \frac{0.05}{0.4} = \frac{5}{40}$
  $= \frac{1}{8} = 0.125$
- Transform the divisor into an integer: $\frac{0.0452}{0.35} = \frac{4.52}{35}$
- Use the repeated subtraction method (long division):

```
35 | 4.52000
   - 3.5          35 × 0.1 = 3.5
     1.02
   - 0.70         35 × 0.02 = 0.7
     0.32
   - 0.315        35 × 0.009 = 0.315
     0.005
   - 0.0035       35 × 0.0001 = 0.0035
     0.0015       Total = 0.1291
```

So $0.0452 \div 0.35 = 0.1291$ (remainder 0.0015)

a 0.129 (to 3 dp) **b** 0.13 (to 2 sf)

The answers compare well with the estimate of 0.125.

You can use a written method to divide difficult fractions. Remember to use the multiplicative inverse.

example

A rectangle has an area of $1\frac{3}{7}$ m^2.
Its width is $\frac{5}{8}$ m.
Find the length of the rectangle, giving your answer as a fraction in its simplest form.

Area $1\frac{3}{7}$ m^2 | $\frac{5}{8}$ m

$1\frac{3}{7} \div \frac{5}{8} = \frac{10}{7} \div \frac{5}{8}$

$= \frac{10}{7} \times \frac{8}{5} \quad = \frac{2 \times 8}{7 \times 1} \quad = \frac{16}{7}$

The length is $2\frac{2}{7}$ m.

You can estimate with fractions as well:
$1\frac{3}{7} \div \frac{5}{8} \cong 1\frac{1}{2} \div \frac{1}{2} = 1\frac{1}{2} \times 2 = 3$

Exercise N2.7

1 Calculate each of these using a written method, giving your answer to 2 decimal places. Remember to estimate first.

a $162 \div 19$
b $189 \div 39$
c $4.99 \div 0.13$
d $20.2 \div 0.24$
e $82.4 \div 0.25$
f $7.4 \div 0.12$
g $^{-}9.21 \div 0.022$
h $^{-}8.29 \div {}^{-}2.4$

2 **a** **i** A jar of coffee holds 0.8 kg. How many jars will be needed to hold 74.3 kg of coffee?
ii How much coffee will be left over?

b 45 kg of washing powder was needed to keep the football kits of Winagain United clean last season. If 0.09 kg is used per wash, how many washes were done?

c A car manufacturer claims that a particular car will travel 783 km on 64 litres of fuel. Find the fuel consumption of the car in km/litre. Give your answer to the nearest km.

Hint: Divide the distance by the amount of fuel.

3 In these pyramids, each brick is the product of the two bricks below. Copy and complete each pyramid.

a

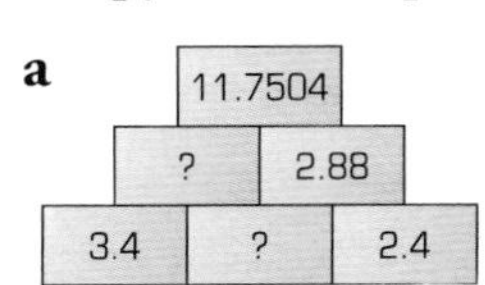

b

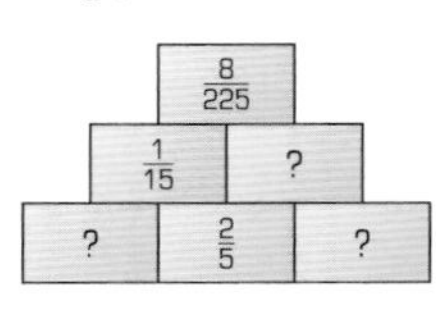

4 **Puzzle**
Using the digits 4, 5, 6, 7 and 8, what are the smallest and largest numbers you can make?

? ? ? ÷ 0 . ? ?

5 The hexagon shown is made up of six equilateral triangles. The area of each triangle is 2.5 m^2 and the perimeter of the hexagon is 14.4 m. Use this information to calculate the vertical height of each triangle (to 2 dp).

6 **Puzzle**
Copy and complete these division puzzles, inserting any of the digits 1, 2, 3 or 4 in the correct place.

a $5.76 \div \frac{?}{??} = 23.04$

b $0.576 \div \frac{?}{??} = 3.744$

7 Work out each of the following, giving your answers correct to:

i 3 decimal places
ii 3 significant figures.

Estimate each answer first.

a $801.5 \div 3.4$
b $0.00716 \div 0.32$
c $3434 \div 33$
d $74.3 \div 0.45$
e $^{-}0.00868 \div 0.00045$
f $^{-}0.0589 \div {}^{-}0.38$

8 **Investigation**
Can an equilateral triangle have a perimeter with the same value as its area? Investigate for other regular shapes.

9 Francesca says it is always easier to turn any division problem into fractions and then use the multiplicative inverse method to solve it.
Do you agree? Show examples to support your comments.

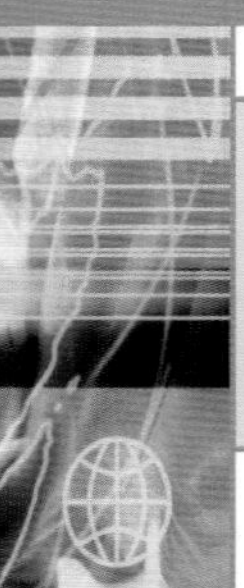

N2.8 Recurring decimals and reciprocals

This spread will show you how to:

- Use algebraic methods to convert a recurring decimal to a fraction.
- Recognise and use reciprocals.
- Use the reciprocal key on a calculator.

KEYWORDS
Recurring decimal
Reciprocal

A recurring decimal has an endlessly repeating sequence of digits.

$0.333\,333\ldots = 0.\dot{3}$ $0.166\,666\ldots = 0.1\dot{6}$ $0.513\,513\ldots = 0.\dot{5}1\dot{3}$

Place the dots above the first and last repeating digits.

- **You can convert any recurring decimal to a fraction.**

example

Convert $0.\dot{3}\dot{4}$ to a fraction.

Let $n = 0.343\,434\,3\ldots$
$100n = 34.343\,43\ldots$

Subtract: $99n = 34.343\,43\ldots - 0.343\,434\,3\ldots$
$99n = 34$
$n = \frac{34}{99}$

Note:
If there are 3 recurring digits you multiply by 1000.

Check on your calculator:

0.34343434

- **The reciprocal of a number is the result of dividing it into 1.**
  The reciprocal of *n* is 1 ÷ *n*.

example

Find the reciprocal of:

a 0.4 **b** 0.27

a $1 \div 0.4 = 10 \div 4$
$= 2.5$

b $1 \div 0.27$
Use a calculator: 

$= 3.703\,703\,704$

The calculator has rounded the last digit up to a 4.

3.703703704

To find the reciprocal of a fraction you turn it upside-down.

The reciprocal of $\frac{2}{5}$ is $\frac{5}{2}$ The reciprocal of $\frac{16}{3}$ is $\frac{3}{16}$

The reciprocal of $\frac{5}{2}$ is $\frac{2}{5}$ The reciprocal of $\frac{3}{16}$ is $\frac{16}{3}$

Remember:
Convert mixed numbers to improper fractions first.

- **The reciprocal of a reciprocal gives the original number.**
- **A number multiplied by its reciprocal equals 1.**

$\frac{2}{5} \times \frac{5}{2} = 1$ $\frac{16}{3} \times \frac{3}{16} = 1$

Exercise N2.8

1 Convert each of these recurring decimals into fractions.

> **Hint:** Count how many recurring digits there are. This will decide whether you multiply by 10, 100, 1000 … .

a 0.232 323 2 … **b** 0.444 444 4 …
c 0.456 456 456 … **d** 0.545 454 5 …
e 0.353 535 … **f** 0.555 555 …
g 0.989 898 989 … **h** 0.373 737 …

2 Copy and complete the table.

Fraction	$\frac{4}{7}$	$\frac{2}{9}$	$\frac{-4}{15}$	$5\frac{1}{6}$	$62\frac{5}{9}$	$^{-}4\frac{6}{7}$
Decimal						

3 Show the key sequence you would use on your calculator to find the reciprocal of 0.39.

> See the second example on page 114.

4 **Puzzle**
Pair up each of these numbers with its reciprocal. Show your working.

13		$0.1\dot{6}$
$0.\dot{0}76\,92\dot{3}$	15	
	6	$0.0\dot{6}$

5 Without using a calculator, work out the fractional equivalent of the reciprocal of each decimal.

a 0.6 **b** 0.56
c 1.2 **d** 5.4
e 43.2 **f** $0.\dot{6}$
g $^{-}0.87$ **h** $^{-}1.67$

6 Convert each of these recurring decimals to a fraction or mixed number.

a $2.\dot{2}\dot{4}$ **b** $3.\dot{6}$
c $4.\dot{4}3\dot{6}$ **d** $7.\dot{8}\dot{4}$
e $78.\dot{6}\dot{5}$ **f** $^{-}4.\dot{5}$
g $^{-}1.\dot{9}\dot{8}$ **h** $100.\dot{3}\dot{7}$

7 **Puzzle**
Each of these numbers, given correct to 4 significant figures, is the reciprocal of a number between 20 and 80. Work out the value of the missing digit in each case.

a 0.015 ?3 **b** 0.03? 71 **c** 0.01? 66

8 **a** Copy and complete the tables below for the functions given.

i $x \rightarrow \frac{1}{x}$

x	1	2	3	4	5
$\frac{1}{x}$					

ii $x \rightarrow \frac{2}{x}$

x	1	2	3	4	5
$\frac{2}{x}$					

iii $x \rightarrow \frac{3}{x}$

x	1	2	3	4	5
$\frac{3}{x}$					

b Draw the graphs of the three functions in part **a**, using the same axes. Comment on your results.

9 **Investigation**
Investigate the graphs produced by these functions:

a $\frac{1}{x}$ **b** $\frac{1}{x+1}$
c $\frac{1}{x+2}$ **d** $\frac{1}{x+3}$

Describe anything you notice and generalise your findings.

10 **Investigation**
The fraction $\frac{1}{7}$ has 7 recurring places of decimals: $\frac{1}{7} = 0.\dot{1}42\,85\dot{7}$.
Investigate whether it is possible to predict the number of recurring decimal places a fraction will have from its numerator and denominator.

Summary

You should know how to ...

1 Know and use the index laws for multiplication and division of positive integer powers.

2 Generate fuller solutions to problems.

3 Recognise limitations on accuracy of data and measurements.

Check out

1 By using the values $t = 4$, $a = 3$ and $b = 2$, show that these statements are true.

   a $(t^a)^b = t^{ab}$

   b $b^t \times b^a = b^{t+a}$

   c $a^t \div a^b = a^{t-b}$

2 **a** Use each of the digits 1, 2, 3 and 4 once only in this multiplication:

   □□ × □□

   to produce the largest solution.

   b Does this arrangement of digits help you to find the largest solution for other sets of four consecutive digits?
   Explain your answer. (You could use the letters a, b, c, d to represent the digits in your explanation.)

   c Investigate further with five consecutive digits in the multiplication:

3 The area of a triangle is 30 cm^2.

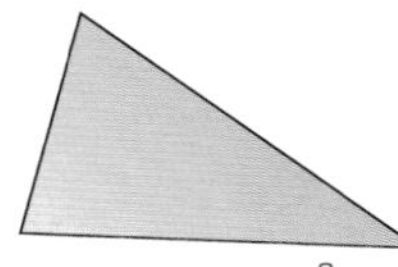

Area = 30 cm^2

If its height is given as 14 cm correct to the nearest cm, calculate the range of values the base of the triangle could actually have.

4 Indices and graphs

This unit will show you how to:

- Simplify or transform algebraic expressions by taking out single-term common factors.
- Multiply a single term over a bracket.
- Derive and use more complex formulae.
- Use index notation for integer powers and simple instances of the index laws.
- Know and use the index laws in generalised form for multiplication and division of positive integer powers.
- Begin to extend understanding of index notation to negative and fractional powers.
- Given values for m and c, find the gradient of lines given by equations of the form $y = mx + c$.
- Investigate the gradients of parallel lines and lines perpendicular to these lines.
- Plot graphs of quadratic and cubic functions.
- Begin to solve inequalities in two variables.
- Square a linear expression, expand the product of two linear expressions of the form $x \pm n$ and simplify the corresponding quadratic expression.
- Generate fuller solutions to problems.

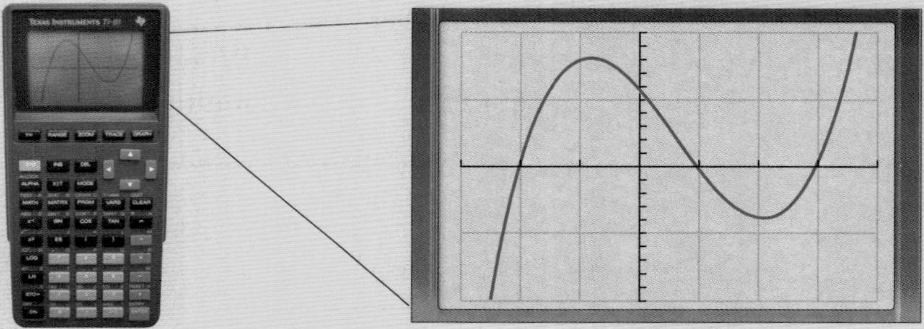

You can use a graphical calculator to explore graphs of functions.

Before you start

You should know how to ...

1 Use index notation.

2 Expand and factorise expressions.

3 Find the gradient of a line segment.

Check in

1 Write these expressions using indices:

a $3 \times 3 \times 3 \times 5 \times 5 \times 5 \times 5 \times 5 \times 5$

b $x \times x \times x \times x \times x$ **c** $a \times b \times a \times a \times b \times a$

d $2 \times c \times d \times 3 \times c \times c \times d \times d$

2 a Which expression does not expand to give the same expression as the others?

b Factorise:

i $4x + 12$ **ii** $10x - 5$ **iii** $xy + xw$ **iv** $x^2 - 2x$

3 Find the gradients of these line segments:

a

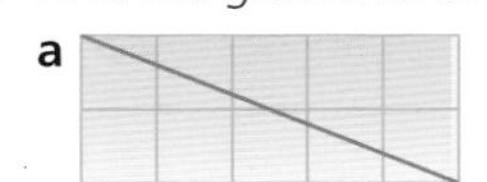

b

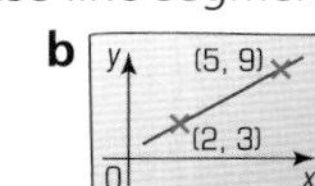

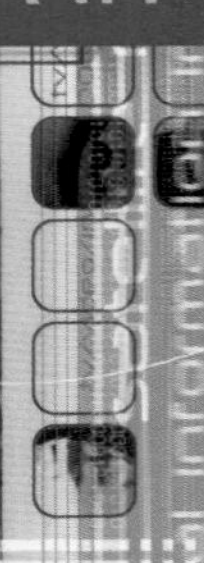

A4.1 Revising indices

This spread will show you how to:

- Use index notation for integer powers and simple instances of the index laws.
- Know and use the index laws in generalised form for multiplication and division of positive integer powers.

KEYWORDS

Index, Exponent, Indices, Base, Simplify, Power

An **index** tells you how many times to multiply a number by itself.

The plural of index is 'indices'. Other words for index are 'exponent' and 'power'.

$3^4 = 3 \times 3 \times 3 \times 3$

In 3^4, 3 is the **base** and 4 is the **index**.

The **laws of indices** can help you to simplify expressions.

Multiplying

- $x^a \times x^b = x^{a+b}$

When you multiply, you add the indices.

Dividing

- $\frac{x^a}{x^b} = x^{a-b}$

When you divide, you subtract the indices.

Brackets

- $(x^a)^b = x^{ab}$

When you have brackets, you multiply the indices.

example

Simplify these expressions.

a $p^6 \times p^7$ **b** $q^{10} \div q^2$ **c** $(z^4)^{10}$

a $p^6 \times p^7 = p^{6+7} = p^{13}$

b $q^{10} \div q^2 = q^{10-2} = q^8$

c $(z^4)^{10} = z^{4 \times 10} = z^{40}$

Note:
You can use the rules of indices in this example, because the base is the same each time.

The laws of indices also help in solving problems.

example

Write an expression for the area of each shape.

a

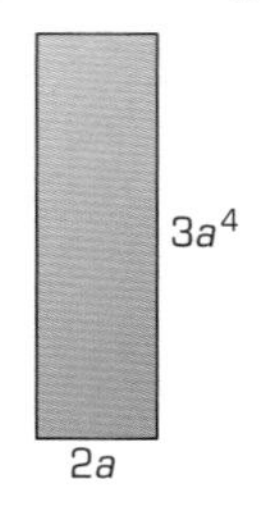

b

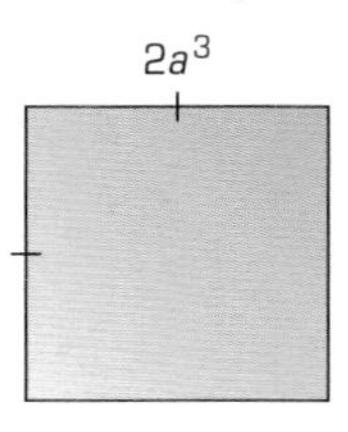

c

3a⁴, 2a³, 2a

a Area $= 2a \times 3a^4 = 6a^5$

b Area $= (2a^3)^2 = 4a^6$

2^2 then $(a^3)^2$

c Area $= 6a^5 + 4a^6$

This cannot be simplified because you are adding unlike terms.

Exercise A4.1

1 Evaluate these without a calculator:

a 10^4 **b** 4^4 **c** 2^8 **d** 1^{50}
e 0^{10} **f** $(^-3)^3$ **g** $(\frac{1}{2})^6$

2 Simplify these expressions, using the laws of indices.

a $x^6 \times x^8$ **b** $y^{12} \div y^4$
c $(z^4)^8$ **d** $y^7 \div y^3$
e $(2^{-2})^3$ **f** $3x^4 \times 4x^3$
g $10y^7 \div 2y^2$ **h** $(3z^2)^3$
i $\frac{(a^6 \times a^8)^2}{a^7}$ **j** $(\frac{8b^4}{2b^{-2}})^3 \times 2b^7$

3 Explain why $a^6 \times b^7$ cannot be simplified beyond $a^6 b^7$.

4 Copy and complete:

a $p^6 \times p^{\square} = p^{14}$ **b** $(2z^2)^{\square} = \square z^6$
c $\frac{10y^8}{2y^{\square}} = \square y^{10}$ **d** $\frac{(p^{\square} \times p^3)^2}{p^{\square}} = p^4$

(How many different answers can you get in part **d**?)

5 Do not use calculators for this question.
x is an unknown number.
Look at the table.

$x^3 = 512$
$x^4 = 4096$
$x^5 = 32\ 768$
$x^6 = 262\ 144$
$x^7 = 2\ 097\ 152$
$x^8 = 16\ 777\ 216$

a Explain how it is possible for the table to show that $512 \times 32\ 768$ is $16\ 777\ 216$.
b Use the table to evaluate $\frac{2\ 097\ 152}{4096}$.
c Use the table to evaluate $(4096)^2$.
d What will the units digit of x^{12} be, given the units digit of x^6 is 4?

6 Find the values of x, y and z without using a calculator if:
$4096 = 64^2 = 8^x = 4^y = 2^z$

7 Find pairs of equivalent expressions:

8 Simplify these expressions.

a $5p^6 \times 4p^2$ **b** $8q^7 \div 2q^4$
c $(4m^3)^4$ **d** $5a^2 \times 4b \times 2a$
e $\frac{3a^4 \times 4b^6}{2b \times 3a^2}$ **f** $(^-4a^3)^3$

9 The expression $3a^2 \times 4b^2 \times 2a^6$ can be simplified to give $24a^8 b^2$. Write as many other expressions as you can that would simplify to give $24a^8 b^2$.

10 True or false? $3^x \times 3^y$ is equivalent to 9^{x+y}.

11 Write a simplified formula for the required quantity in each case.

a

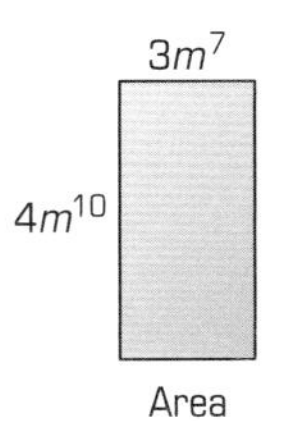

b

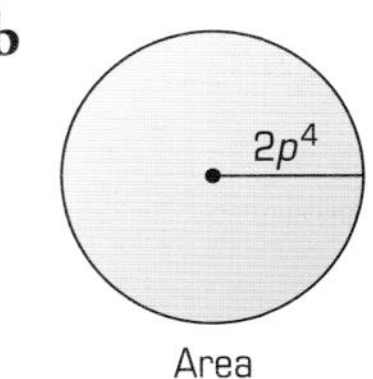

c

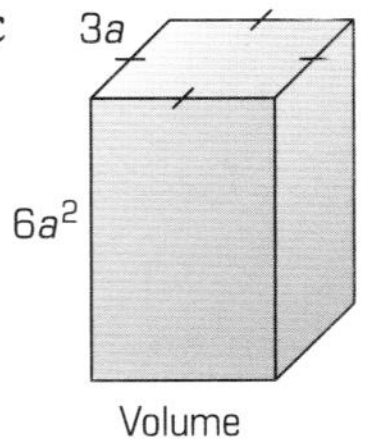

d

12 Show that the volume of this solid is $168x^6$ cubic units.

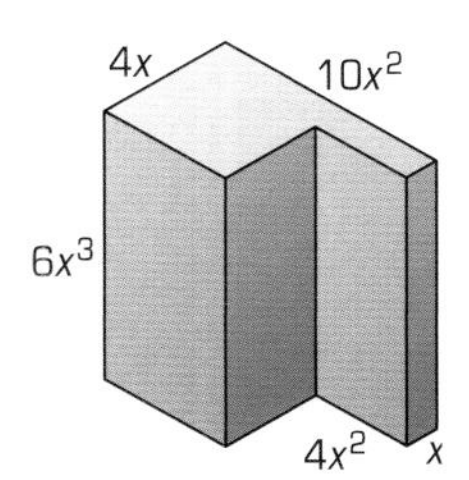

Negative and fractional indices

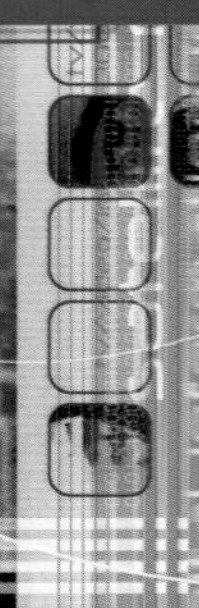

This spread will show you how to:

- Know and use the index laws in generalised form for multiplication and division of integer powers.
- Begin to extend understanding of index notation to negative and fractional powers, and recognise that the index laws can be applied to these as well.

KEYWORDS

Index, indices Reciprocal
Evaluate Index laws

You should be familiar with positive indices. $2^3 = 8$ $(^-3)^5 = {}^-243$

Check this by working out $^-3 \times {}^-3 \times {}^-3 \times {}^-3 \times {}^-3$

You should also know about zero, negative and fractional indices.

Zero indices
Any number to the power of zero is 1.
$17^0 = 1$, $(^-3)^0 = 1$

Negative indices
To evaluate 5^{-2}:
- work it out as if it were a positive index: $5^2 = 25$
- write the reciprocal: $\frac{1}{25}$

Fractional indices
The power of $\frac{1}{2}$ is a square root.
$\Rightarrow$ $36^{\frac{1}{2}} = 6$
The power of $\frac{1}{3}$ is a cube root.
$\Rightarrow$ $125^{\frac{1}{3}} = 5$

You can summarise these facts using algebra.

- $a^0 = 1$
- $a^{-n} = \frac{1}{a^n}$
- $a^{\frac{1}{n}} = \sqrt[n]{a}$

example

Here is a formula with **P** as the subject.
$P = x^0 + x^{\frac{1}{2}} + x^{-2}$
Evaluate P when $x = 4$.

$$P = 4^0 + 4^{\frac{1}{2}} + 4^{-2}$$
$$= 1 + 2 + \frac{1}{16} = 3\frac{1}{16}$$

You should remember the index laws.

$x^a \times x^b = x^{a+b}$ $(x^a)^b = x^{ab}$
$x^a \div x^b = x^{a-b}$

You can solve simple equations involving negative and fractional indices.

example

Solve the equations:

a $16y^{-3} = 2$ **b** $2x^{\frac{1}{2}} = 12$ **c** $(x^{\frac{1}{4}})^2 = 2$

a $y^{-3} = \frac{1}{y^3}$, so $16y^{-3} = \frac{16}{y^3}$
$\frac{16}{y^3} = 2$
$16 = 2y^3$
$8 = y^3$
$2 = y$

b $2x^{\frac{1}{2}} = 12$
$x^{\frac{1}{2}} = 6$
$\sqrt{x} = 6$
$x = 6^2$
$= 36$

c $(x^{\frac{1}{4}})^2 = 2$
$x^{\frac{2}{4}} = 2$
$x^{\frac{1}{2}} = 2$
$x = 2^2$
$x = 4$

Remember: Multiply the indices.

Exercise A4.2

1 **a** If $x = 4$ and $y = 8$, match each expression with its value.

Expression

$x^{\frac{1}{2}}$ x^{-3} x^0 y^{-2} $y^{\frac{1}{3}}$

Value

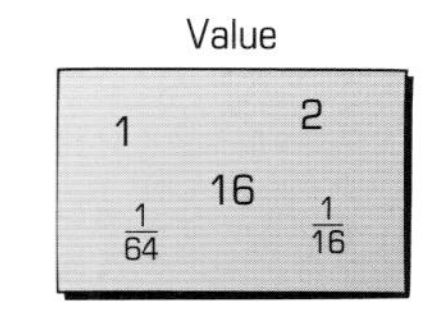

b For any unmatched values, write an expression in terms of x or y with this value.

2 **a** Simplify each expression using the laws of indices.

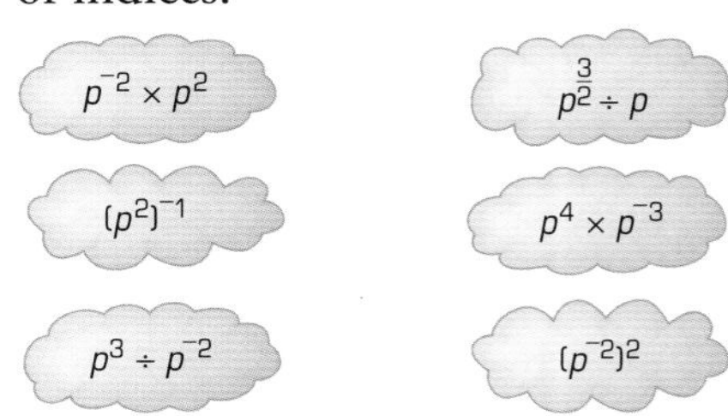

b If $p = 9$, put your simplified expressions in ascending order.

3 Simplify these.

a a^0 **b** $x^6 \times x^{-8}$

c $(x^{-3})^{-2}$ **d** $\frac{x^4}{x^{-2}}$

e $x^3 \times x^{-3}$ **f** $(x^{\frac{1}{4}})^2 \times x$

4 A different way of writing x^{-6}, for example, is to use its reciprocal $(\frac{1}{x^6})$. This looks neater and is easier to use in further problem solving. Simplify the following, writing your answer in the form $\frac{1}{y^n}$, where n is an integer.

a $y^4 \times y^{-7}$ **b** $(y^3)^{-2}$

c $\frac{y^8}{y^{10}}$ **d** $y^{\frac{1}{2}} \times y^{-1\frac{1}{2}}$

e $y \div y^{1\frac{1}{2}}$ **f** $(y^{\frac{-3}{2}})^4$

5 **a** Write a simplified expression for the area of this rectangle.

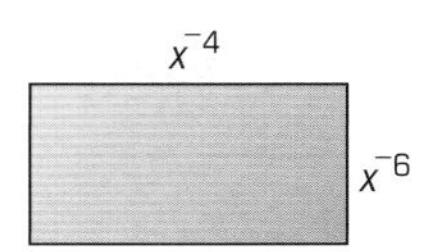

b If $x = 2$, what is the value of the area?

c If the area is 1, what is the value of x?

6 True or false? The area of this triangle is always $\frac{1}{2}$, regardless of the value of y.

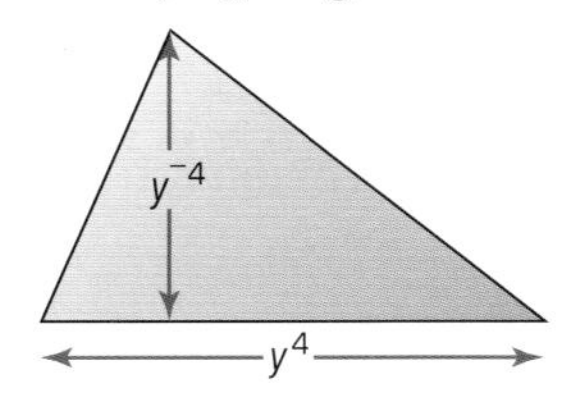

7 Write an equation using the given information.
Solve it to find the unknown in each case.

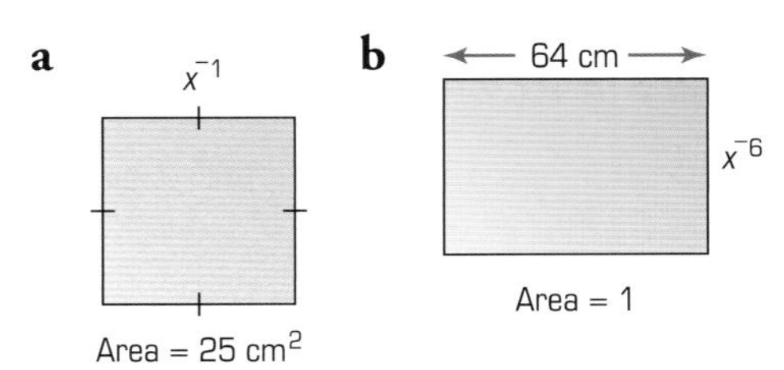

c $\sqrt{x}$ $\frac{1}{x^2}$ x Product = $\frac{1}{4}$

8 If $x = \frac{1}{8}$, put these expressions in descending order
Repeat for $x = {}^{-}8$.

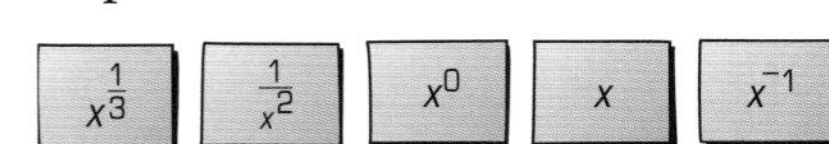

A4.3 Simplifying expressions

This spread will show you how to:

- Simplify or transform algebraic expressions.
- Derive a formula.

KEYWORDS

Expression, Variable, Simplify, Like terms, Operation, Coefficient

When you simplify an expression, you should look carefully at the operations involved. Here are some helpful hints.

Multiplication and division

- In algebra you write expressions without × or ÷ signs.

  $3p \times q = 3pq \qquad 15k \div 4m = \frac{15k}{4m}$

- It helps to deal with the coefficients first, then the variables.

  $4k \times 3m \times 2k = 24k^2 m \qquad \frac{15m}{5n} = \frac{3m}{n}$

- You can use the laws of indices to help with × and ÷.

  $3m^6 \times 8m^7 \times 3m^{^-2} = 72m^{11} \qquad 12m^3 \div 4m^7 = 3m^{^-4}$

- Think of brackets as meaning 'all × by' and fractions as 'all ÷ by'.

  $4x(x^2 + 3xy) = 4x^3 + 12x^2y \qquad \frac{3x+15}{3} = x + 5$

Remember:
A **variable** is a letter that can take different values.
A **coefficient** is the number in front of a variable.

Note:
The sign to the left of a term moves with the term.
$2p + 3q - 3p$
$= 2p - 3p + 3q$
$= {}^-p + 3q$ or $3q - p$

Addition and subtraction

- Only like terms can be collected together – simply count how many of each you have.

  $3x + 2y + 4x - 5y = 3x + 4x + 2y - 5y$
  $= 7x - 3y$

An expression like $3x + 2x^2$ cannot be simplified.

- Some terms may appear unlike, but they are really like terms.

  $3ab + 6ba = 9ab$

By convention, write letters in alphabetical order.

- Brackets can sometimes appear in addition and subtraction expressions.

  $3m + ({}^-2m) = 3m - 2m = m \qquad 3 - (2x + 4y) = 3 - 2x - 4y$

Be careful with negatives.

Some problems involve a mixture of operations, so extra care is needed.

example

Write an expression for the area of this trapezium. Simplify your expression.

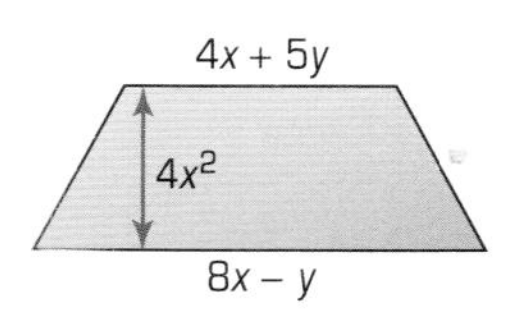

$\text{Area} = \frac{\text{sum of parallel sides}}{2} \times \text{height}$
$= \left(\frac{4x + 5y + 8x - y}{2}\right)4x^2$
$= \left(\frac{12x + 4y}{2}\right)4x^2$
$= (6x + 2y)4x^2$
$= 24x^3 + 8x^2y$

Exercise A4.3

1 Copy each grid. Replace the expression in each cell with its simplified form.

a

$3 \times x$	$5 \times a \times b$	$2 \times a \times 3 \times b$
$3a \times 4b$	$2a \times 5b \times 4c$	$3x \times 2x \times x$
$2b^2 \times 5b^3$	$3b^3 \times 6b^{-2}$	$5x^7 \times 2x^9 \times 3x^6$

b

$15 \div m$	$3a \div 4$	$6b \div 13c$
$\frac{15ab}{3a}$	$\frac{24a^2}{6a^2}$	$\frac{10xyz}{2xw}$
$\frac{12x^7}{6w^4}$	$\frac{8p^2}{4p^{-3}}$	$\frac{20x^6y^5}{4x^2y^3}$

c

$3a + 9a$	$2x - 4x + 7x$	$3a + 4b - 6a - 2b$
$3xy + 4xy$	$5ab - 6ba$	$10xy + 11yz$
$3x^2 + 4x - 5x^2 + xy$	$pqrs + qsrp$	$3ab + 6$

2 Decide if each statement is true or false.

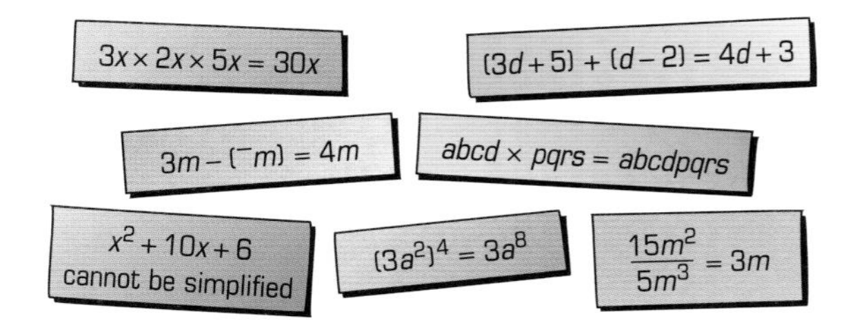

3 **a** Write a simplified expression for the:
i area **ii** perimeter of this rectangle.

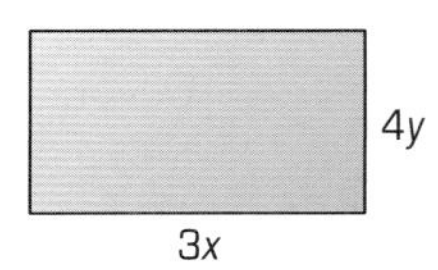

b A different rectangle has area $12y^2$ and perimeter $14y$. What are the dimensions of this rectangle?

4 Write an expression for the missing lengths in each rectangle, in simplified form.

a

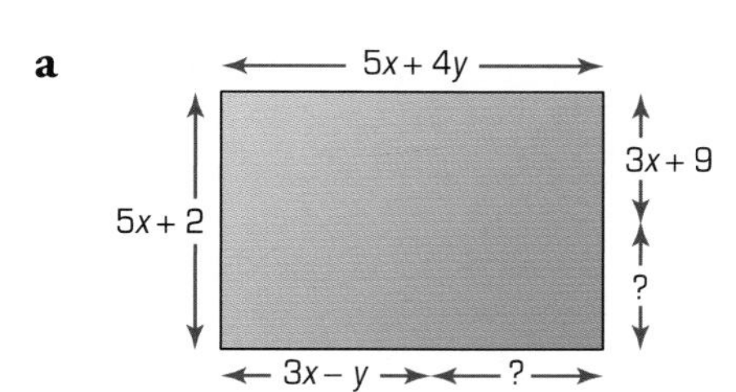

b

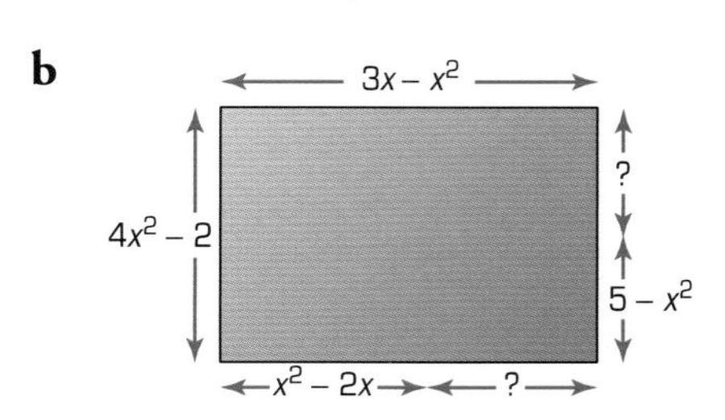

5 Use a formula or background knowledge to find a simplified expression for the given quantity.

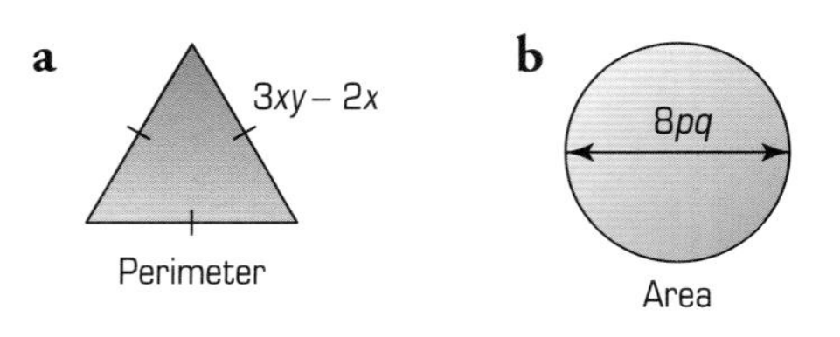

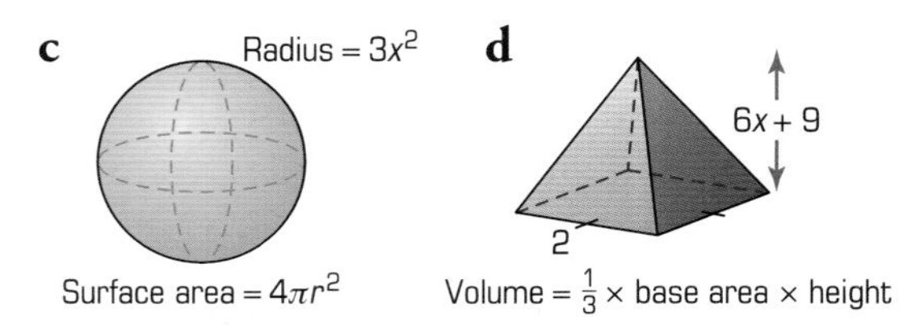

e

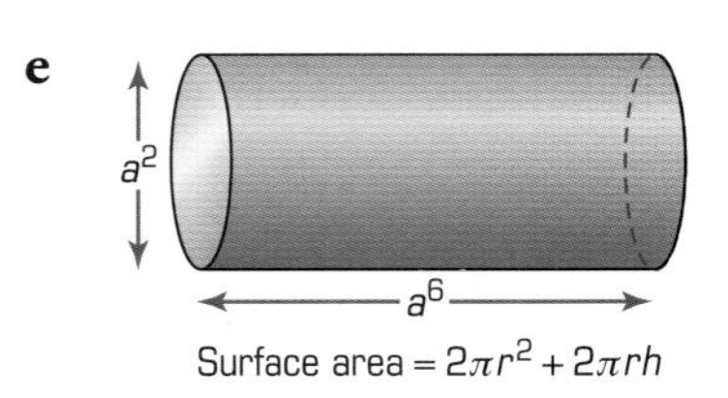

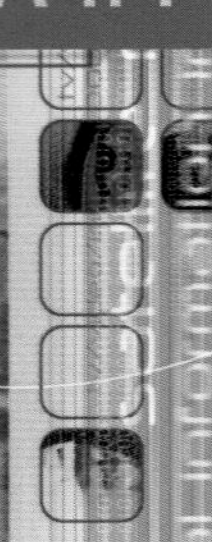

A4.4 Expanding and factorising

This spread will show you how to:

- Simplify or transform algebraic expressions by taking out single-term common factors.
- Multiply a single term over a bracket.

KEYWORDS

Expand, HCF, Factorise, Expression, Common factor

You should know how to expand and factorise expressions.

Expanding
You remove brackets by **multiplying** all terms inside by the term outside.

expand: $3x(x + 2) \to 3x^2 + 6x$; factorise: $3x^2 + 6x \to 3x(x + 2)$

Factorising
You insert brackets by **dividing** each term by a common factor. This is called 'taking out a common factor'.

You need to take care with negative terms outside a bracket.

example

Write an expression for:

a the total area of this shape

b the difference in height between rectangle A and rectangle B.

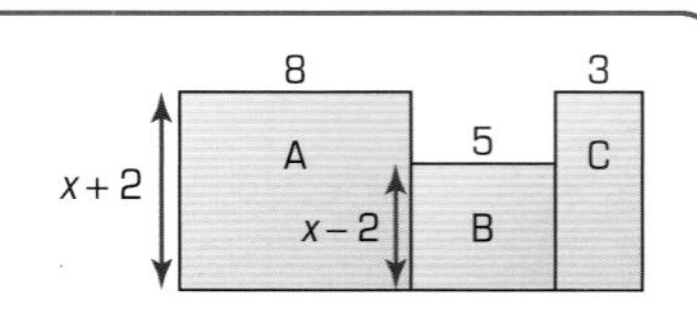

a Total area
$= 8(x + 2) + 5(x - 2) + 3(x + 2)$
$= 8x + 16 + 5x - 10 + 3x + 6$
$= 16x + 12$

b Height of A – height of B
$= (x + 2) - (x - 2)$
$= x + 2 - x + 2$
$= x - x + 2 + 2$
$= 4$

Insert brackets as you are subtracting x **and** $^-2$.

minus × minus = plus

You can sometimes factorise an expression in more than one way.

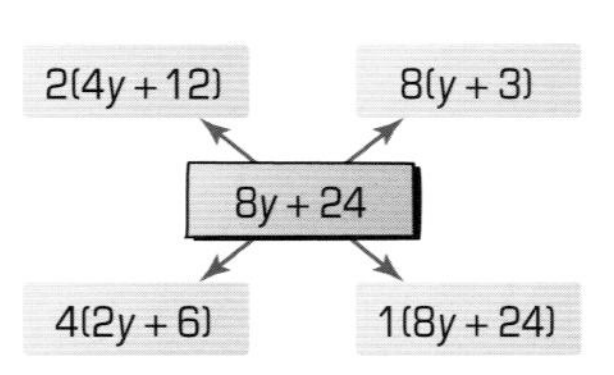

In this case you should aim to use the HCF of $8y$ and 24, which is 8, as a factor.
$8y + 24 = 8(y + 3)$

Remember:
HCF means 'highest common factor'.

- You **fully factorise** an expression by using the HCF of the terms as a common factor.

Here are some more expressions that have been fully factorised.

$10x + 15 = 5(2x + 3)$
$3x^2 + 2x = x(3x + 2)$

10 and 15 have an HCF of 5.

$3x^2$ and $2x$ have an HCF of x.

$5ab - 10b = 5b(a - 2)$
$2p + 4pq = 2p(1 + 2q)$

$2p \div 2p = 1$

Expressions with a minus work in the same way.

You can check that you have factorised correctly by expanding again. You should get the expression that you started with.

Exercise A4.4

1 Find pairs of equivalent expressions. For the odd one out, write an expression equivalent to it.

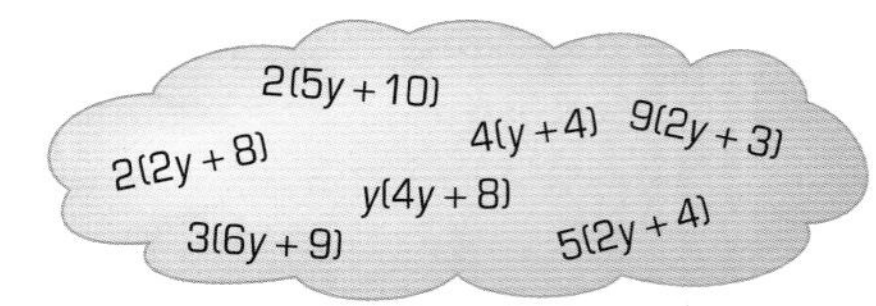

2 Expand and simplify these expressions.

a $4(x + 8) + 2(x - 3)$
b $5(2y - 4) + 3(6y - 2)$
c $2x(x + 3) + 3x(2x - 2)$
d $2(m - 3) - 3(m - 2)$
e $2p(p^2 + p) - p(p - p^2)$
f $(p + 4) - (4 - 2p)$

3 Write expressions involving brackets for the required quantities. Expand and simplify your answers.

a

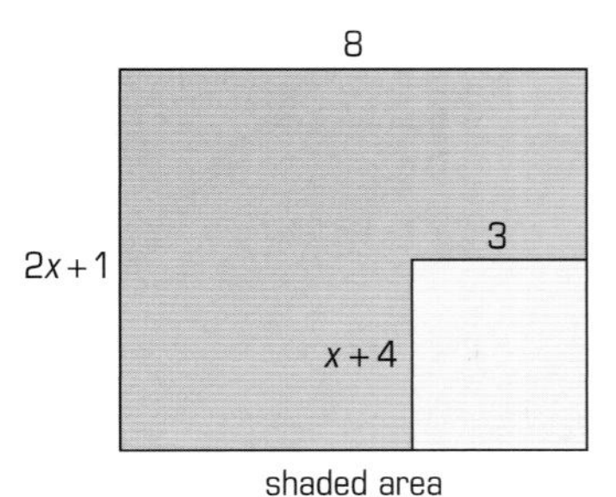

shaded area

b

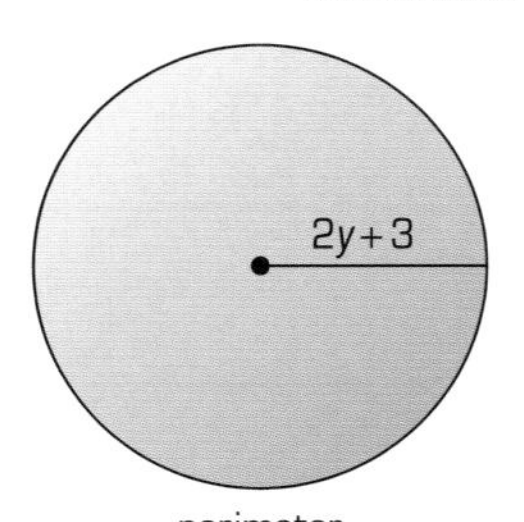

perimeter

c

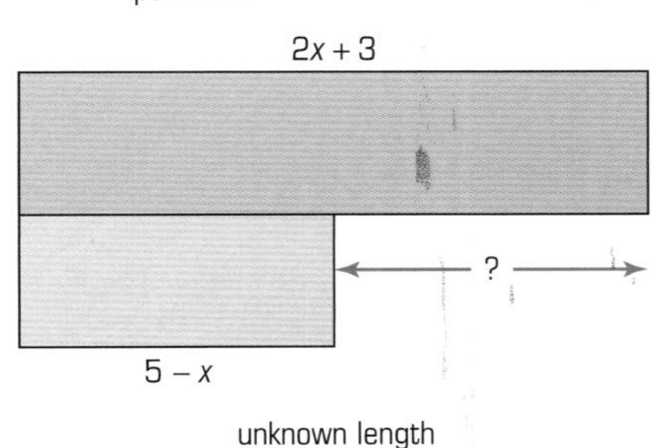

unknown length

4 Which expression is the 'odd one out' because it does not simplify as the others do?

a $\left(\frac{2(3x + 4) - 2}{2}\right) - 2$ **b** $\frac{3(2x + 4)}{2} - 5$

c $\frac{6 - 2(1 - 3x)}{2} - 1$ **d** $\frac{5(2x + 4)}{5} + x$

5 Factorise these expressions by completing the missing boxes.

a $3x + 9 = \square(x + 3)$
b $12x - 36 = \square(x - \square)$
c $x^2 + 2x = x(\square + \square)$
d $5x + 10 - 15x^2 = 5(\square + \square - 3\square)$
e $3x + 6xy = \square(1 + 2\square)$

6 Factorise these expressions fully.

a $6x + 18$ **b** $10x^2 - 20$
c $3xy + 6x$ **d** $12ab + 24b - 6b^2$
e $5z^2 - 10z$ **f** $3mn + 6m^2 - 9m^2n$

7 By expanding, check if each student has completed the factorisation correctly. If not, correct it yourself.

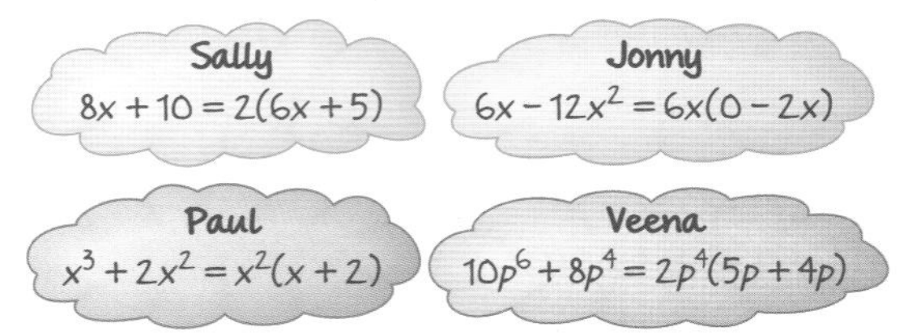

8 The identity shows that multiplying a number by 8 and adding 10 is equivalent to multiplying it by 4, adding 5 and doubling the result.

$8x + 10 \equiv 2(4x + 5)$

Find equivalent operations for each of these.

a Multiply a number by 3 and subtract 9.
b Multiply a number by itself and add double this number.
c Subtract a number from 4 and double the result.

A4.5 Expanding double brackets

This spread will show you how to:

- Square a linear expression.
- Expand the product of two linear expressions.
- Simplify the corresponding quadratic expression.

KEYWORDS

Product	Square
Quadratic	Expand
Expression	Simplify

These two rectangles are the same.

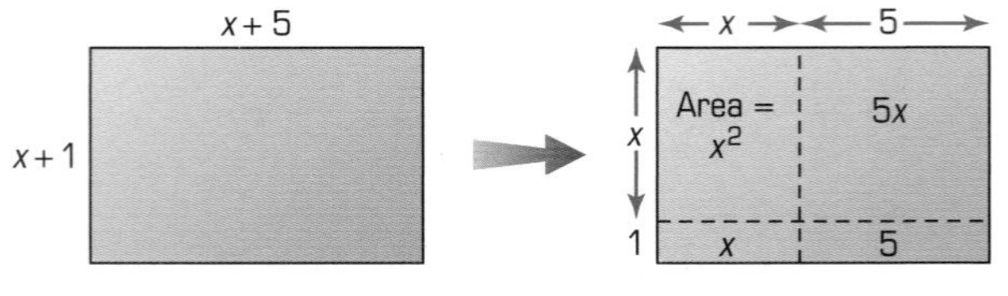

Area = $(x + 1)(x + 5)$

$\Rightarrow \quad (x + 1)(x + 5) = x^2 + 6x + 5$

Area = $x^2 + 5x + 1x + 5 = x^2 + 6x + 5$

Note:
A **product** is the result of a multiplication.

$(x + 1)$ and $(x + 5)$ are linear expressions.
$x^2 + 6x + 5$ is a quadratic expression.

You can expand a pair of brackets without having to draw a rectangle.
Multiply each term in the first bracket by each term in the second bracket:

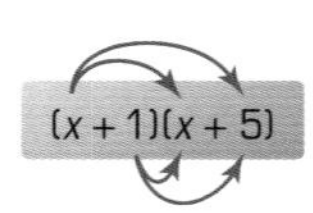

F	O	I	L
$x \times x = x^2$	$x \times 5 = 5x$	$1 \times x = x$	$1 \times 5 = 5$

Collect the terms: $x^2 + 5x + x + 5 = x^2 + 6x + 5$

FOIL is a useful acronym.
You multiply:
Firsts, **O**uters, **I**nners, **L**asts.

Remember:
When you **square** an expression, you multiply it by itself.

example

Expand these expressions, and simplify your answer.

a $(x + 3)(x + 4)$ **b** $(x - 4)(x - 6)$ **c** $(x + 6)^2$

a $(x + 3)(x + 4)$
$= x^2 + 4x + 3x + 12$
$= x^2 + 7x + 12$

b $(x - 4)(x - 6)$
$= x^2 - 6x - 4x + 24$
$= x^2 - 10x + 24$

c $(x + 6)^2 = (x + 6)(x + 6)$
$= x^2 + 6x + 6x + 36$
$= x^2 + 12x + 36$

Double brackets often appear in problems involving area.

example

Find the difference in area between the rectangle and the square.
Simplify your answer.

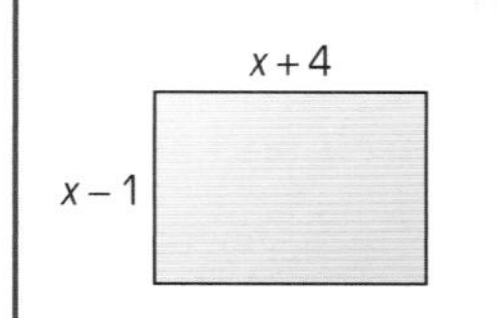

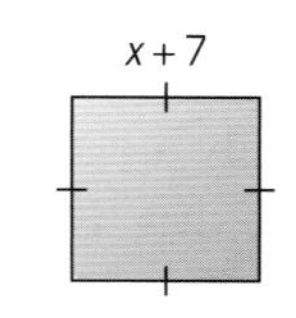

Area of rectangle $= (x + 4)(x - 1)$
$= x^2 - x + 4x - 4$
$= x^2 + 3x - 4$

Area of square $= (x + 7)^2$
$= (x + 7)(x + 7)$
$= x^2 + 7x + 7x + 49$
$= x^2 + 14x + 49$

Difference in area $= (x^2 + 14x + 49) - (x^2 + 3x - 4)$
$= x^2 + 14x + 49 - x^2 - 3x + 4$
$= x^2 - x^2 + 14x - 3x + 49 + 4$
$= 11x + 53$

Exercise A4.5

1 a The diagrams show different rectangles split into four smaller rectangles. Copy the diagrams and fill in the areas of the small rectangles.

i

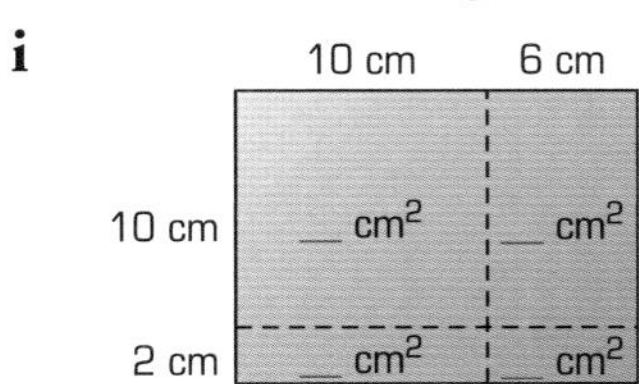

ii

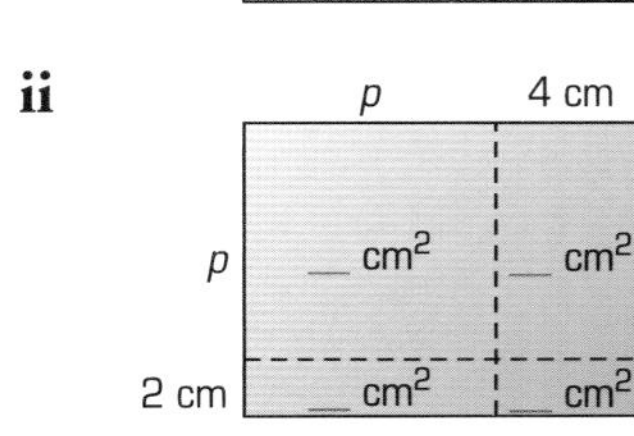

b Using your answers from part **a**

i evaluate 12×16 (no calculators)

ii multiply out $(p + 4)(p + 2)$.

2 Multiply out each pair of brackets in the first box and match with an expression from the second box.

a	$(x + 4)(x + 5)$	**b**	$(x + 10)(x + 2)$
c	$(x - 2)(x - 5)$	**d**	$(x + 3)(x - 5)$
e	$(x - 4)^2$	**f**	$(x + 3)^2$

i	$x^2 - 2x - 15$	**ii**	$x^2 + 6x + 9$
iii	$x^2 + 12x + 20$	**iv**	$x^2 - 7x + 10$
v	$x^2 + 9x + 20$	**vi**	$x^2 - 8x + 16$

3 Remove the brackets and simplify:

a $(z + 8)(z + 1)$ **b** $(a + 4)(a - 3)$

c $(b + 4)^2$ **d** $(c - 1)(c + 5)$

e $(m + 2)(m - 5)$ **f** $(n - 4)(n + 1)$

g $(p + 11)(p - 8)$ **h** $(q - 3)(q - 2)$

i $(t - 3)^2$ **j** $(x - 9)(x - 11)$

4 Explain how you know, using algebra, that adding 3 to a number and then squaring the result is not the same as squaring the number and then adding 3.

5 Write an expression for each area or volume. Expand and simplify your answer.

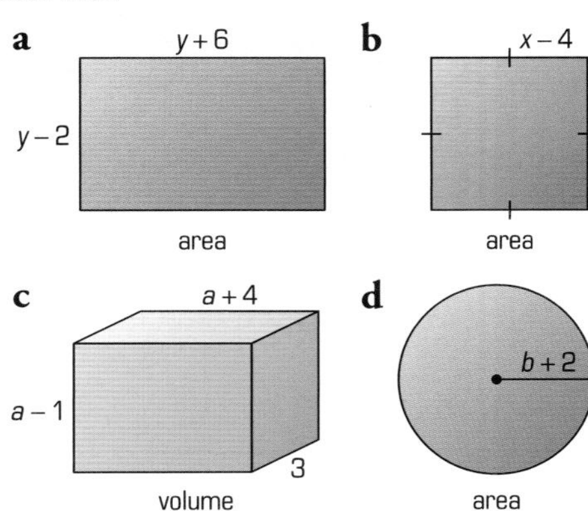

6 Remove all brackets and simplify:

a $(a - 3)(a + 1) + (a + 2)(a + 4)$

b $(b - 6)(b + 2) + (b + 3)(b + 2)$

c $(c + 3)^2 + (c - 3)(c - 5)$

d $d(d + 4) + (d - 6)(d + 1)$

e $(2e + 1)^2 + (e + 1)^2$

f $(f + 2)^2 - (f + 3)(f + 4)$

g $(2g + 1)(3g - 1) + (2g + 2)^2$

h $(2h + 3)(3h - 1) - (2h + 3)^2$

7 True or false?

$(x + 3)(x + 2) = x^2 + 5x + 6$

Explain your answer.

8 a Expand $(n - 1)^2$, simplifying your answer.

b Use the answer to part **a** to work out

i 99^2 **ii** 49^2 without a calculator.

9 Prove that the product of any two odd numbers is always odd.

10 Challenge

Show that $(x + 2)(x + 3)(x + 4)$ is identical to $x^3 + 9x^2 + 26x + 24$.

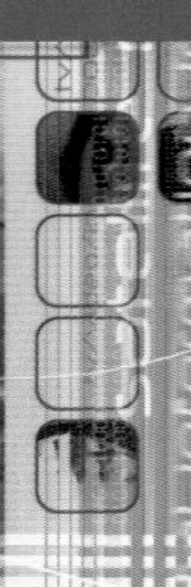

A4.6 Gradients of parallel and perpendicular lines

This spread will show you how to:

- Find the gradient of lines given by equations of the form $y = mx + c$.
- Investigate the gradients of parallel lines and perpendicular lines.

KEYWORDS
Gradient
Parallel
Perpendicular

- **The gradient of a straight line = $\frac{\text{change in vertical distance}}{\text{change in horizontal distance}}$**

Or $m = \frac{y_2 - y_1}{x_2 - x_1}$

- **The equation of a straight line can be written in the form $y = mx + c$, where m is the gradient.**

(x_2, y_2)

(x_1, y_1) $y_1 - y_2$

$x_1 - x_2$

example

a Find the gradient of the line segment shown in the diagram.
b Write down the gradient of the straight line with equation $y = 5x - 3$.
c Find the gradient of the straight line joining $(^-1, 2)$ and $(1, ^-1)$.

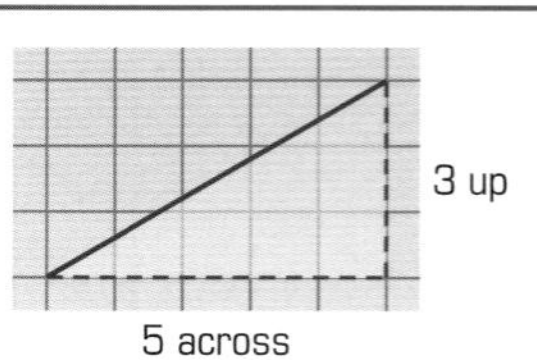

a Gradient $= \frac{\text{change in vertical distance}}{\text{change in horizontal distance}} = \frac{3}{5}$

b $m = 5$, so gradient is 5.

c $m = \frac{y_2 - y_1}{x_2 - x_1} = \frac{^-1 - 2}{1 - (^-1)} = \frac{^-3}{2}$ (or $^-1\frac{1}{2}$)

Remember:
Lines with negative gradients slope downwards left to right.

- **The gradients of parallel lines are equal.**

example

Give the gradient of a line parallel to $y = 4x - 2$.

$m = 4 \Rightarrow$ gradient of $y = 4x - 2$ is 4.
$\Rightarrow$ gradient of a parallel line is 4.

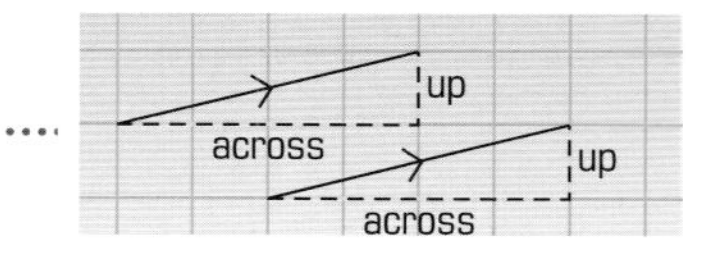

The gradients of perpendicular lines are also related.

This line has gradient m.

Rotate it by 90° clockwise.

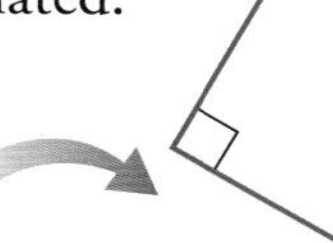

The new line has gradient $^-\frac{1}{m}$.

- **If the gradient of a straight line is m, then a line perpendicular to it will have gradient $^-\frac{1}{m}$.**

example

Give the gradient of a line perpendicular to $2y = x + 8$.

$2y = x + 8 \Rightarrow y = \frac{1}{2}x + 4$, so $m = \frac{1}{2}$

A perpendicular line will have gradient $\frac{^-2}{1} = ^-2$.

Exercise A4.6

1 **a** Write down the gradients of these lines.

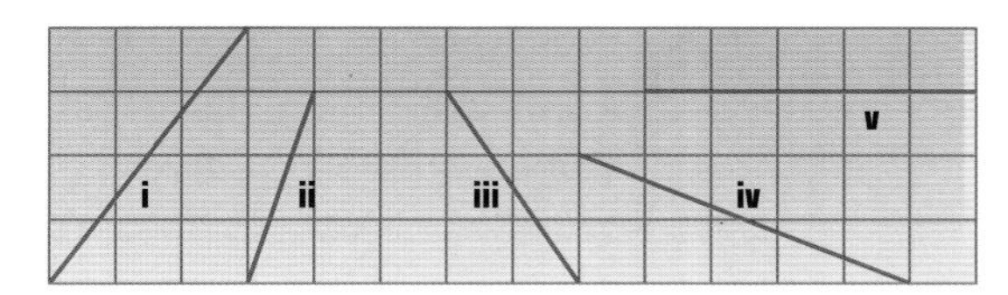

b Write down the gradients of the lines perpendicular to lines **i**–**iv** in part **a**. Why can't you write the gradient of a line perpendicular to line **v**?

2 Find the gradients of the lines joining:

a (5, 6) to (10, 16)
b (2, 1) to (5, 2)
c ($^-3$, 6) to (10, 7)
d (4, $^-1$) to ($^-2$, $^-7$)
e (a, b) to (c, d).

3 True or false?

a $y = 3x + 2$ has a gradient of 3.
b $y = 3x + 2$ and $2y = 6x + 1$ are parallel.
c $y = 3x + 2$ and $y = \frac{1}{3}x + 2$ are perpendicular.
d Lines with gradient $\frac{5}{2}$ and $2\frac{1}{2}$ are parallel.
e Line with gradient $\frac{^-5}{6}$ and 1.2 are perpendicular.

4 Find the equations of these lines.

a

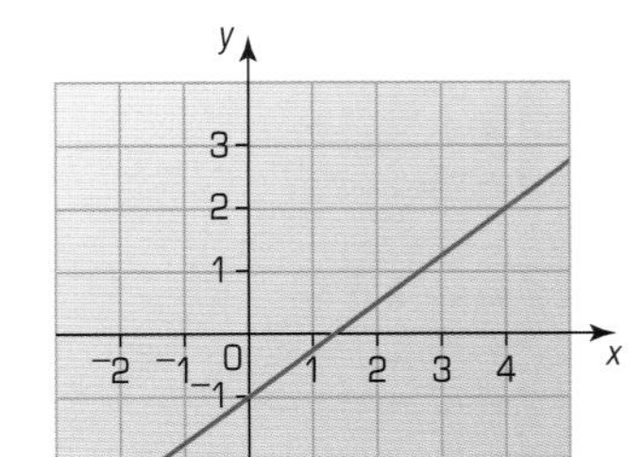

b

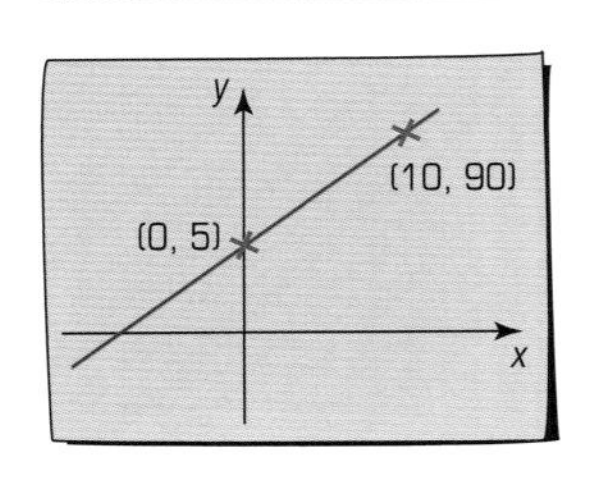

5 **a** Sandeep solved a pair of simultaneous equations graphically.

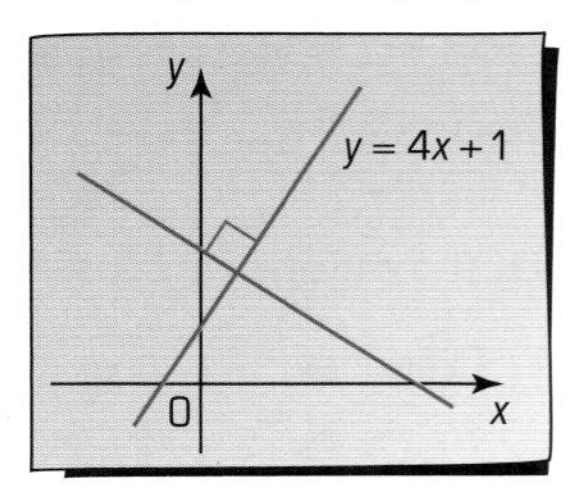

The solution was $x = 1$, $y = 5$.
Use this solution and the information in the diagram to give the pair of simultaneous equations that she solved.

b Jon found the solution $x = 2$, $y = {}^-1$ to his simultaneous equations. Here is his graph.

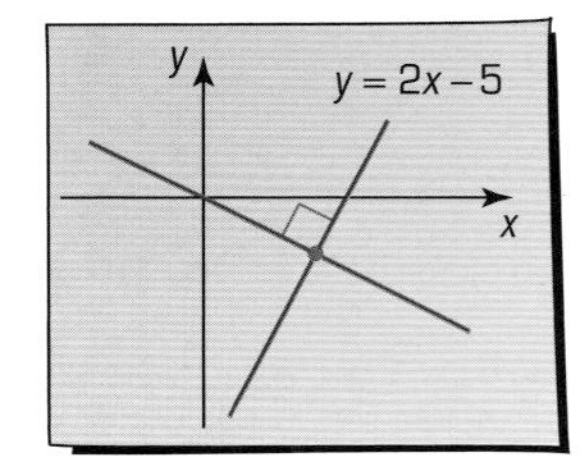

Use his solution and the information in the diagram to give his pair of simultaneous equations.

6 Write an expression for:

a The gradient of a line joining (x, y) to $(x^2, 3y)$
b The gradient of a line perpendicular to the one joining $(t, 4)$ to $(2t, 5)$
c The gradient of a line perpendicular to the one joining (m, n) to (v, w).

7 Find the value of p if the gradient of the line joining $(p, 3p)$ to $(2p, 20)$ is 1.

8 True or false?
The triangle formed by joining (5, 12), ($^-12$, 5) and ($^-7$, 17) is a right-angled triangle.

Draw a diagram.

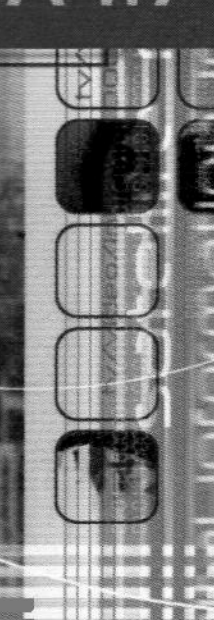

A4.7 Further curved graphs

This spread will show you how to:

- Plot graphs of simple quadratic and cubic functions.
- Know simple properties of quadratic functions.

KEYWORDS

Quadratic	Reciprocal
Parabola	Hyperbola
Cubic equation	Curve

Here are some graphs that you should know.

Horizontal and vertical

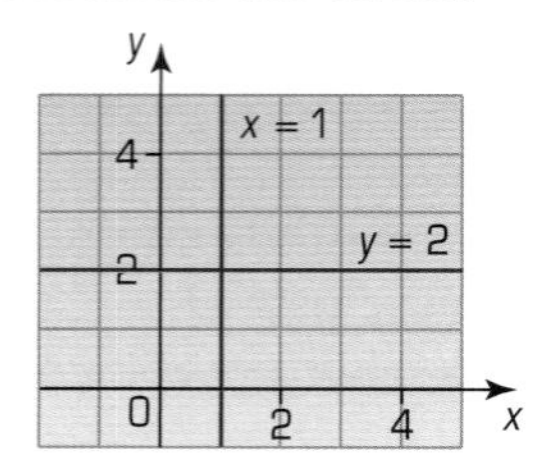

$x = a$ is vertical
$y = b$ is horizontal
(a and b are constants)

Diagonal

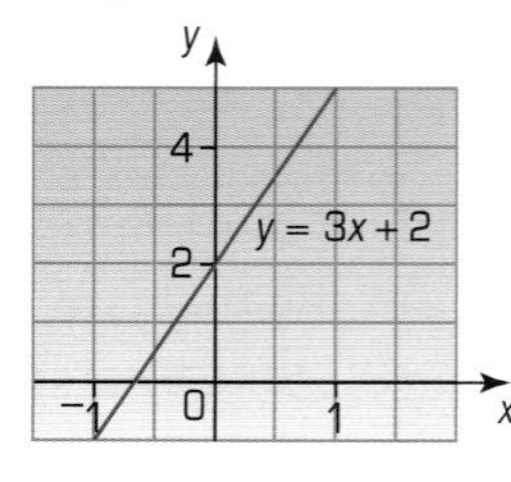

$y = mx + c$ is a straight diagonal line with gradient m, cutting the y-axis at $(0, c)$

Quadratic equation

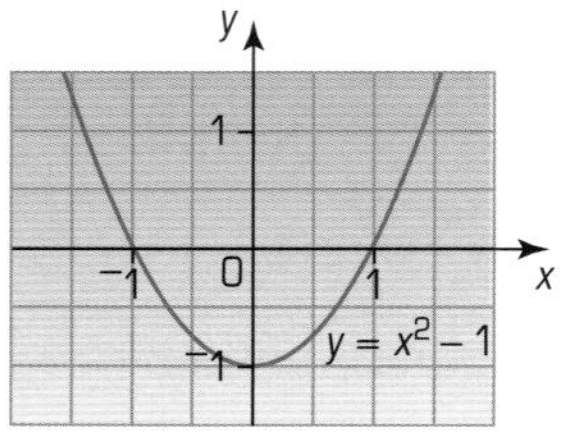

A parabola
Formed from equations with highest power of $x = x^2$

Cubic equation

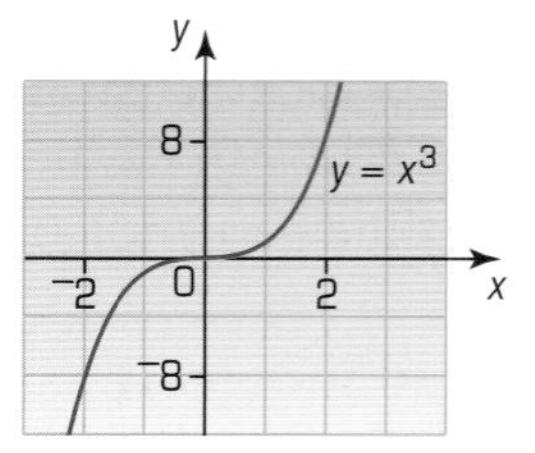

Graph is a curve
Formed from equations with highest power of $x = x^3$

Reciprocal graphs

These are graphs containing the reciprocal of x in their equation, such as $y = \frac{1}{x}$.

Reciprocal graphs have a characteristic shape.

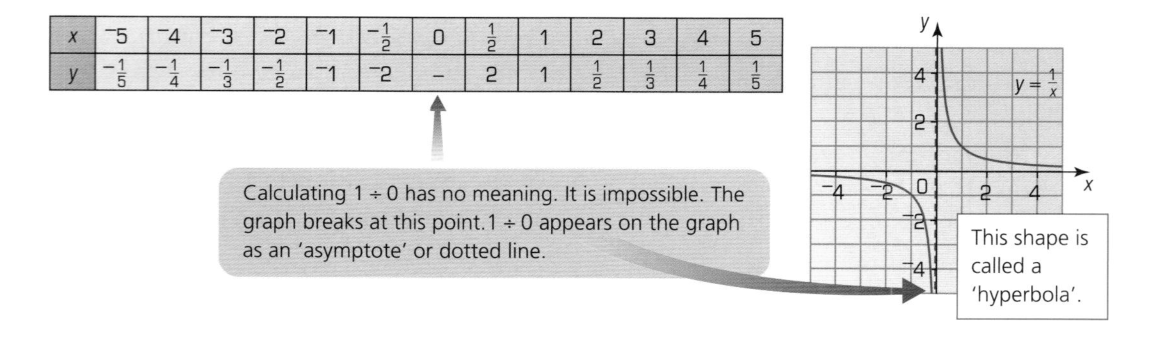

x	$^-5$	$^-4$	$^-3$	$^-2$	$^-1$	$^-\frac{1}{2}$	0	$\frac{1}{2}$	1	2	3	4	5
y	$^-\frac{1}{5}$	$^-\frac{1}{4}$	$^-\frac{1}{3}$	$^-\frac{1}{2}$	$^-1$	$^-2$	–	2	1	$\frac{1}{2}$	$\frac{1}{3}$	$\frac{1}{4}$	$\frac{1}{5}$

By altering an equation, a graph will keep its shape but change in some way.

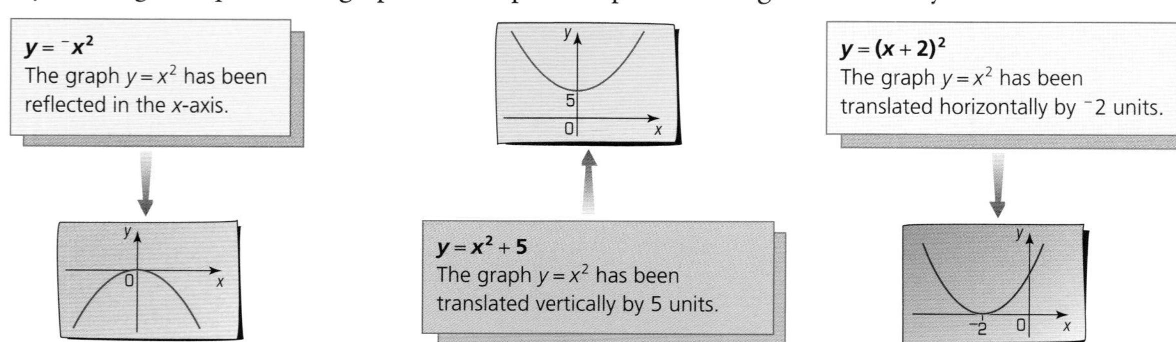

Exercise A4.7

1 Look at the six equations of graphs A to F.

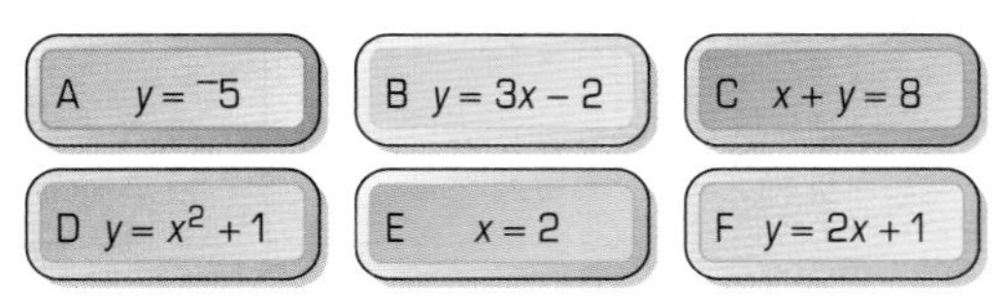

a Which graphs pass through (0, 1)?
b Which graph is parallel to the x-axis?
c Which graph is not a straight line?
d Which graphs intersect at $(2\frac{1}{2}, 5\frac{1}{2})$?

2 **a** Draw a pair of axes labelled 0 to 12. Plot at least six points whose x-coordinate and y-coordinate multiply together to make 12. Join these to complete the curve $xy = 12$.
b What is the name of this curve? Why is $xy = 12$ this shape?

3 **a** Predict the shape of each graph.
i $y = (x + 3)^2$
ii $y = x^3 + x + 1$

b Copy and complete these coordinate tables for the graphs in part **a**.

i

x	$^{-}5$	$^{-}4$	$^{-}3$	$^{-}2$	$^{-}1$	0	1	2
$x + 3$								
$y = (x + 3)^2$								

ii

x	$^{-}2$	$^{-}1$	0	1	2	3	4
x^3							
1	1	1	1	1	1	1	1
$y = x^3 + x + 1$							

c Choose appropriate axes and plot the graphs. Use your diagrams to check your predictions in part **a**.

4 **Investigation**
Using a graphical calculator, appropriate computer package or by drawing graphs by hand, investigate these graph groups.

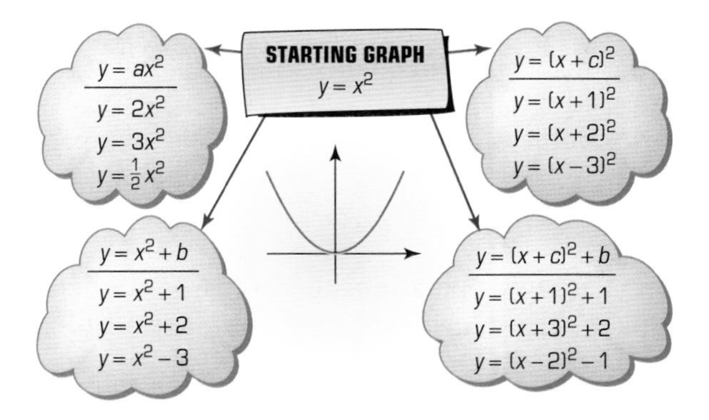

In each case, explain how altering the equations changes the graph.

5 **a** Using your results from question 4, sketch these graphs.
i $y = x^2 + 9$ **ii** $y = 7x^2$
iii $y = (x - 1)^2$ **iv** $y = (x + 1)^2 + 4$

b Give the equations of these graphs.

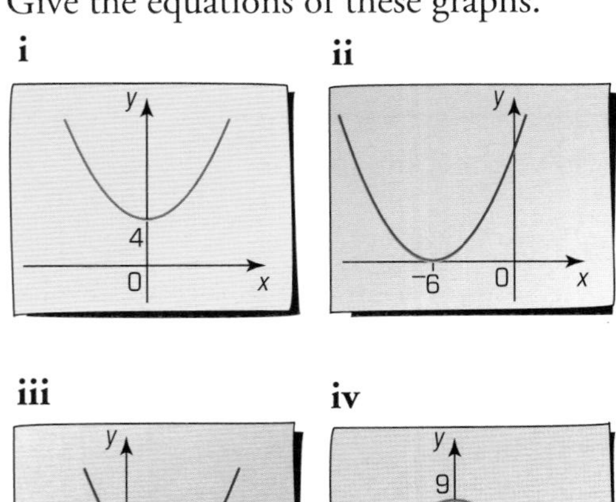

Solving linear inequalities

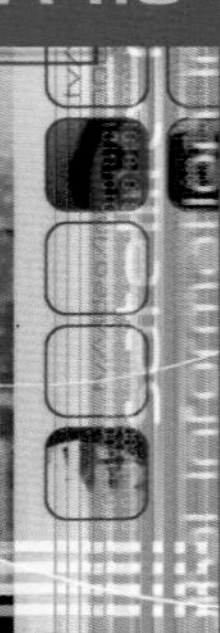

This spread will show you how to:

- Solve linear inequalities in one variable, and represent the solution set on a number line.

KEYWORDS
Inequality
Solve

You can solve an inequality in a similar way to solving an equation.

For example, $3x + 2 \geqslant 5$ → compare with $3x + 2 = 5$

$3x \geqslant 3$ → $3x = 3$

$x \geqslant 1$ → $x = 1$

On a number line, the solution of $3x + 2 \geqslant 5$ looks like this:

−1 0 1 2 3

There are two operations to look out for when solving inequalities. Look at what happens to the statement in the box:

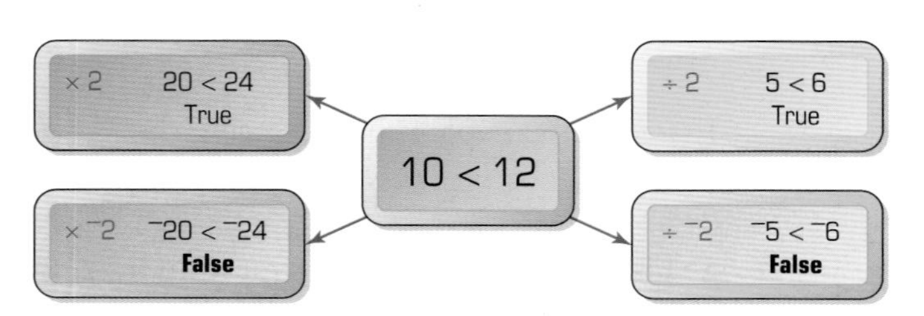

- If you multiply or divide by a negative number, an inequality becomes false. You must reverse the inequality to keep it true.

example

Solve these inequalities, and represent them on a number line.

a $8x + 4 \leqslant 5x - 2$ **b** $^{-}4x < 16$ **c** $\frac{2(3y + 6)}{5} < \frac{5y + 1}{4}$

a $8x + 4 \leqslant 5x - 2$

$3x + 4 \leqslant {}^{-}2$

$3x \leqslant {}^{-}6$

$x \leqslant {}^{-}2$

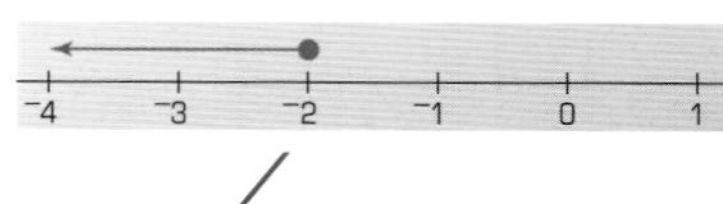

If the circle is filled in then x can take the initial value.

b $^{-}4x < 16$

$x > {}^{-}4$

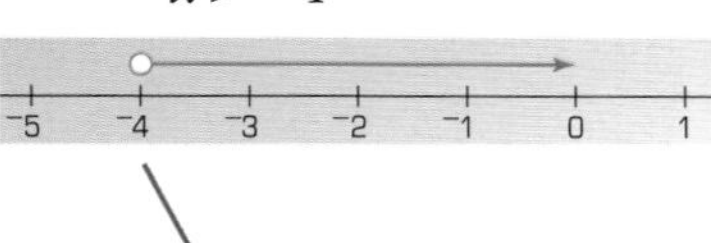

If the circle is empty, then x must be greater or less than the initial value

c $\frac{2(3y + 6)}{5} < \frac{5y + 1}{4}$

$8(3y + 6) < 5(5y + 1)$

$24y + 48 < 25y + 5$

$48 < y + 5$

$43 < y$

This is the same as $y > 43$.

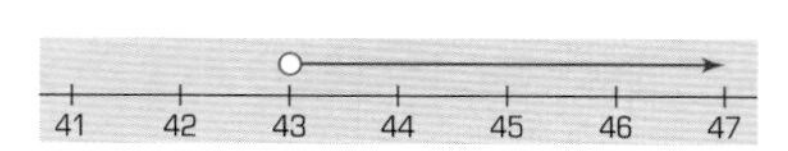

Exercise A4.8

1 Solve these inequalities. Illustrate your solutions on a number line.

- **a** $5x \leqslant 15$
- **b** $2x + 1 > 18$
- **c** $3x - 2 < 10$
- **d** $4(x + 2) \leqslant 16$
- **e** $3x - 5 > 2x + 7$
- **f** $4z - 3 \leqslant 3(z - 2)$
- **g** $^{-}6x < 30$
- **h** $\frac{x}{^{-}5} \geqslant 15$
- **i** $10 - 3x < 15$

2 In each question, list the integers that satisfy the given inequality.

- **a** $28 < 7z \leqslant 49$
- **b** $^{-}8 \leqslant 4x \leqslant 16$
- **c** $10 \leqslant {}^{-}2n \leqslant 20$
- **d** $2p < 18$ and $3p > 9$
- **e** $3x + 2 \leqslant 16$ and $^{-}2x \leqslant 4$

3 Explain why it is not possible to find a value of n such that $3n \leqslant 21$ and $2n + 1 > 15$.

4 x and y are both integers. It is known that $3 < x \leqslant 6$ and that $9 < y < 12$.

- **a** If xy is 40, what could x and y be?
- **b** What is the largest possible value of xy?
- **c** What is the smallest possible value of $x - y$?

5 Solve these inequalities. Illustrate your solution set on a number line.

- **a** $5x + 7 \leqslant 8x - 4$
- **b** $3(x - 2) \leqslant 5(x + 6)$
- **c** $\frac{x + 5}{4} \geqslant \frac{x - 3}{7}$
- **d** $\frac{x + 3}{^{-}5} \leqslant 10$
- **e** $3(y + 7) \leqslant \frac{y - 4}{^{-}2}$
- **f** $\frac{5(3 - 2y)}{12} > 1$

6 Indicate, on a number line, the range of values that satisfy:

- **a** $3x + 6 \leqslant 18$ and $^{-}2x > 2$
- **b** $4y - 2 \leqslant 0$ and $^{-}2y \leqslant 4$
- **c** $5 - 2z \leqslant 13$ and $4z + 6 \leqslant 10$
- **d** $8 \leqslant 2x \leqslant x + 7$
- **e** $y < 3y + 2 < 2y + 6$
- **f** $4z + 1 < 8z < 3(z + 2)$

Hint: In **d** solve the two inequalities separately.

7 In each case, write an inequality and solve it to find the range of values of the unknown.

a Area is greater than perimeter.

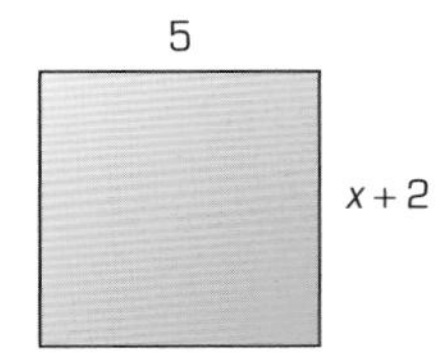

b Mean of first group is smaller than mean of second group.

c ? must be greater than 6. Solve for z.

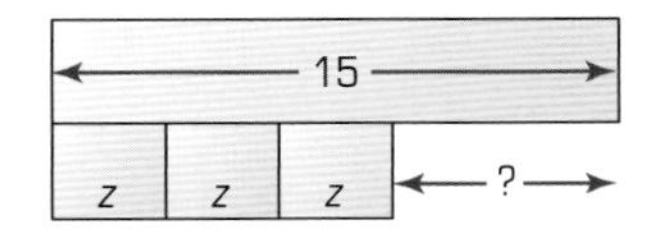

d The area of the rectangle exceeds that of the triangle.

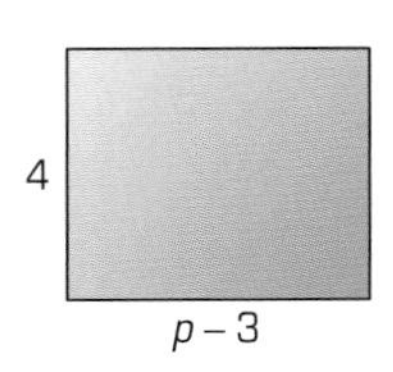

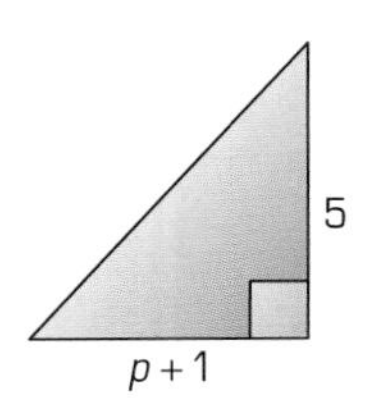

A4.9 Graphical inequalities

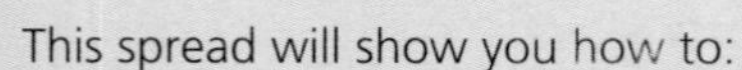

This spread will show you how to:

- Begin to solve inequalities in two variables.

KEYWORDS

Inequality　Region

Greater than or equal to (⩾)

Less than or equal to (⩽)

An **inequality** is a mathematical statement involving one of the signs $<$, $>$, $\leqslant$ or $\geqslant$.

This minibus holds up to 12 people.

Let p = the number of people on the minibus. Then $p \leqslant 12$.

Instead of $p \leqslant 12$, you could write $12 \geqslant p$.

You can represent inequalities in two dimensions on a graph.

This diagram shows some of the points that satisfy the inequality $x \geqslant 3$.

(4, 1)

(1, ⁻1)

This point has an x-coordinate greater than 3.

This point has an x-coordinate less than 3.

In this diagram, the region that satisfies $x \geqslant 3$ is unshaded.

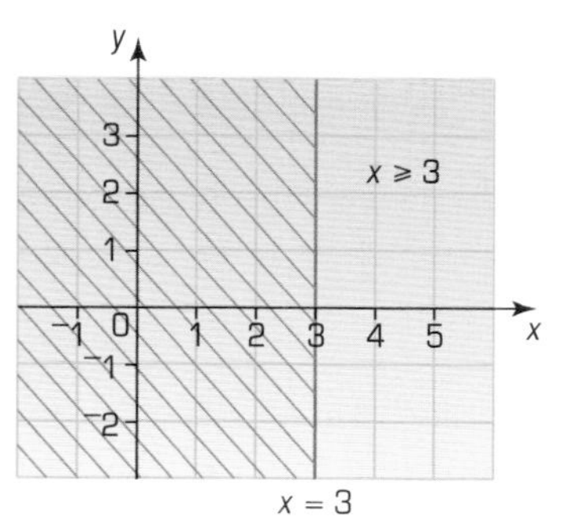

example

Leave unshaded the regions represented by these inequalities on a graph.

a $x < 2$　**b** $y \geqslant {}^-1$　**c** $x > 1$ and $y \leqslant 2$

a $x < 2$

x < 2

x = 2

Points on the dotted line are not counted in the region.

b $y \geqslant {}^-1$

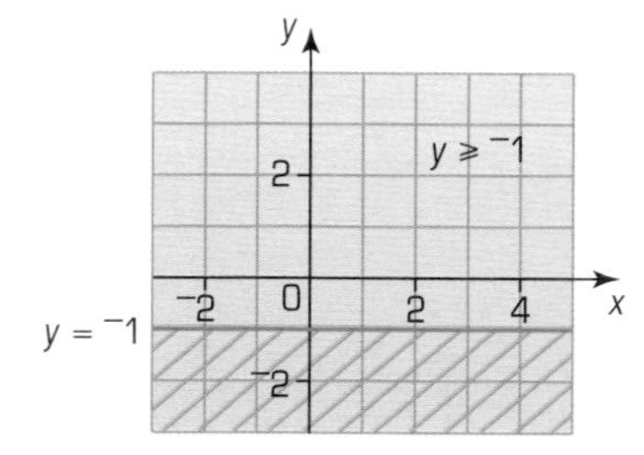

c $x > 1$ and $y \leqslant 2$

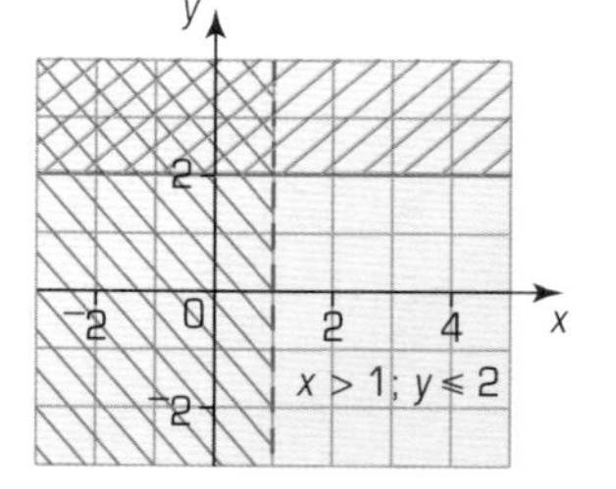

Exercise A4.9

1 What inequalities do these number lines show?

a **b**

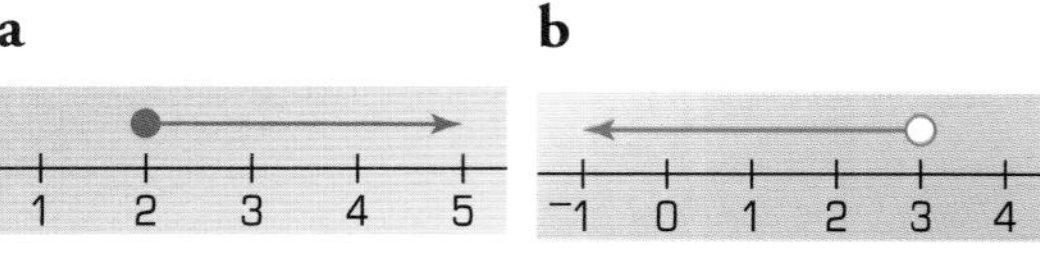

c **d**

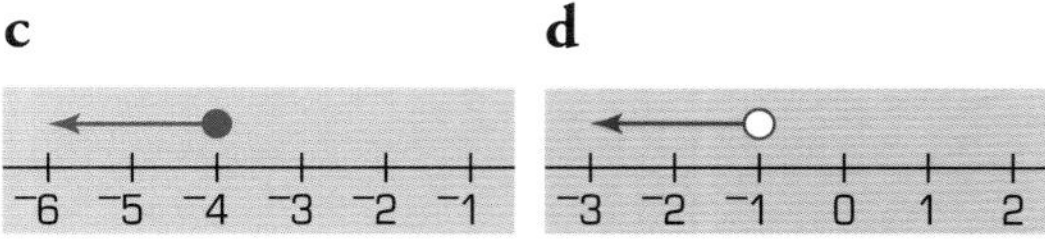

e **f**

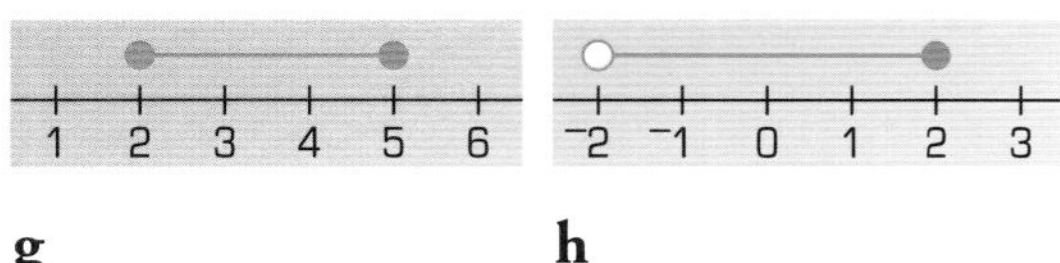

g **h**

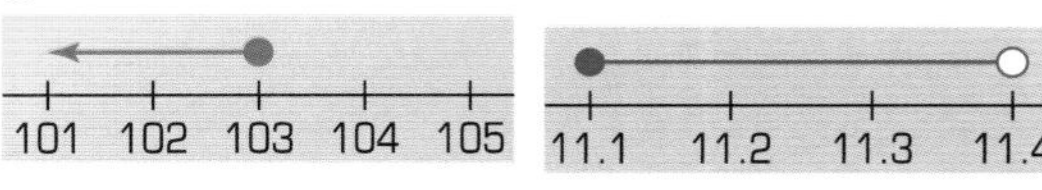

2 Draw a number line to represent:

a $x \geqslant 8$ **b** $p \leqslant {}^{-}2$
c $m > 7$ **d** $g < {}^{-}2$
e ${}^{-}3 \leqslant h \leqslant 5$ **f** $10 < p < 17$

3 True or false? $4 \leqslant x$ is shown by:

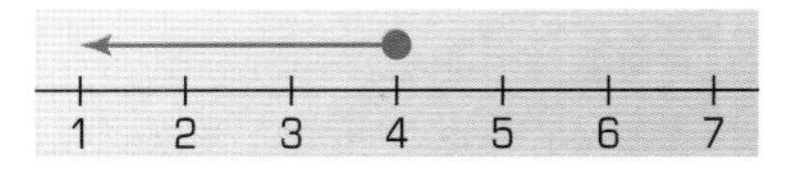

4 What inequality is represented by the unshaded region in each case?

a

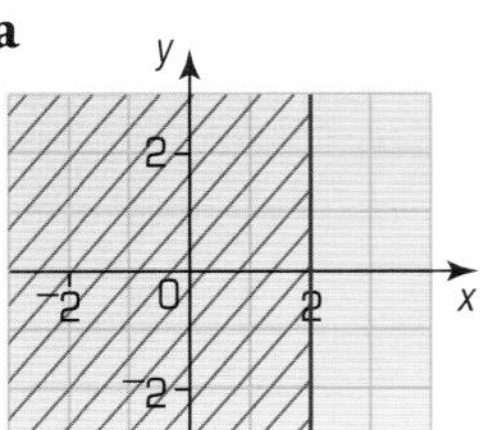

b

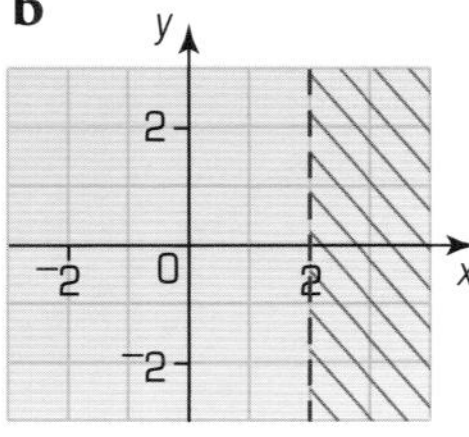

5 Draw inequality regions to represent:

a $x \geqslant 1$ **b** $y < 3$
c $x > {}^{-}2$ **d** $y \leqslant {}^{-}4$
e ${}^{-}2 \leqslant x \leqslant 5$ **f** $1 < y < 3$
g $x \geqslant 4$ and $y > 1$ **h** $x \leqslant 2$ and $x \geqslant 8$
i $x \leqslant 2$, $y \leqslant 2$ and $y > {}^{-}3$

6 What inequalities describe the following unshaded regions?

a

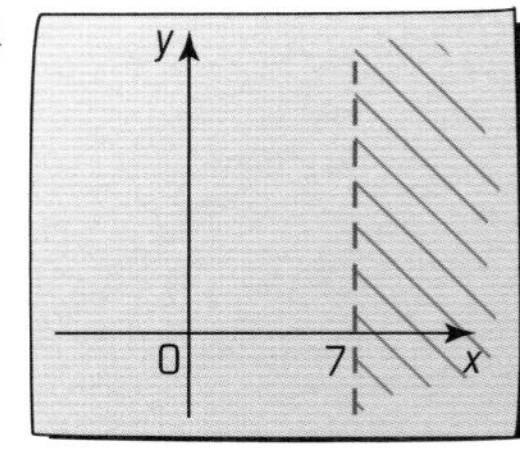

b

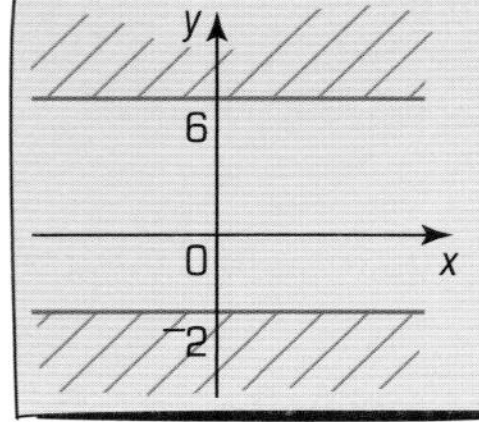

c

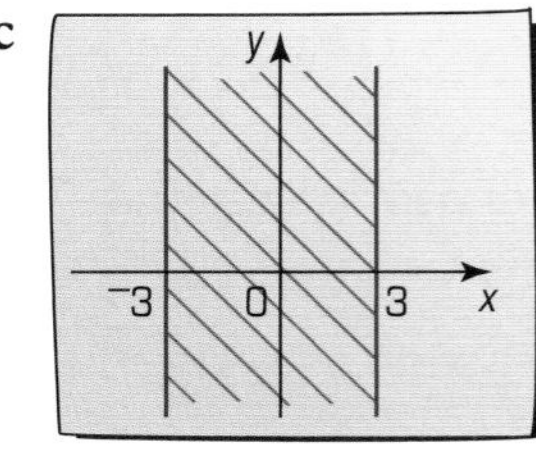

d

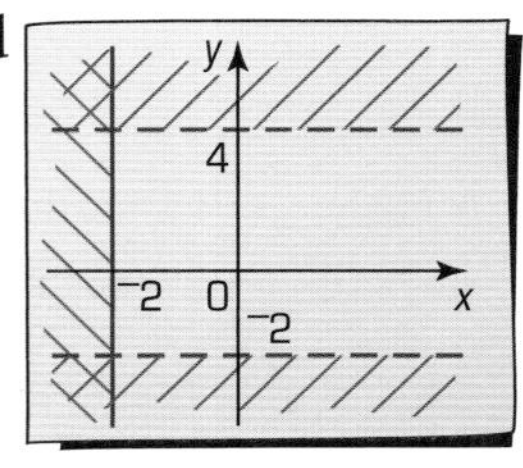

7 Write down inequalities to enclose:

a an unshaded square
b an unshaded rectangle.

How do you think you could give instructions to construct an unshaded right-angled triangle?

8 Draw a pair of axes. Represent each clue on the axes. What are the possible integers that I may be thinking of?

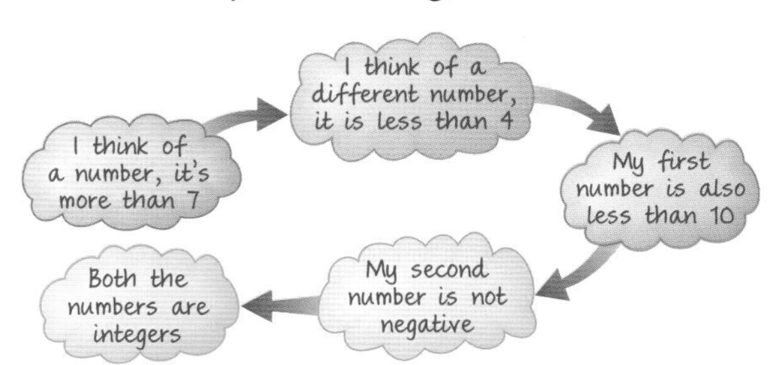

A4 Summary

You should know how to …

1 Know and use the index laws in generalised form for multiplication and division of positive integer powers.

2 Square a linear expression, expand the product of two linear expressions of the form $x \pm n$ and simplify the corresponding quadratic expression.

3 Investigate the gradients of parallel lines and lines perpendicular to these lines.

4 Generate fuller solutions to problems.

Check out

1 a Which expression is the odd one out? Explain why.

$2x^2 \times 18x^4$ $\quad (6x^3)^2 \quad$ $\frac{72x^4}{2x^{-3}}$

b Which expression has the greatest value when $x = 16$?

2 a True or false? $(y + 3)^2$ is $y^2 + 9$.

b Expand:

i $(x + 3)(x + 4)$

ii $(x - 7)(x + 4)$

iii $(x - 5)^2$

3 a

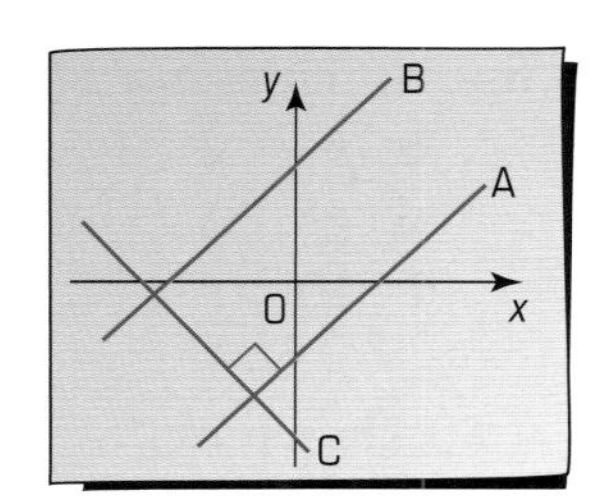

If line A has equation $y = 3x - 2$, give possible equations of lines B and C, explaining your choice.

b Repeat for line A with equation $4y + 8 = x$.

4 Draw a pair of axes labelled $^{-}10$ to 10. Cross out any points not satisfied by the given clues. Hence, locate the integer coordinates at which the treasure lies.

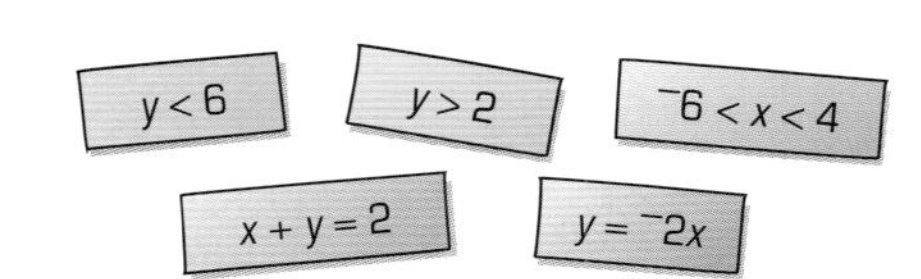

2 Probability

This unit will show you how to:

- Use the vocabulary of probability in interpreting results involving uncertainty and prediction.
- Identify all the mutually exclusive outcomes of an experiment.
- Know that the sum of probabilities of all mutually exclusive outcomes is 1 and use this when solving problems.
- Estimate probabilities from experimental data.
- Understand relative frequency as an estimate of probability and use this to compare outcomes of experiments.
- Justify generalisations, arguments or solutions.

You can estimate expected results using experimental probability.

Before you start

You should know how to ...	Check in
1 Use efficient methods to calculate with fractions and decimals.	**1** Find: **a** $\frac{1}{5}+\frac{2}{3}$ **b** $\frac{1}{5}\times\frac{2}{3}$ **c** $0.3+0.15$ **d** 0.3×0.15
2 Calculate a proportion as a fraction or a decimal.	**2** Find: **a** $\frac{2}{3}$ of 45 **b** 0.35 of 60
3 Draw a line graph.	**3** Every Monday morning Angela sits a multiplication test. Her results for half a term are: 18 17 16 19 15 17 16 Draw a line graph for this data.

D2.1 Probability and expected frequency

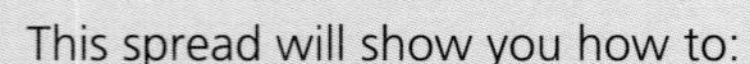

This spread will show you how to:

- Use the vocabulary of probability in interpreting results involving uncertainty and prediction.

KEYWORDS

Expected frequency
Exhaustive
Event
Independent
Trial
Mutually exclusive
p(n)

Two events are **exhaustive** if it is certain that at least one of them will occur.

Event A: number 1, 2, 3 or 5
Event B: even number
A and B are exhaustive events.

Events are **mutually exclusive** if they cannot happen at the same time.

Event Y: yellow marble
Event G: green marble
Y and G are mutually exclusive events.

Events are **independent** if the outcome of one event has no effect on the outcome of the other event.

Event D: number 6
Event M: yellow marble
The number on the dice will not affect the colour of the marble.
D and M are independent events.

- The sum of probabilities of mutually exclusive outcomes = 1

example

A bag contains green and yellow marbles. The probability of choosing a green marble is $\frac{3}{4}$.

a What is the probability of choosing a yellow marble?

Joel thinks there are seven green marbles.

b Explain why he must be wrong.

Tegan takes out one marble from the bag. It is yellow.

c What is the smallest number of green marbles there could be in the bag?

a $p(\text{green}) = \frac{3}{4} \Longrightarrow p(\text{yellow}) = 1 - \frac{3}{4} = \frac{1}{4}$

Remember: p(green) is a shorter way of writing 'the probability of choosing a green marble'.

b The total number of marbles is an integer.
$\frac{3}{4}$ of an integer can never be 7.

c There is at least 1 yellow marble.
1 out of 4 marbles are yellow $\Longrightarrow$
There are at least 3 green marbles.

- Expected frequency = number of trials × probability

example

Huw rolls a dice 40 times.
How many times should he expect an even number?

Remember: In statistics, a trial is a repetition of an experiment.

Number of trials = 40 $\qquad$ $p(\text{even}) = \frac{1}{2}$
Expected frequency = $40 \times \frac{1}{2} = 20$
Huw should expect the dice to land on an even number 20 times.

Exercise D2.1

1 Stuart rolls an ordinary fair dice.
He defines the events A to E as follows:
Event A: prime number
Event B: number less than 5
Event C: square number
Event D: 4 or greater
Event E: even number

a Name two events that are:
  i exhaustive
  ii mutually exclusive.

b Stuart rolls the dice 15 times.
What is the expected number of times that he will get a square number?

2 Each member of Tony's football club gets one raffle ticket.

	Ticket colour	Number used
Under 8	Buff	1 to 16
Under 10	Red	1 to 20
Under 13	Blue	1 to 24

Tony will select one ticket at random.
The table shows information about the tickets.

a How many members of the football club are there?

b What is the probability that the winning ticket will be buff coloured?

c What is the probability that the winning ticket will show number 15?

d Tony does pick ticket number 15.
What is the probability that the ticket is red?

3 Emma rolls a dice and tosses a coin.
What is the probability that she gets:

a a six and a head

b an odd number and a tail?

4 Harriet has two bags of marbles.

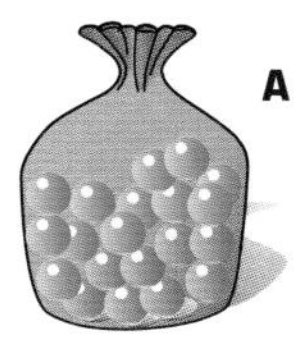

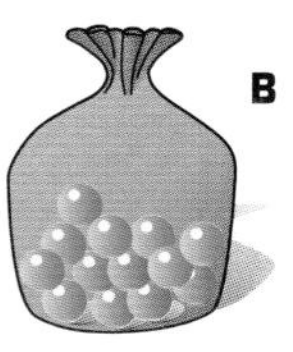

Bag A contains 8 blue and 12 green marbles.
Bag B contains 5 blue and 8 green marbles.
Harriet takes one marble at random from one of the bags.
She wants to choose a green marble.
Which bag should Harriet choose from?
Explain how you decided on your answer.

5 Joel has a bag containing orange, red, green and yellow marbles.
Joel takes one marble out of the bag.
The table shows the probability of each colour being chosen.

Colour	Orange	Red	Green	Yellow
Probability	0.4	0.05	0.25	0.3

a Explain why the number of orange marbles in the bag cannot be 20.

b What is the smallest possible number of each colour of marble in the bag?

6 Becca has a bag containing purple, orange and white bricks.
There is at least one of each of these colours in the bag.
Becca takes one brick out of the bag at random.
The probability that she chooses an orange brick is $\frac{3}{8}$.
There are 40 bricks in the bag.
What is the greatest number of purple bricks there could be in the bag?

D2.2 Probability and independent events

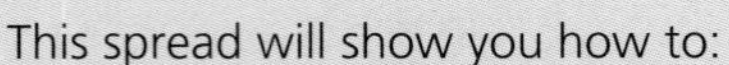

This spread will show you how to:

- Identify all the mutually exclusive outcomes of an experiment.
- Know that the sum of probabilities of all mutually exclusive outcomes is 1 and use this when solving problems.

KEYWORDS
Tree diagram
Independent events
Outcome
Event
Mutually exclusive

Oscar is making two glass vases.
Each vase is made independently.
The probability that a vase cracks while it is being made is 0.08.

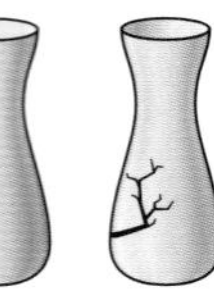

Oscar draws a tree diagram to show all the outcomes.

'Crack' and 'no crack' are mutually exclusive events.
The probability of 'no crack' is $1 - 0.08 = 0.92$.

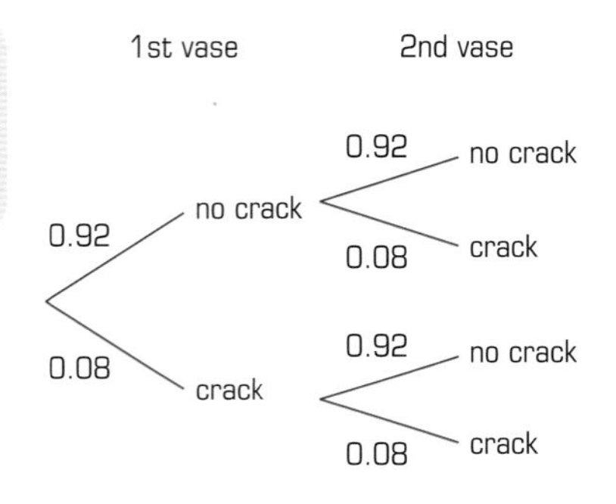

Remember:

- You write the outcomes at the end of the branches.
- You write the probabilities on the branches.

The tree diagram shows that there are four possible outcomes:

(no crack, no crack) (crack, no crack)
(no crack, crack) (crack, crack)

Oscar can use the tree diagram to calculate probabilities.

The probability that **both** vases crack while being made is:
$0.08 \times 0.08 = 0.0064$

The probability that **only one** of the vases cracks while being made is:

$(0.92 \times 0.08) + (0.08 \times 0.92)$
$= 0.0736 + 0.0736$
$= 0.1472$

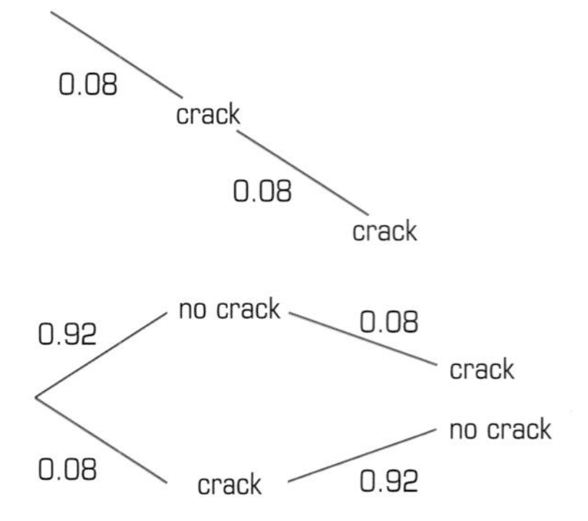

- You can calculate the probability of independent events on a tree diagram by **multiplying** along the branches.
- If you can take more than one route on a tree diagram, **add** the resultant probabilities.

Exercise D2.2

1 A drawing pin is thrown twice.
The probability that it lands point up is 0.7.
  a Write down the probability that it lands point down.
  b Draw a tree diagram to show all the different possible outcomes.
  c Calculate the probability that the drawing pin lands point up both times.
  d Calculate the probability that it lands point up once only.

2 A triangular spinner, with equal sectors numbered 1, 2, 3, is spun twice.
The scores are added.

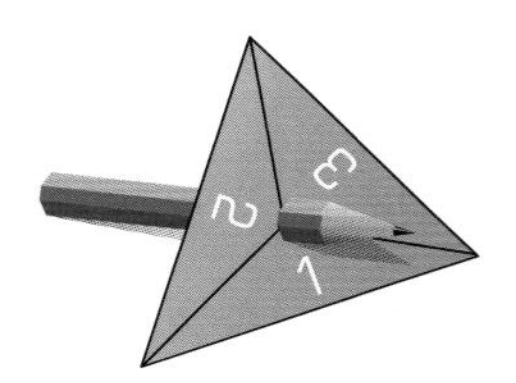

  a Draw a diagram to show all the possible outcomes.
  b What is the probability of getting a score of 2?
  c What is the probability of getting a score of 3?
  d What is the probability of getting a score greater than 4?

3 This spinner is spun twice.

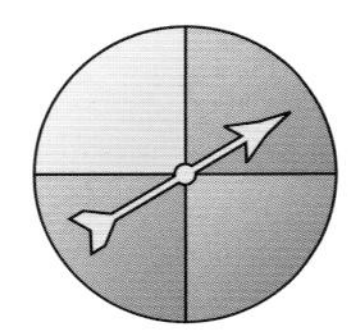

  a Draw a tree diagram to show all the possible outcomes.
  b What is the probability that it lands on red both times?
  c The spinner is spun for a third time.
  What is the probability that it shows red all three times?

4 A fair coin is thrown three times.
  a Show that the probability that it lands tails twice and heads once is $\frac{3}{8}$.
  b What is the probability that it lands tails every time?

5 Joe makes two clay models.
The probability that a model is broken within one day of its being made is 0.02.
  a Calculate the probability that both models are broken within one day of being made.
  b Calculate the probability that only one of the models breaks within one day of being made.
  c Joe has enough clay to make 50 models.
  If he makes 50 models, how many will not be broken within one day?
  Explain your answer.

6 Henry makes two paper planes.

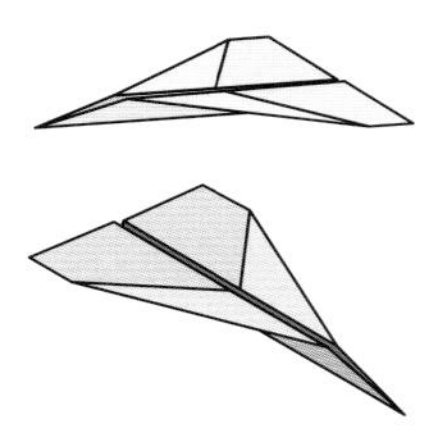

The probability that a plane can still fly one hour after being made is 0.68.
  a Calculate the probability that both planes can still fly one hour after they are made.
  b Calculate the probability that only one of the planes can still fly one hour after being made.
  c Henry made 25 paper planes.
  How many do you expect could not fly one hour after being made?

Experimental probability

This spread will show you how to:
- Understand relative frequency as an estimate of probability.
- Estimate probabilities from experimental data.

KEYWORDS
Relative frequency
Estimate
Proportion
Bias
Trial

You can estimate probability by carrying out an experiment.

- Experimental probability = $\frac{\text{number of successful trials}}{\text{total number of trials}}$

Experimental probability is also called **relative frequency**.

Louise suspects that a particular coin is biased towards heads.

The first 10 spins of the coin give:
H T H H H T H T H H

Louise calculates the relative frequency of obtaining a head:
$\frac{7}{10} = 0.7$

The next 10 spins of the coin give:
T T H T H T H T H T

The relative frequency of obtaining a head from all 20 results is: $\frac{11}{20} = 0.55$

Louise concludes that she needs to carry out more trials to find out if the coin is biased.

Louise spins the coin 40 more times.
She records the results of each set of 10 spins.

Set of 10 spins	1	2	3	4	5	6
Number of heads	7	4	6	5	6	5
Relative frequency	0.7	0.55	0.567	0.55	0.56	0.55

The relative frequency is settling down to around 0.55.

0.55 is Louise's estimated probability of obtaining a head with her coin.

Louise needs to carry out even more trials if her estimate is to be reliable.

- Relative frequency is the proportion of successful trials in an experiment.
- Relative frequency can give an estimate of probability.
- The more trials you carry out, the more reliable your estimate of probability will be.

Exercise D2.3

1 Lewis spins a coin 10 times.

a Copy and complete the table to show the relative frequency of obtaining a head after each spin.

Spin number	1	2	3	4	5	6	7	8	9	10
Head or tail	T	H	H	T	H	H	T	T	H	H
Relative frequency										

b Do you think the coin is fair? Give a reason for your answer.

c How could you get a more accurate estimate of the probability of this coin landing heads?

2 An equilateral spinner with three sectors coloured brown (B), orange (O) and green (G) is spun.
The colours from the first 20 spins are:

B B G O G O O B B O
G B B G O B G G O O

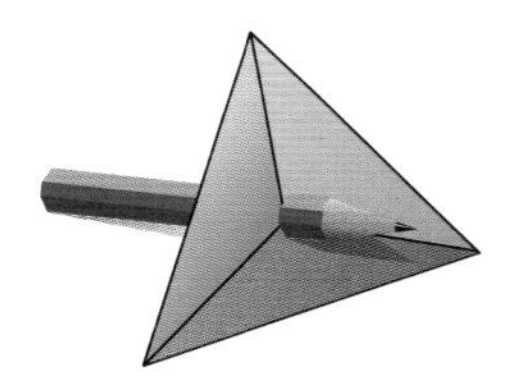

a Draw a relative frequency table to show how the probability of getting green changes with each spin.

b Use these 20 spins to estimate the probability of getting each colour.

c Do you think the spinner is fair? Give a reason for your answer.

d If the spinner were spun 3000 times, how many of each colour would you expect to get?

3 Clara spins a coin 50 times.
She calculates the relative frequency of obtaining a tail after each group of 10 spins.

Number of spins	10	20	30	40	50
Relative frequency	0.6	0.55	0.5	0.55	0.54

a How many times did the coin land tails in:
  i the first 10 spins
  ii the second 10 spins?

b In which group of 10 spins did Clara get:
  i the smallest number of tails
  ii the largest number of tails?

c If Clara spins the coin a further 10 times, exactly how many tails does she need so that the relative frequency is 0.5?

4 When a dog's toy is thrown it can land blue side up or blue side down.
Terri threw the dog's toy 320 times.
It landed blue side up 64 times.
Use these figures to estimate the probability that the next time it is thrown the dog's toy lands:

a blue side up

b blue side down.

Terri throws the dog's toy 10 more times.
It lands blue side up 4 times.

c Estimate the probability of the toy landing blue side up for the new 10 trials only.

d Add the trial results together and estimate the new probability of 'blue side up' from the total number of trials.

e What is the change in the total estimated probability when the new results are added?

D2.4 Relative frequency and experiments

This spread will show you how to:

- Understand relative frequency as an estimate of probability and use this to compare outcomes of experiments.

KEYWORDS
Limit
Estimate
Trial
Relative frequency diagram

Experimental probability is useful if you do not know the theoretical probability of an event.

Fred plays tic-tac-toe against his computer.
He wants to estimate the probability that he can win against the computer.

These are the results of the first 12 games.

Game number	1	2	3	4	5	6	7	8	9	10	11	12
Win or lose	W	L	L	W	L	L	L	W	L	W	W	L
Relative frequency	1	0.5	0.33	0.5	0.4	0.33	0.29	0.375	0.33	0.4	0.45	0.42

Fred draws a relative frequency diagram to show how the probability changes.

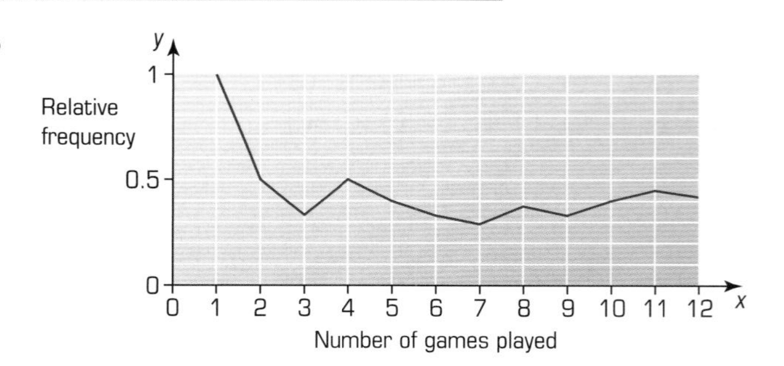

As Fred plays more games the relative frequency becomes less changeable.
It is tending towards a **limiting value** of 0.4.
Fred estimates that p(win) = 0.4.

▶ Relative frequency is the only realistic way of giving an estimate of probability.

Out of the next eight games, Fred only wins two.

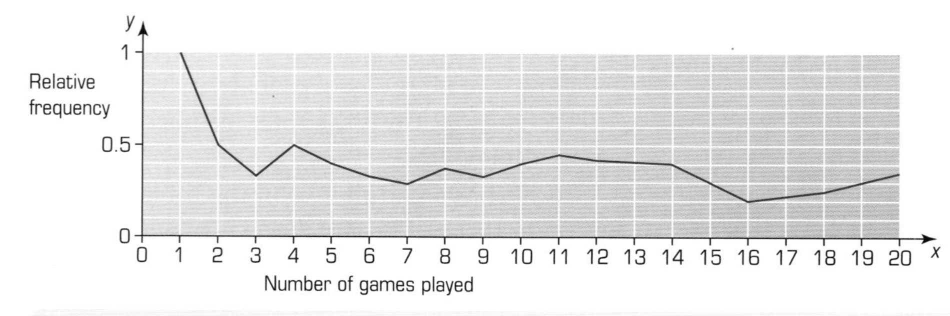

The relative frequency is now 0.35.

This is different to Fred's earlier estimate of p(win).

▶ The more trials you carry out, the more reliable your estimate of probability will be.

Exercise D2.4

1 Rachael has some seeds of a particular plant.
She wants to find the probability that a seed will germinate.
The seeds come in packs of ten. She buys eight packs.

a Copy and complete the table to show the relative frequency of the number of seeds that germinate.

Pack number	1	2	3	4	5	6	7	8
Number of seeds that germinated	8	7	9	7	8	6	9	6
Relative frequency								

b Draw a relative frequency graph to show how the probability of germinating seeds changes.

c Rachael has another pack of 10 of the same seeds.
Estimate how many of these seeds will germinate.
Give a reason for your answer.

d A nursery usually sells 9000 of these plants in one season.
What is the minimum number of seeds they should sow to ensure they have enough to sell? Explain your answer.

2 Callum and Peter play a game 20 times.
The graph shows the relative frequency of games that Peter won.

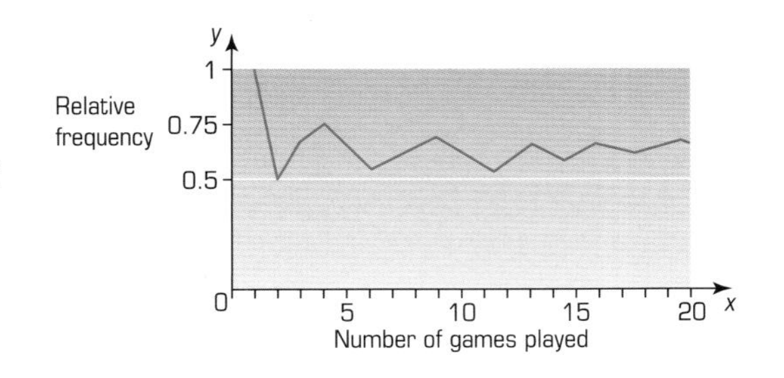

a Use the graph to decide if Peter won or lost each of the first four games.

b What percentage of the 20 games did Peter win?

c Callum and Peter play the game again.
Estimate the probability that Callum wins.

d Callum and Peter actually play a further five games.
Callum wins all of them.
Copy the graph and continue the line to show how the graph changes as these next five games are played.

3 A bag of white grains of rice had got mixed with a bag of brown grains of rice.
Jenny wanted to estimate how much of this mixed bag was white grains of rice.
She picked 20 grains of rice, counted the number of white grains and replaced them in the bag. Jenny repeated this several times.

Trial number	1	2	3	4	5	6
Number of white grains of rice	14	11	12	13	9	13
Relative frequency						

a Copy and complete the table to show the relative frequency of the number of white grains of rice in the bag.

b Draw a relative frequency graph to show how the probability of the number of white grains of rice changes.

c Use your results to estimate the number of white grains of rice that Jenny may have in a handful of 150 rice grains from the bag.

Summary

You should know how to …

1 Understand relative frequency as an estimate of probability and use this to compare outcomes of experiments.

2 Generate fuller solutions to problems.

Check out

1 Lesley and Duncan were typing reports.
Each report was five pages long.
Each page contained 100 words.
The table shows the number of errors found on each of the first five pages.

Page	1	2	3	4	5
Lesley	7	5	5	4	5
Duncan	5	9	6	4	7

Use relative frequency to compare and comment on the reliability of Lesley's and Duncan's typing skills.

2 Dervla and Emma were each playing in a brass band. During one concert, the number of wrong notes they played per piece was recorded.

Piece	1	2	3	4	5	6	7	8
Dervla	9	7	5	3	5	2	1	2
Emma	8	4	5	4	4	1	1	2

a Use relative frequency to compare and comment on Dervla and Emma's playing.

b Draw a graph to show how the number of wrong notes varied per piece. Suggest a reason for any features of the graph.

c Estimate the number of wrong notes in piece 9 for

i Dervla

ii Emma

How will this estimate change if you leave out the value for piece 1?

3 Transformations and congruence

This unit will show you how to:

- Transform 2-D shapes by combinations of translations, rotations and reflections.
- Know that translations, rotations and reflections map objects on to congruent images.
- Apply the conditions SSS, SAS, ASA or RHS to establish the congruence of triangles.
- Know that if 2-D shapes are similar, corresponding angles are equal and corresponding sides are in the same ratio.
- Enlarge 2-D shapes by a fractional scale factor.
- Understand the implications of enlargement for area and volume.
- Use and interpret maps and scale drawings.
- Use proportional reasoning to solve a problem.
- Know underlying assumptions, recognising their importance and limitations.
- Generate fuller solutions to problems.

Rangoli patterns are made by repeated reflection.

Before you start

You should know how to …

1 Apply simple transformations to 2-D shapes.

2 Apply Pythagoras' theorem in simple cases.

3 Simplify a ratio.

Check in

1 Copy the diagram on to squared paper. Extend both axes to $^-6$.

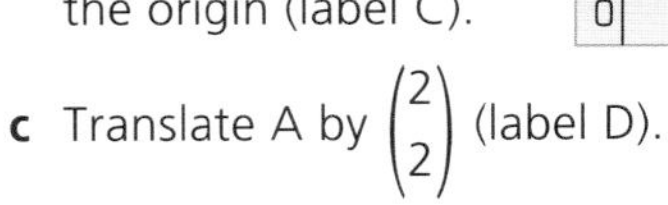

a Reflect A in the x-axis (label B).

b Rotate A 180° about the origin (label C).

c Translate A by $\begin{pmatrix} 2 \\ 2 \end{pmatrix}$ (label D).

2 Find the unknown lengths:

a

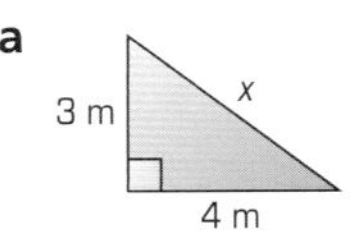

b

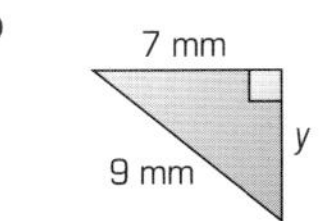

3 Simplify these ratios:

a 15 : 3 : 6 **b** 150 m : 5 m **c** 200 m : 1 km

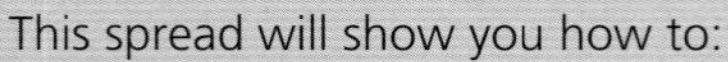

S3.1 Transformations review

This spread will show you how to:

- Transform 2-D shapes by combinations of translations, rotations and reflections.
- Know that translations, rotations and reflections preserve length and angle and map objects on to congruent images.

KEYWORDS

Vector	Mirror line
Object	Congruent
Reflection	Rotation
Image	Translation

In these diagrams, the **object** △ABC has been transformed to the **image** △A′B′C′ by a

... **reflection** in the line $y = 2$

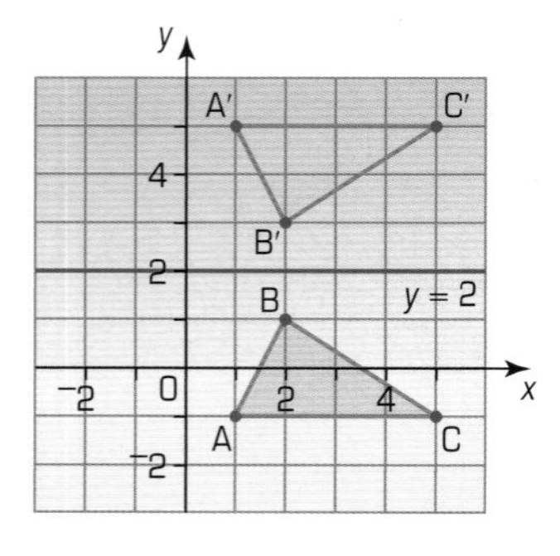

... **translation** by the vector $\binom{4}{-2}$

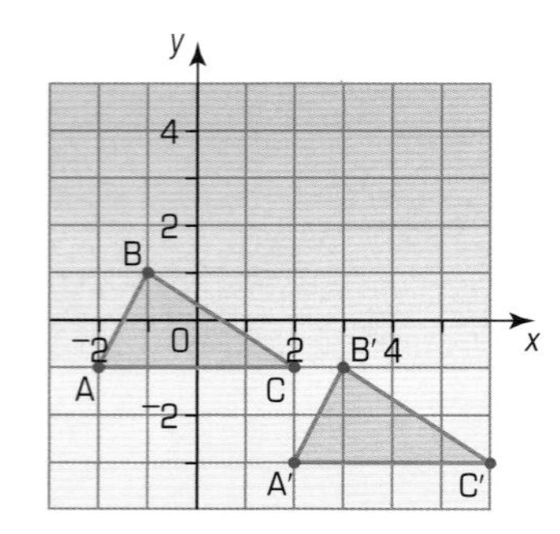

...**rotation** by 90° clockwise about (2, ⁻1).

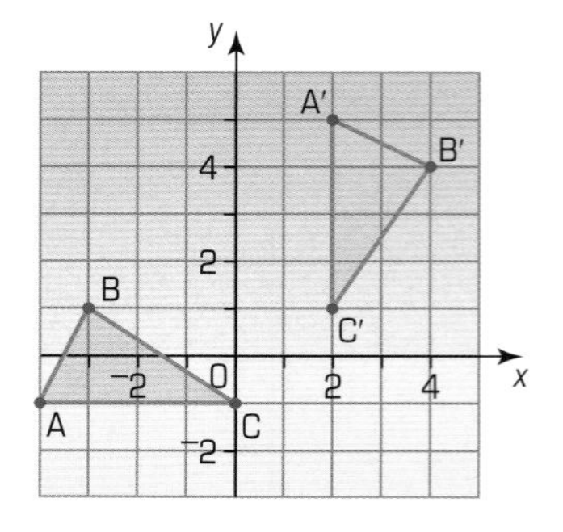

- You specify a ...
  - Reflection by a **mirror line**
  - Translation by a **vector**
  - Rotation by an **angle**, direction and **centre of rotation**.

- You can map congruent shapes onto each other by a reflection, a rotation or a translation.

example

The shapes P, Q, R and S are congruent.

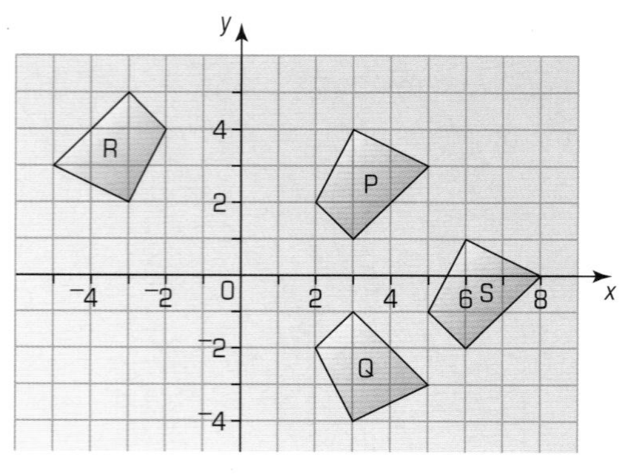

Describe the transformations that map P onto Q, R and S.

P → Q Reflection in the x-axis

P → R Rotation through 180° about (0, 3).

P → S Translation by $\binom{3}{-3}$

Remember:
Two shapes are **congruent** if they are identical. This means that their corresponding sides and angles are equal.

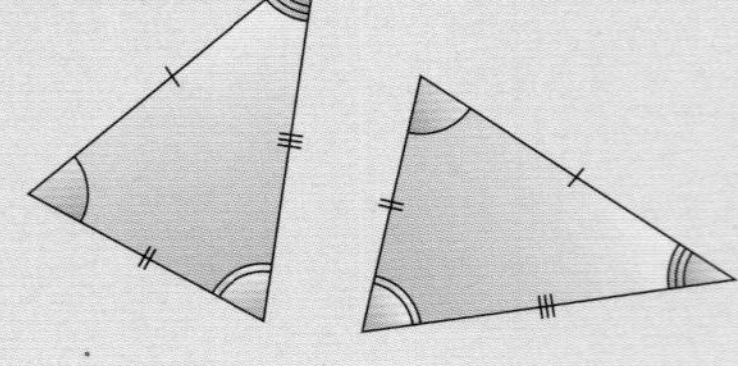

Exercise S3.1

1 On separate copies of the diagram, reflect the T-shape in each of the mirror lines:

a $x = 1$ **b** the x-axis

c the y-axis **d** $y = 1$ **e** $x = {}^{-}2$

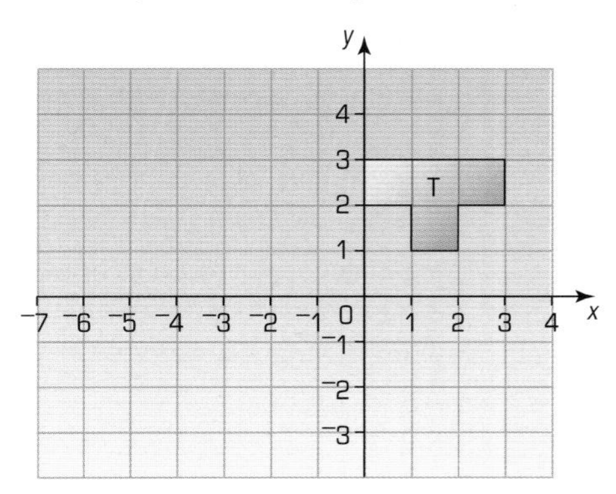

2 Use a separate copy of the diagram for each of **a** – **i**.

Rotate the L-shape:

a 90° clockwise about (0, 1)

b $\frac{1}{2}$ turn anticlockwise about (⁻1, ⁻2)

c $\frac{1}{2}$ turn clockwise about (3, ⁻2)

d 90° anticlockwise about (1, 2)

e $\frac{1}{4}$ turn clockwise about the origin.

Translate the L-shape using these vectors:

f $\binom{0}{1}$ **g** $\binom{2}{-3}$ **h** $\binom{3}{0}$ **i** $\binom{-2}{-3}$

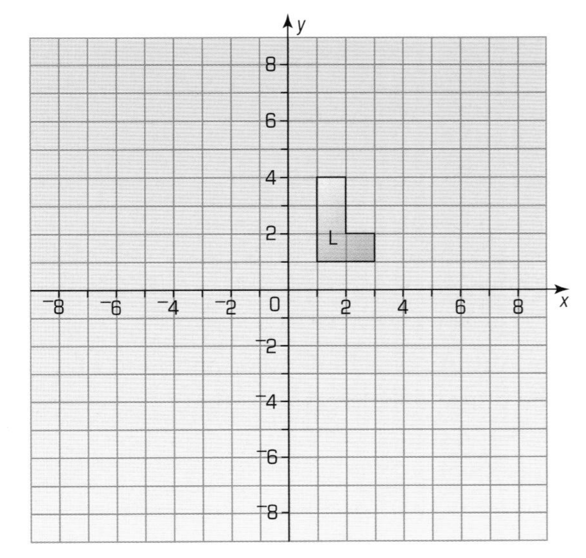

3 **a** On a copy of the diagram, reflect the triangle in the line $x = 1$, then translate through $\binom{2}{-3}$.

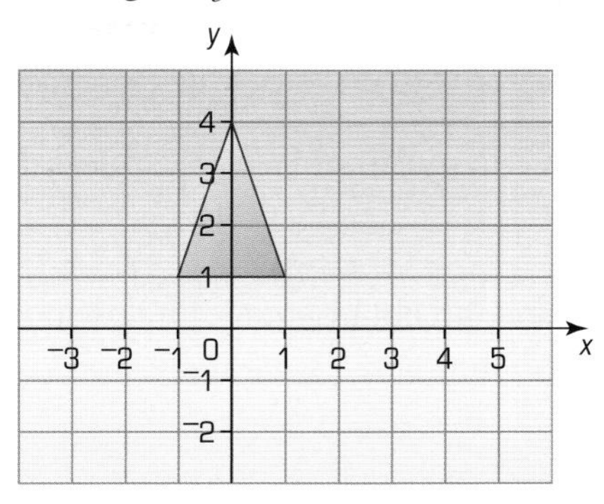

b What single transformation is equivalent to the combined transformations in **a**?

4 The diagram shows six congruent shapes. Describe these transformations:

a A to B **b** A to C **c** D to E

d C to F **e** E to F **f** B to F

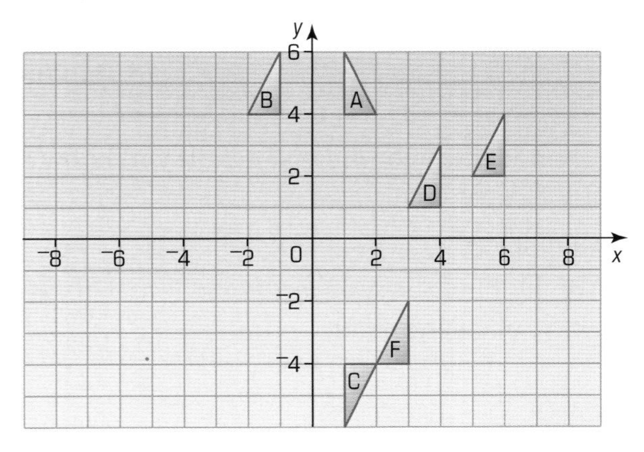

5 What shape do the combined image and object form when:

a a right-angled triangle is reflected in one of its shorter sides

b a square is rotated three times in succession through a quarter turn about the same corner

c an equilateral triangle is rotated through 180° about the midpoint of one side

d an isosceles triangle is reflected in one of its equal sides?

S3.2 Enlargements review

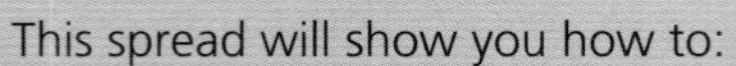

This spread will show you how to:

- Enlarge 2-D shapes, given a centre of enlargement and a whole-number scale factor.

KEYWORDS
Enlargement
Scale factor
Ratio
Similar

When you enlarge an object, the object and its image are **similar**.

- To specify an enlargement you need a centre of enlargement and a scale factor.

Enlargement preserves angles. Each length on the shape is enlarged by the same amount.

example

Enlarge ABCD by scale factor 2 about (1, 2).

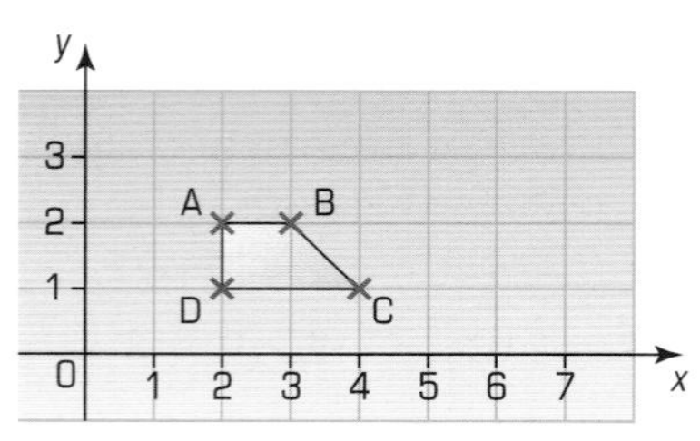

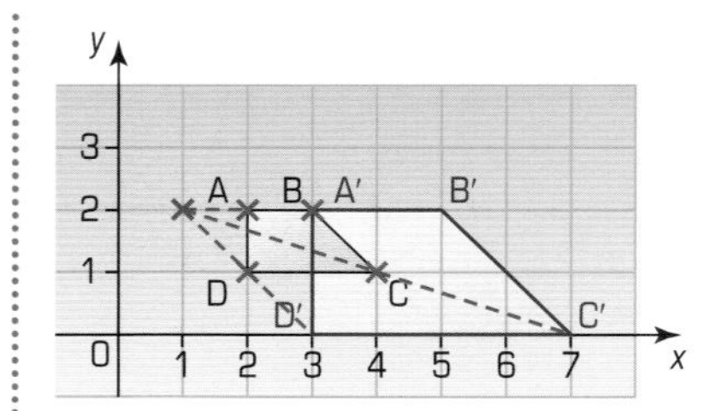

Draw ray lines from (1, 2) to each vertex of ABCD.
Extend these ray lines to twice their original length.
Join the extended lines to make the image A′B′C′D′.

You can identify the enlargement if you know the object and its image.

example

A′B′C′D′ is an enlargement of ABCD. Find the scale factor and centre of enlargement.

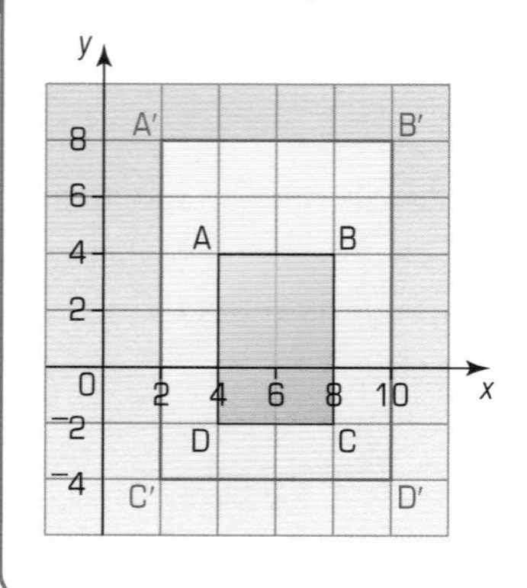

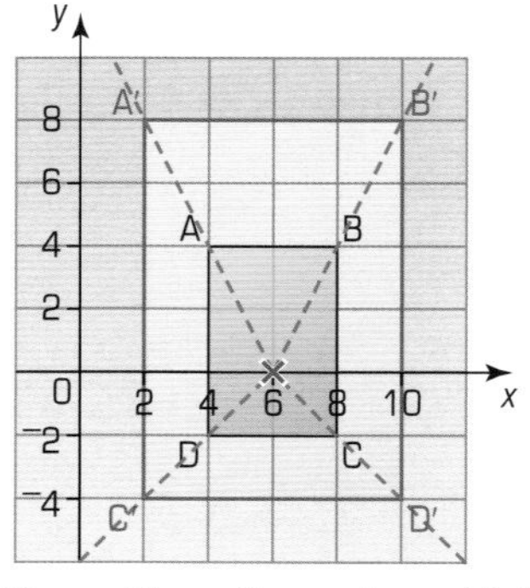

Draw lines from A to A′, B to B′, C to C′ and D to D′.
They cross at the centre of enlargement (6, 0).
AB = 4 units, A′B′ = 8 units.
The scale factor is $8 \div 4 = 2$.

- The ratio of any two corresponding line segments is the scale factor of the enlargement.

Exercise S3.2

1 Copy these diagrams on to squared paper and draw enlargements with centre C and the given scale factor.

a

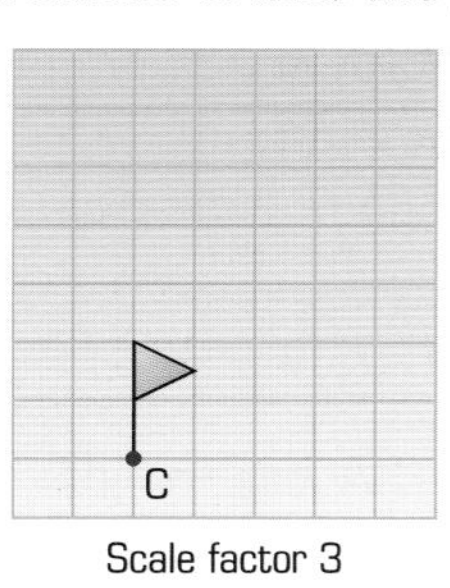

Scale factor 3

b

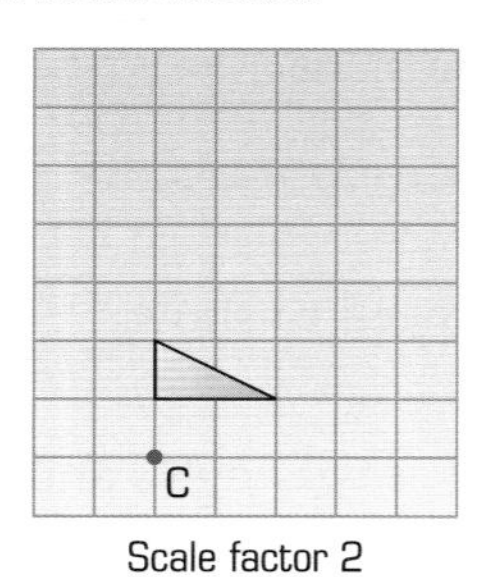

Scale factor 2

c

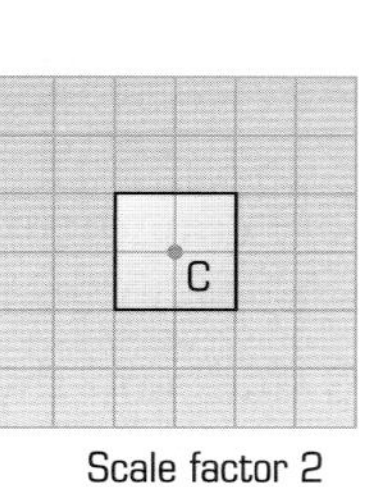

Scale factor 2

d

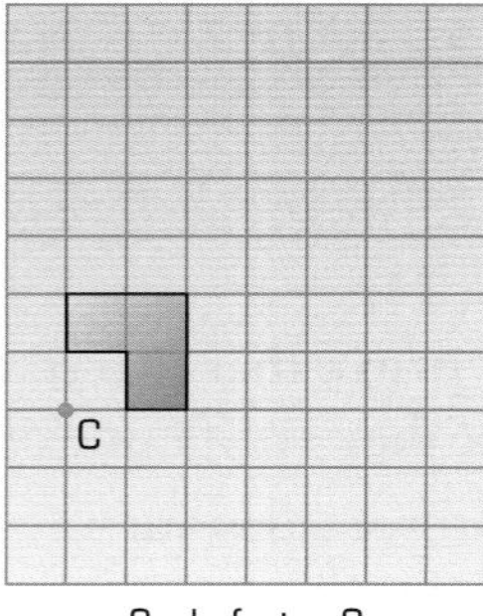

Scale factor 3

2 Copy this grid on to squared paper four times.
Draw each of these enlargements of R on a new grid.

a scale factor 2, centre (0, 0)
b scale factor 2, centre (1, 1)
c scale factor 3, centre (2, ⁻1)
d scale factor 3, centre (⁻1, 0)

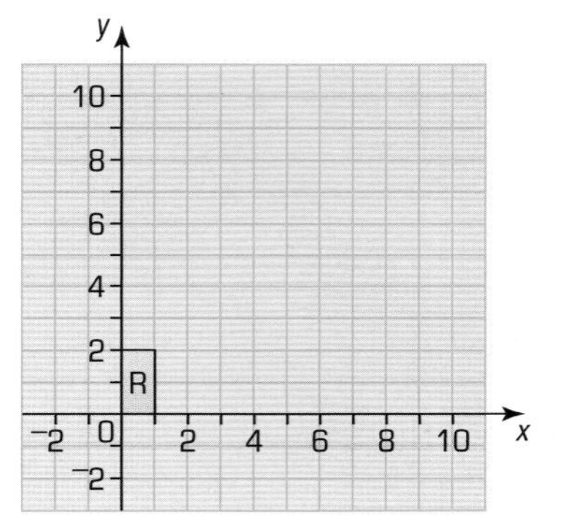

3 Find the scale factor and centre of enlargement for these pairs of shapes.

a

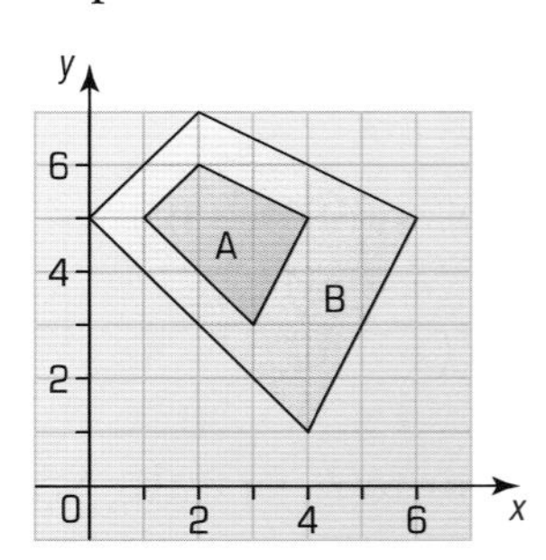

b

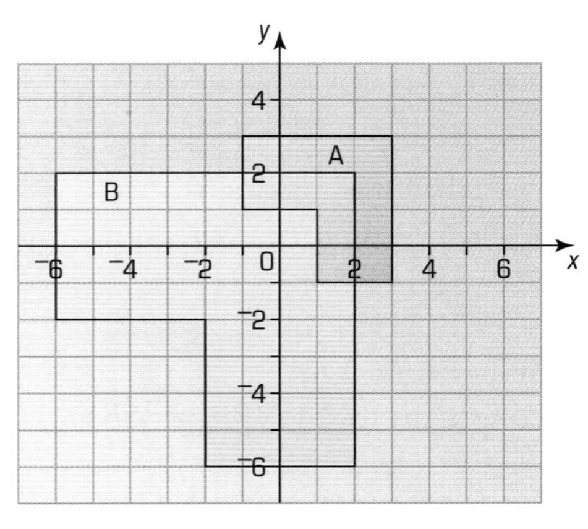

4 Work out the perimeter of shapes A and B in question 3b.

a Calculate the ratio of perimeter A to perimeter B.
b Calculate the ratio of area A to area B.
c What do you notice about your results to parts **a** and **b**?
d Investigate the relationship between area, length and volume of enlargements.

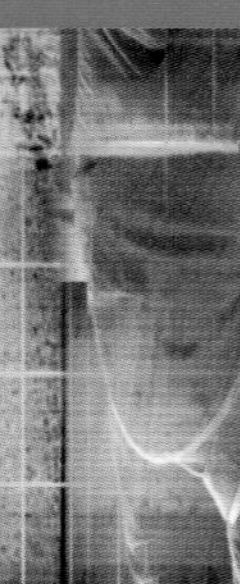

S3.3 Congruent triangles

This spread will show you how to:

- Apply the conditions SSS, SAS, ASA or RHS to establish the congruence of triangles.

KEYWORDS
Congruent
Hypotenuse
Pythagoras' theorem
SSS
SAS
ASA
RHS

To **prove** that two triangles are congruent you need to show that they satisfy one of these four conditions:

All three sides are equal (SSS).

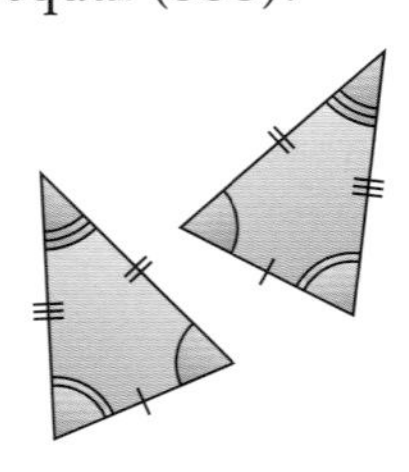

Two sides and the included angle are equal (SAS).

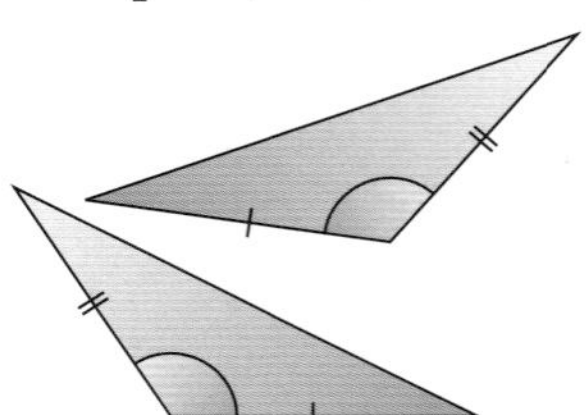

Two angles and a corresponding side are equal (ASA).

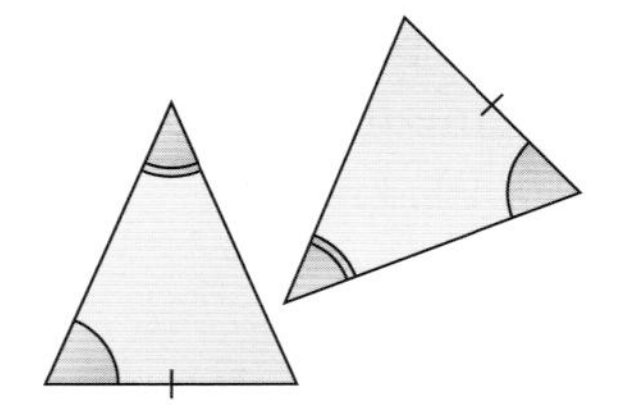

A right angle, hypotenuse and side are equal (RHS).

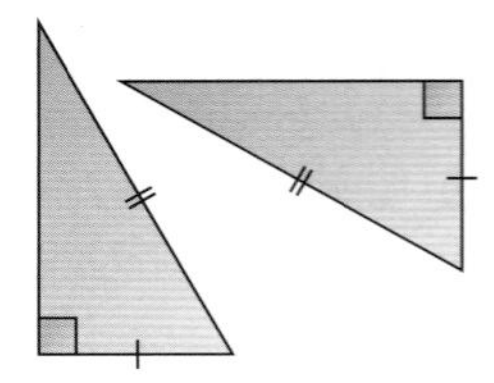

example

Show that the triangles in each pair are congruent.

a

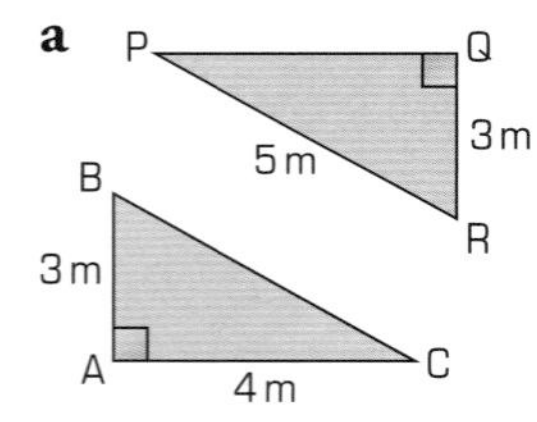

b

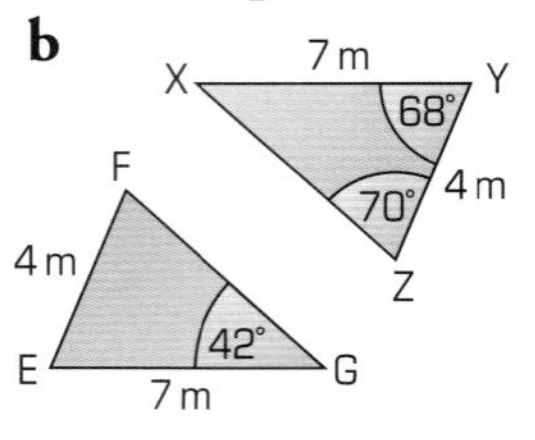

c

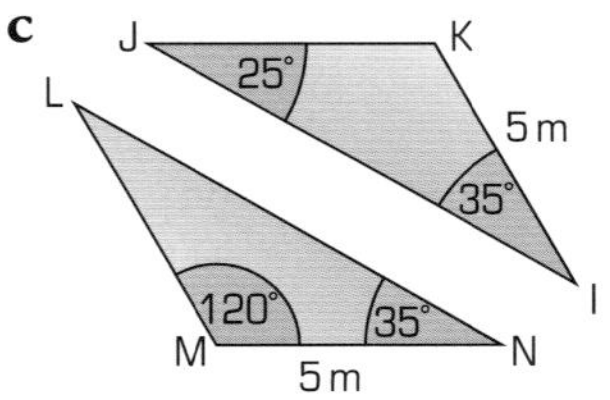

d

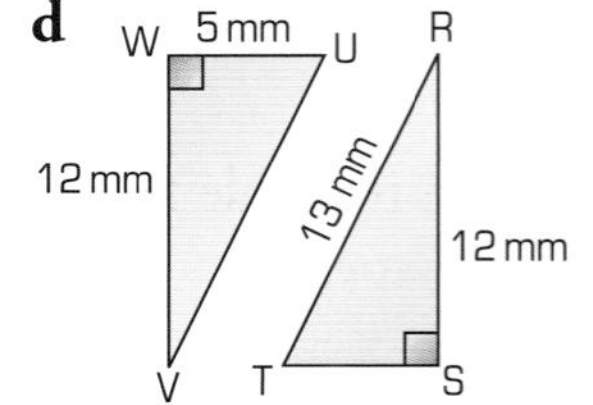

a $\triangle ABC$: $BC^2 = 3^2 + 4^2 = 25$, $BC = 5$ m (by Pythagoras' theorem)
$\triangle PQR$: $PQ^2 = 5^2 - 3^2 = 16$, $PQ = 4$ m
So $AB = QR$ (= 3 m)
$AC = PQ$ (= 4 m)
$BC = PR$ (= 5 m)
$\triangle ABC$ is congruent to $\triangle PQR$ (by SSS).

b $\triangle XYZ$: $\angle X = 180° - (68° + 70°) = 42°$
$EF = YZ$ (= 4 m)
$EG = XY$ (= 7 m)
$\angle G = \angle X$ (= 42°)
$\triangle EFG$ is congruent to $\triangle XYZ$ (by SAS).

c $\triangle IJK$: $\angle K = 180° - (25° + 35°) = 120°$
So $\angle M = \angle K$ (= 120°)
$MN = KI$ (= 5 m)
$\angle N = \angle I$ (= 35°)
$\triangle LMN$ is congruent to $\triangle IJK$ (by ASA).

d $\triangle UVW$: $UV^2 = 5^2 + 12^2 = 169$
$UV = 13$ mm
$\angle W = \angle S$ (= 90°)
$UV = RT$ (= 13 mm)
$WV = RS = 12$ mm
$\triangle UVW \equiv \triangle TRS$ (by RHS).

This is the hypotenuse.

Exercise S3.3

In questions **1–6**:

a Copy the diagrams.

b Decide whether or not the triangles are congruent. If they are explain your answer using one of the four rules. If they're not, explain why.

1

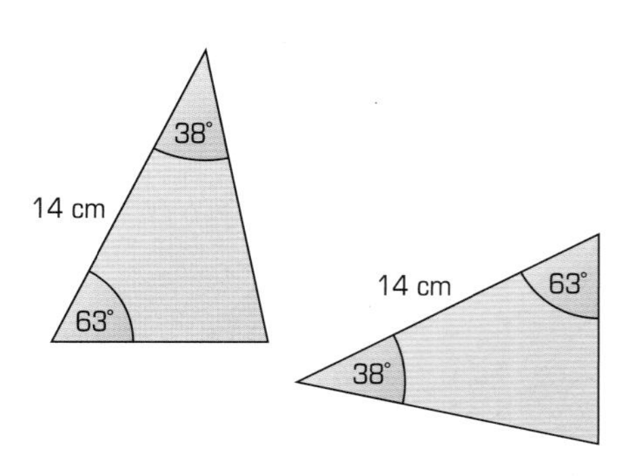

2

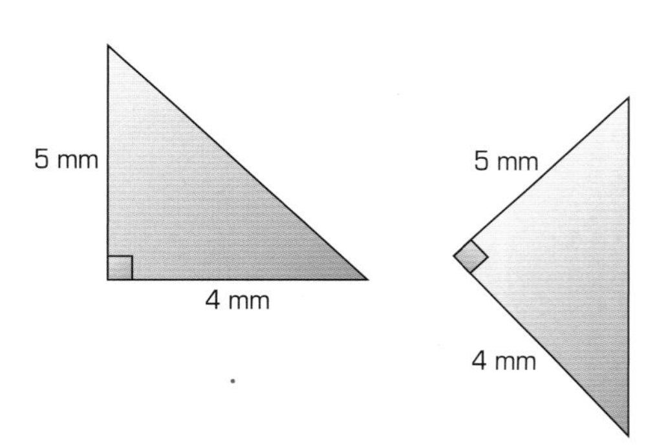

3

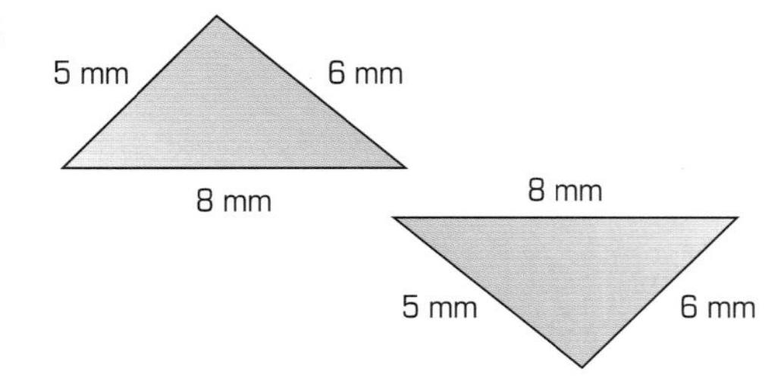

4

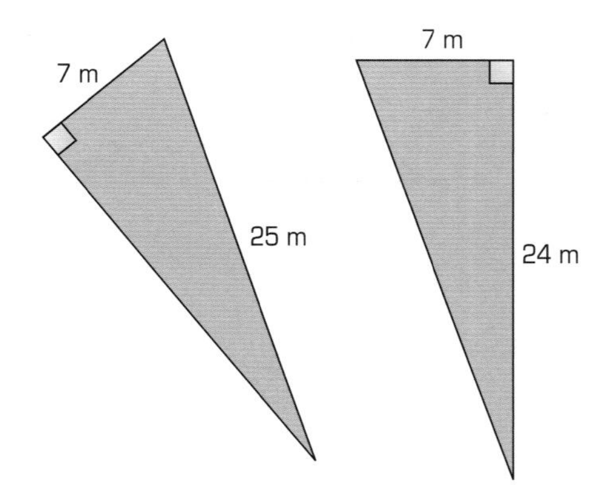

5

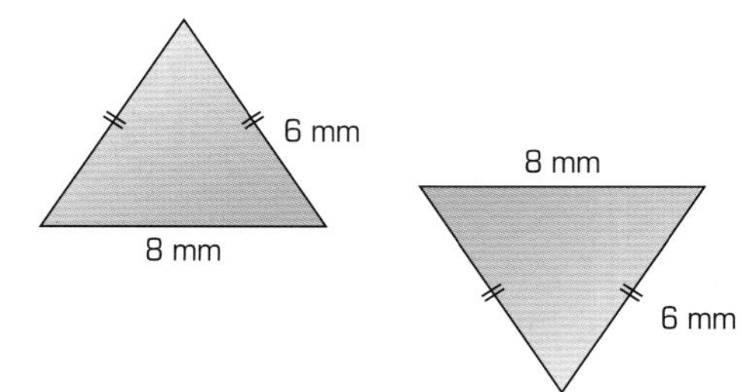

6

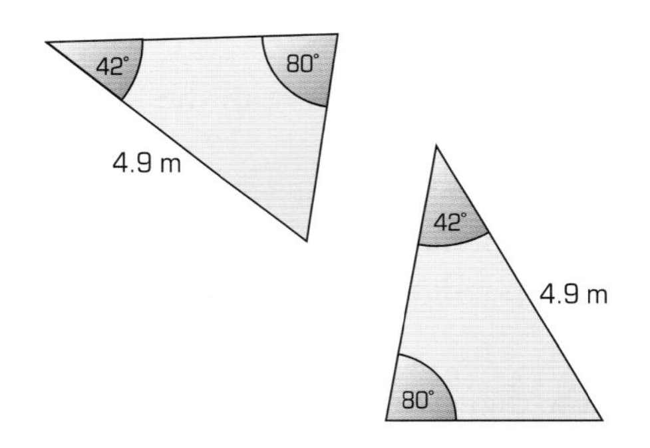

7 Draw an isosceles triangle PQR, with PQ = PR. Draw the perpendicular from P to QR to meet QR at a point S. Show that the two triangles PQS and PRS are congruent. Hence deduce that the base angles of an isosceles triangle are equal.

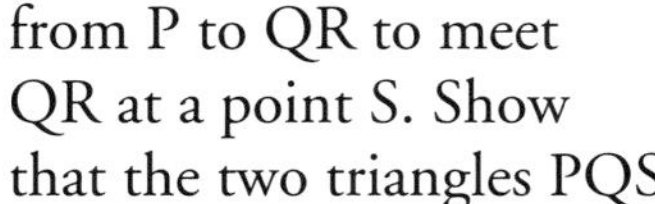

8 Here is a rhombus ABCD with its two diagonals (AC and BD) drawn in. The diagonals cross at the point X.

a Show that the triangles ABX, ADX, BCX and CDX are congruent.

b Explain how you know that the diagonals bisect each other at right angles.

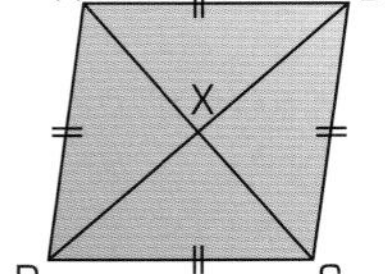

S3.4 Similar shapes

This spread will show you how to:

- Know that if 2-D shapes are similar, corresponding angles are equal and corresponding sides are in the same ratio.

KEYWORDS
Similar
Ratio
Corresponding

These two triangles are **similar**.
They have the same shape.

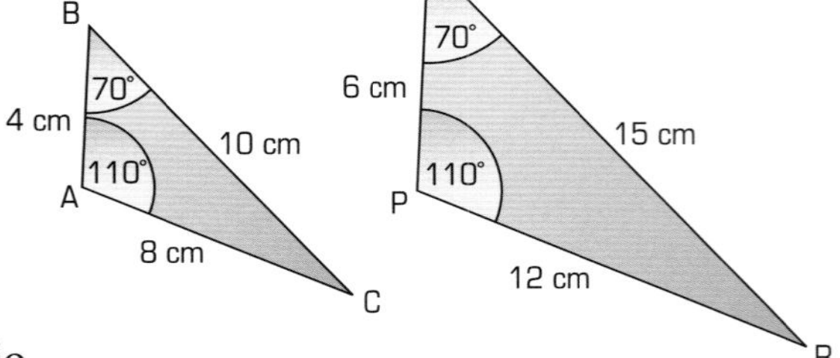

$\angle A = \angle P \qquad \angle B = \angle Q \qquad \angle C = \angle R$
$\Rightarrow$ Corresponding angles are equal.
$AB : PQ = 4 : 6 = 2 : 3$
$AC : PR = 8 : 12 = 2 : 3$
$BC : QR = 10 : 15 = 2 : 3$
$\Rightarrow$ Corresponding sides are in the same ratio.

- **Two 2-D shapes are similar if:**
  - **corresponding angles are equal**
  - **corresponding sides are in the same ratio.**

If two 2-D shapes have:
- equal angles
- corresponding sides in the same ratio

the shapes are similar.

If you know that two shapes are similar, you can find unknown lengths and angles.

example

$\triangle ABC$ and $\triangle PQR$ are similar.
Find $\angle R$ and $\angle Q$.

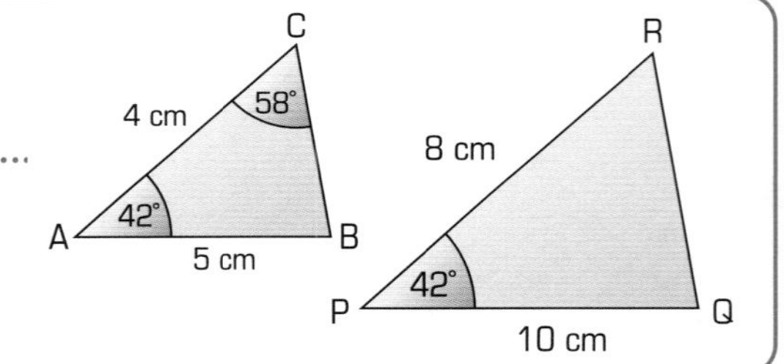

$\angle R = \angle C = 58°$ (corresponding angles)
$\angle Q = 180° - (58° + 42°)$ (angles in a triangle)
$= 80°$

If you have sufficient information, you can show that two shapes are similar.

example

a Show that $\triangle ABE$ is similar to $\triangle ACD$.
b Hence find ED and CD.

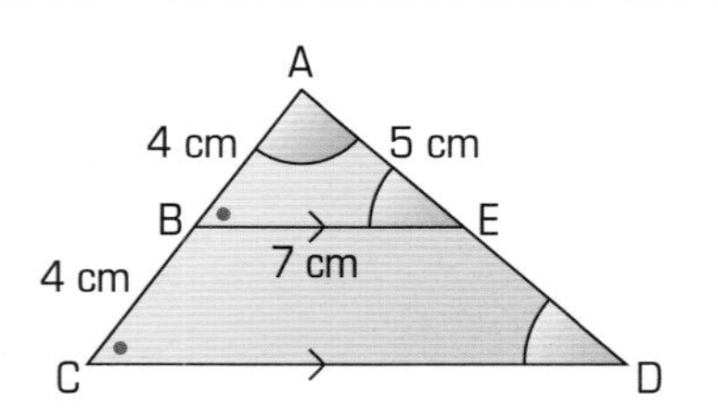

a $\angle A$ is common to both triangles.
$\angle B = \angle C$ (corresponding angles)
$\angle E = \angle D$ (corresponding angles)
So $\triangle ABE$ is similar to $\triangle ACD$.

b Since the triangles are similar, AB and AC are corresponding sides.
$AB : AC = 4 : 8 = 1 : 2$
So $AE : AD = 1 : 2 = 5 : 10 \Rightarrow ED = 10\text{ cm} - 5\text{ cm} = 5\text{ cm}$
$BE : CD = 1 : 2 = 7 : 14 \Rightarrow CD = 14\text{ cm}$

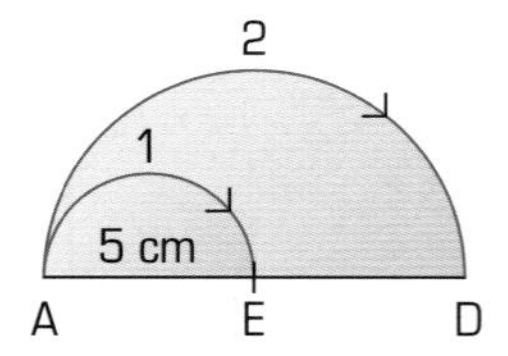

Exercise S3.4

1 **a** Copy each diagram.
b Calculate the missing angles.
c Decide if the triangles are similar by checking the angles.
d If they are similar, colour corresponding sides and angles the same (you will need six differently coloured crayons).

i

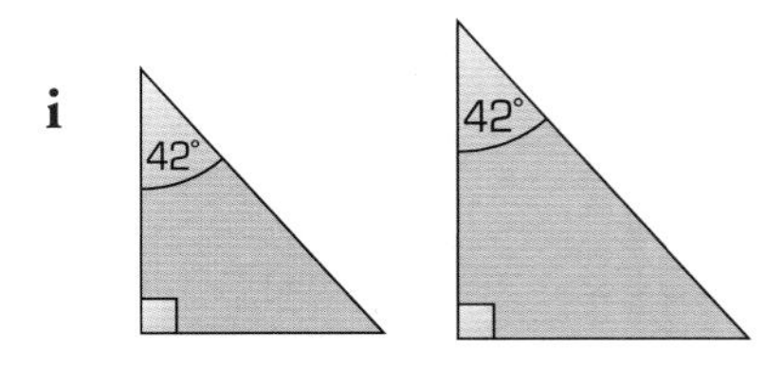

ii

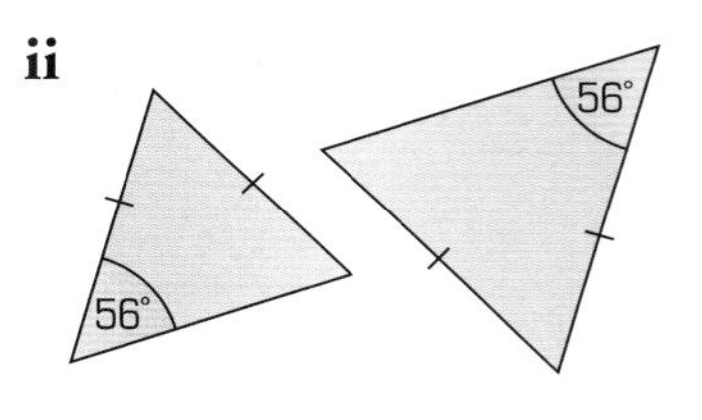

iii

iv

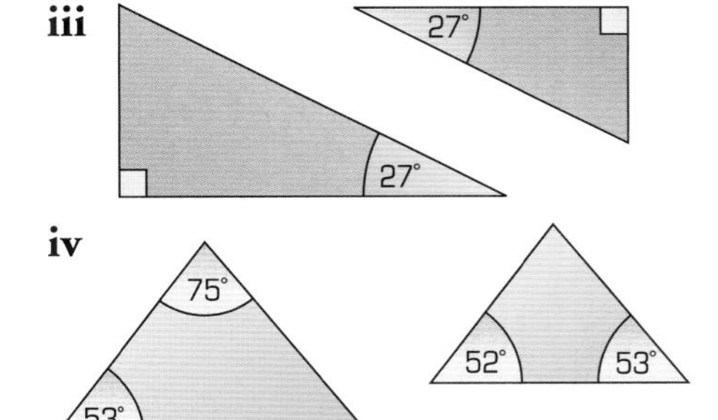

2 These two triangles are similar.

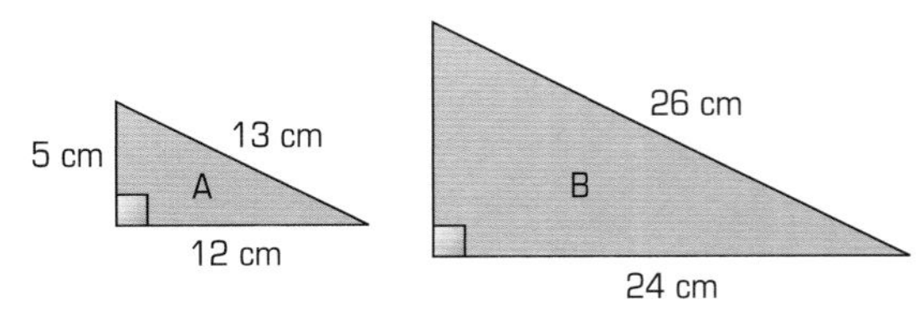

a Find **i** the perimeter and **ii** the area of each triangle.
b What is the ratio of perimeter A to perimeter B?
c What is the ratio of area A to area B?
What do you notice about your results in parts **b** and **c**?

3 Copy each diagram. Using angle properties of parallel lines, find the pairs of equal angles.
Use colour to show equal angles.

a

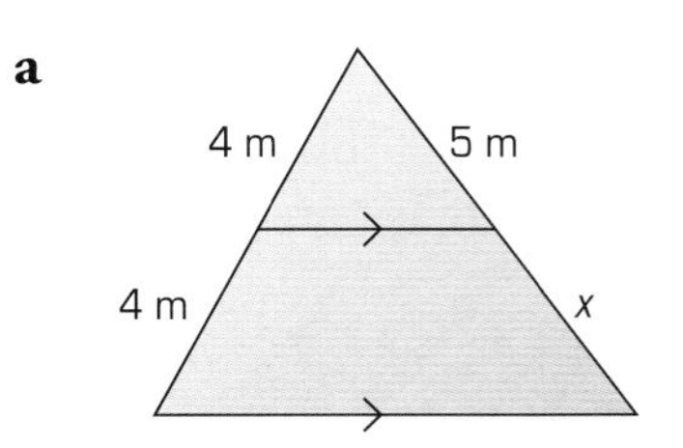

b

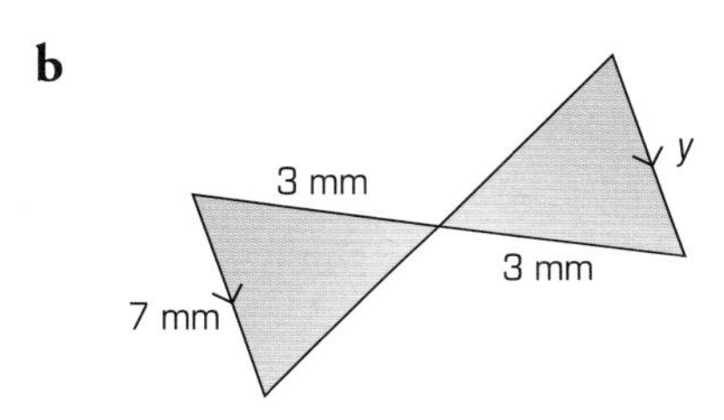

c

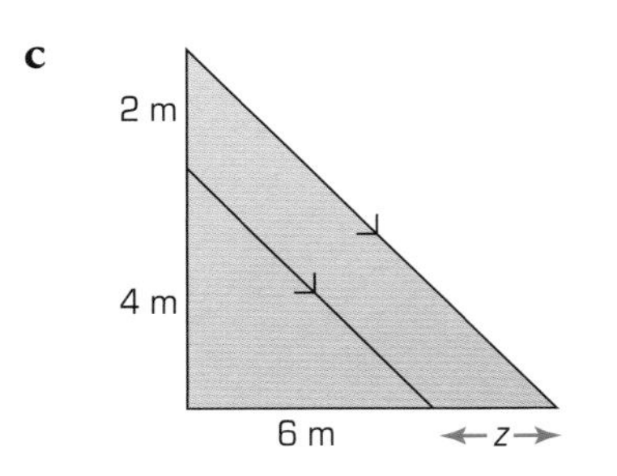

d

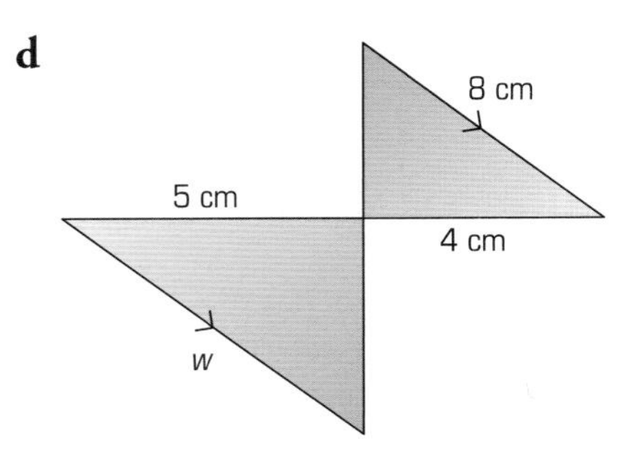

4 Find the unknown sides in each shape in question 3. Explain your reasoning.

Enlargement and similarity

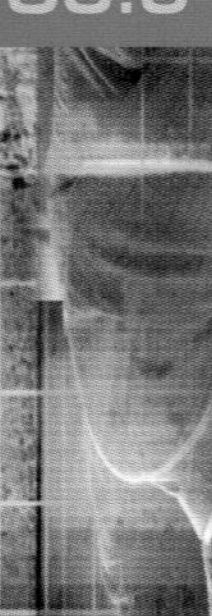

This spread will show you how to:

- Enlarge 2-D shapes, given a fractional scale factor.
- Understand the implications of enlargement for area and volume.

KEYWORDS

Enlargement, Similar, Scale factor, Ratio, Centre of enlargement

When you enlarge a shape, the original shape and its enlarged image are **similar**.

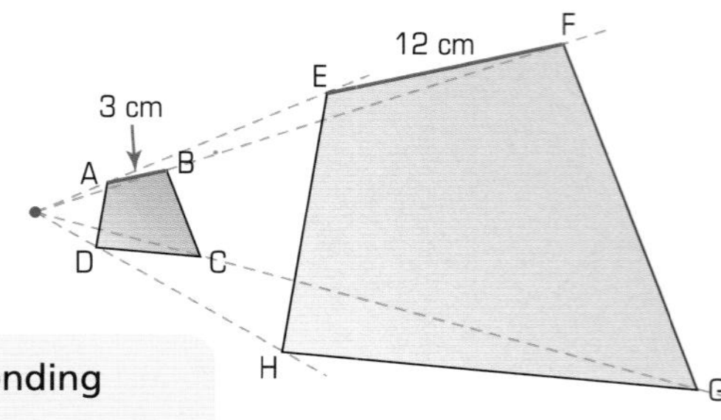

AB : EF = 3 : 12 = 1 : 4

The scale factor of the enlargement is 4.

► **In an enlargement, the ratio of any two corresponding line segments is equal to the scale factor.**

An enlargement can have a fractional scale factor.

example

a Enlarge the parallelogram ABCD by scale factor $\frac{1}{3}$ with centre of enlargement ($^{-}1$, 3).
Label the image A′B′C′D′.

b Write down the ratio of the lengths A′B′ : AB.

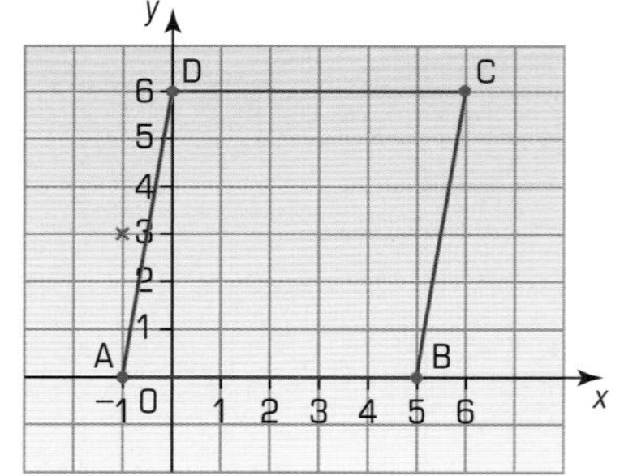

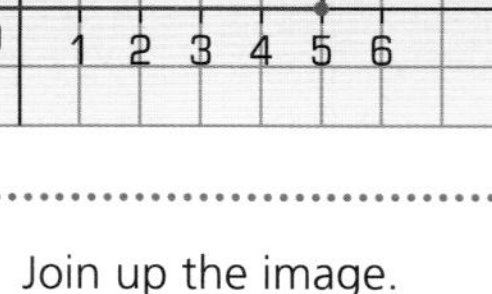

a Draw lines from ($^{-}1$, 3) to each of the vertices.

Measure $\frac{1}{3}$ of the distance along each line.

Join up the image.

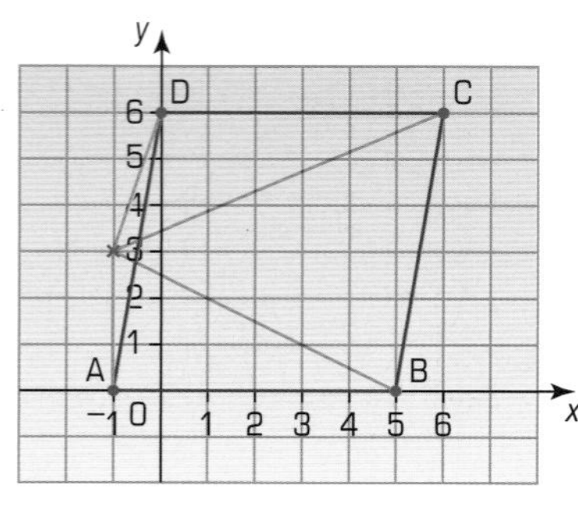

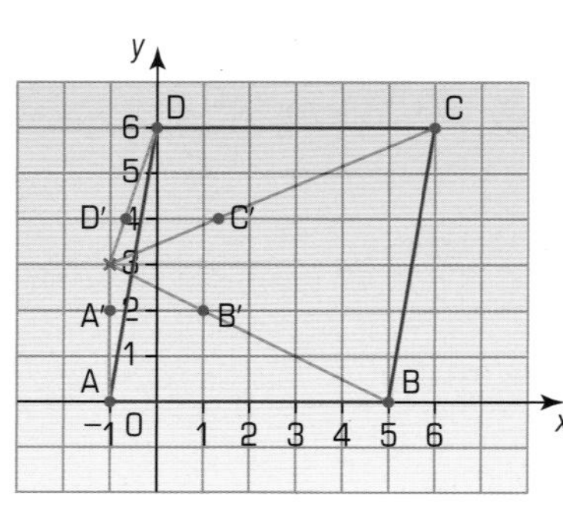

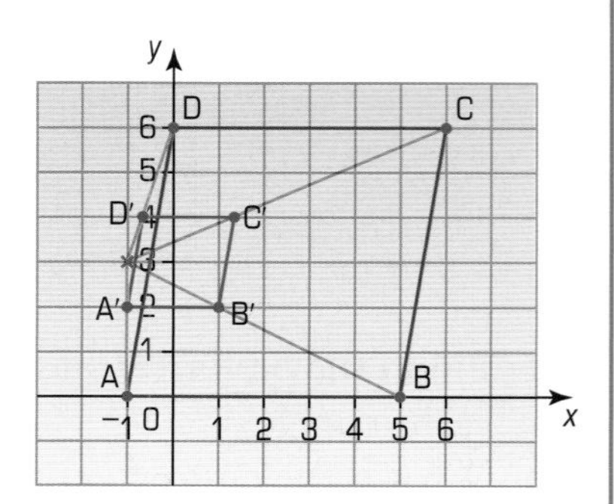

b Scale factor is $\frac{1}{3} \Longrightarrow$ A′B′ : AB = $\frac{1}{3}$: 1 = 1 : 3

It is called an enlargement even though the shape gets smaller.

Exercise S3.5

1 Copy the diagram and draw an enlargement, scale factor 2, centre C.

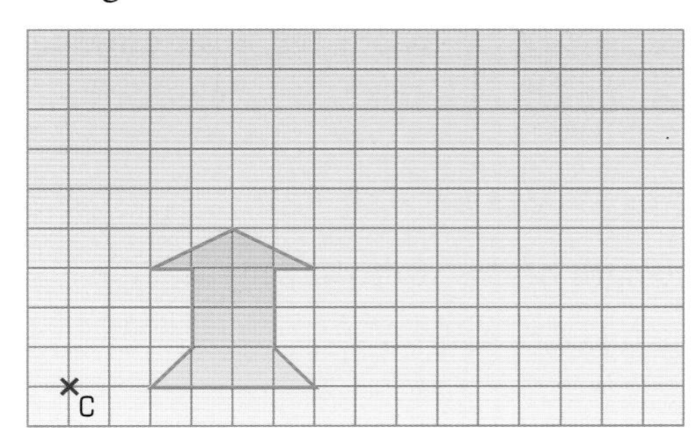

2 **a** Find the centre and scale factor of this enlargement.

b Find the ratio of the areas of rectangle ABCD and A′B′C′D′.

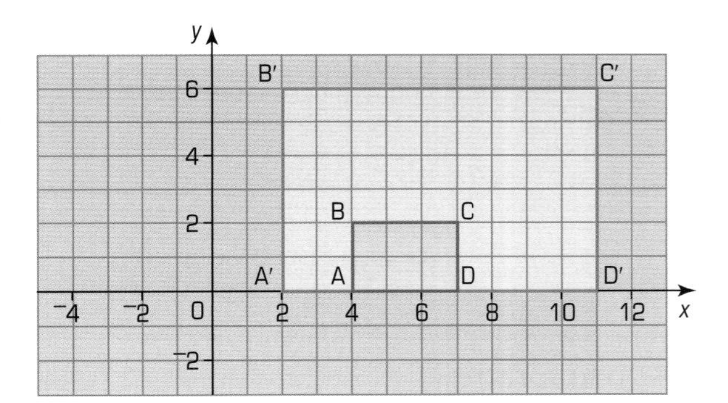

3 On separate copies of this diagram, draw enlargements of shape R with:

a scale factor 2, centre (0, 0)

b scale factor 3, centre (1, 2)

c scale factor 3, centre ($^-1$, 0).

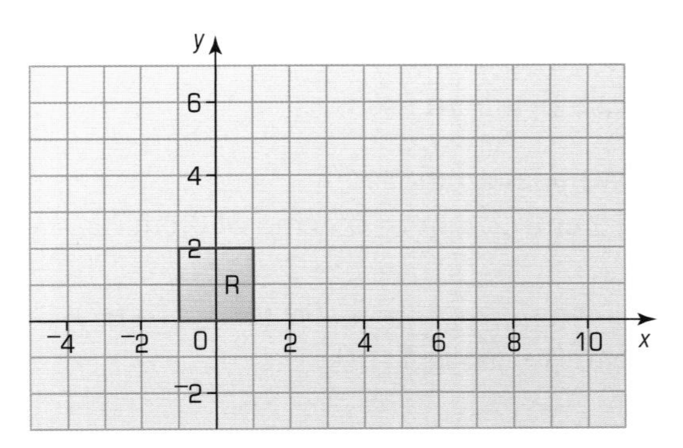

4 On two separate copies of this diagram, draw enlargements of the T-shape with:

a scale factor $\frac{1}{2}$, centre ($^-3$, $^-2$)

b scale factor $\frac{1}{3}$, centre (0, 1).

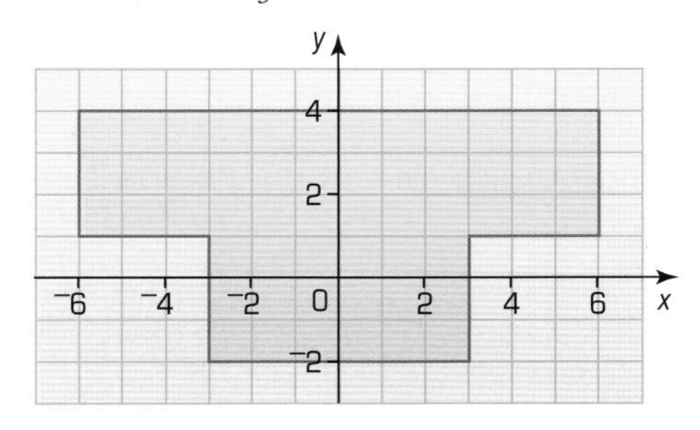

5 Rectangle P′Q′R′S′ is an enlargement of rectangle PQRS. Find:

a the centre of enlargement

b the scale factor.

The ray lines are shown to help you.

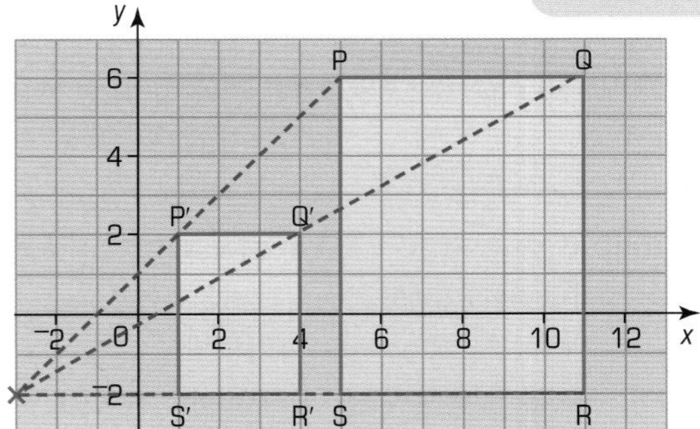

6 **a** Draw a square on a grid. Choose a centre of enlargement and use different scale factors such as 2, $\frac{1}{2}$, 3, $\frac{1}{3}$, 4 and $\frac{1}{4}$. Construct a table of measurements from the images of your square.

Scale factor	1	2	$\frac{1}{2}$	3	$\frac{1}{3}$	4	$\frac{1}{4}$
Length of side							
Area							

b Imagine that your square is one face of a cube. Calculate the volume of each cube and add the volumes to your table. What do you notice?

S3.6 Scale and proportion

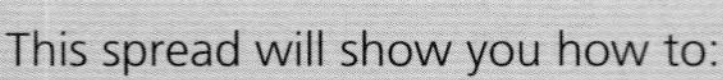

This spread will show you how to:
- Use and interpret maps and scale drawings.
- Use proportional reasoning to solve a problem.

KEYWORDS

Map, Scale drawing, Scale, Ratio

Maps and scale drawings are examples of enlargement by a fractional scale factor.

The length of this swimming pool is 25 m.

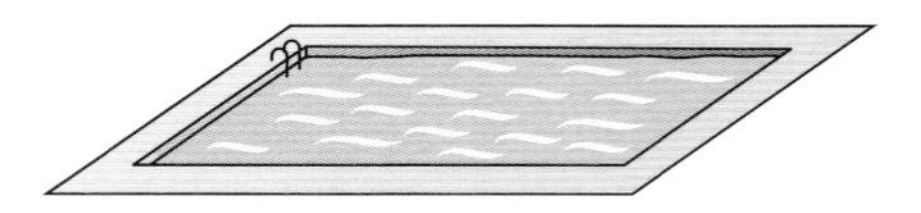

The length of this scale drawing is 5 cm.

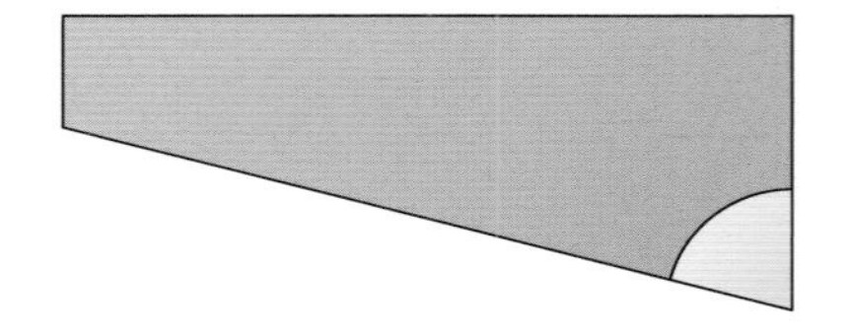

For the scale drawing of the swimming pool:

5 cm represents 25 m or 2500 cm

1 cm represents 500 cm

The **scale** of the drawing is 1 : 500.

Note: You could also write the scale as '1 cm to 5 m'.

► **You can express the scale of a map or a drawing as a ratio.**

The quantities in the ratio must first be written in the same units.

example

A map is drawn to a scale of 1 cm to 2 km.
Write this as a ratio.

Note: Choose the **smaller** unit to convert to.

First convert to the same units: 2 km = 2000 m = 2000 × 100 cm
= 200 000 cm

The scale of the map is 1 cm : 200 000 cm, or 1 : 200 000.

You can use proportional reasoning to solve problems with scale.

example

a The scale of a map is 2 cm to 8 km. The real distance between two villages is 30 km. Find the distance between the villages on the map.

b Sam draws a scale diagram of a box. The box has length 2.5 m. The scale drawing has length 15 cm. Find the scale that Sam used.

a

	Map		Real	
	2 cm	→	8 km	
÷8	0.25 cm	→	1 km	÷8
×30	7.5 cm	→	30 km	×30

So 30 km is represented by 7.5 cm on the map.

b Write as a ratio:

15 cm : 2.5 m = 15 cm : 250 cm
= 15 : 250
= 3 : 50 (or 1 : 16.67)

Sam's scale is 3 : 50.

Exercise S3.6

1 We can write scales in two ways: type 1 – 1 cm to 5 km and type 2 – 1 : 1 000 000.

a Change these scales so that they are written like type 1.

i 1 : 10 000 **ii** 1 : 2 000 000
iii 1 : 50 000 000 **iv** 2 : 30 000 000

b Change these scales so that they are written like type 2.

i 1 cm to 50 km
ii 1 cm to 15 000 m
iii 1 cm to 10 km
iv 2 cm to 5 km.

2 My bedroom is rectangular in shape and measures 3.6 m by 4.2 m. I want to fit this furniture into my bedroom:

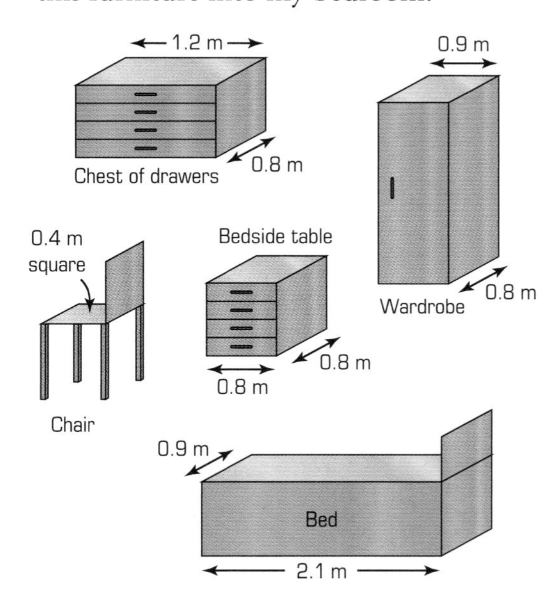

a Draw the outline of the bedroom to a scale of 1 : 50.
Mark the position of the door.

b Draw a plan view of each item of furniture to the same scale and cut out the shapes (make sure that you label them carefully).

c Arrange them in the bedroom plan (make sure that all drawers and doors will open!).

d How much floor space is left?

3 A groundsman marks out a football pitch.

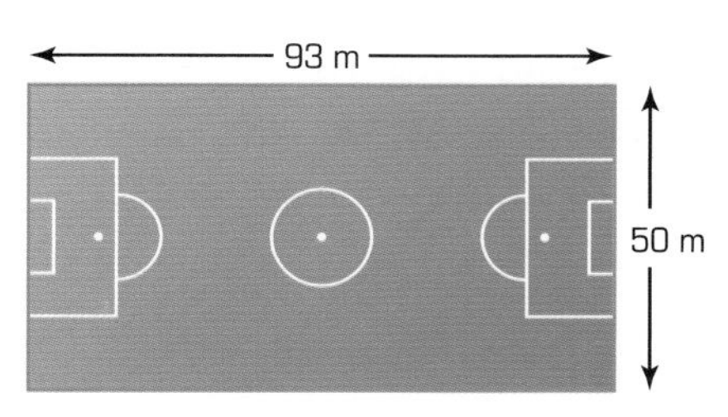

Draw a scale diagram of the pitch using a scale of 1 cm to 1 m.
On the diagram, mark the centre circle and the area outside each goal.

4 Here is part of a map of Lodge Moor.
The scale is 2 cm : 1 km.
Find the distance from:

a The Sportsman Inn to Malin Bridge Café

b Lodge Moor Hospital to Rowel Bridge.

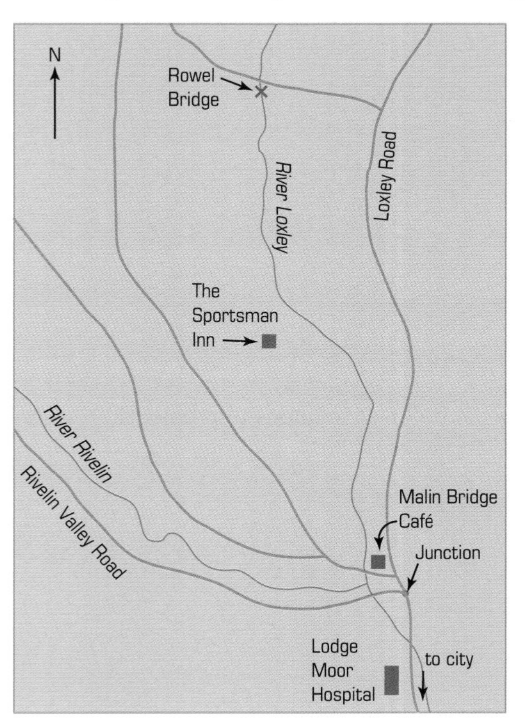

Summary

You should know how to ...

1 Apply the conditions SSS, SAS, ASA or RHS to establish the congruence of triangles.

2 Know that if 2-D shapes are similar, corresponding angles are equal and corresponding sides are in the same ratio.

3 Generate fuller solutions to problems.

Check out

1 Are these triangles congruent?

a

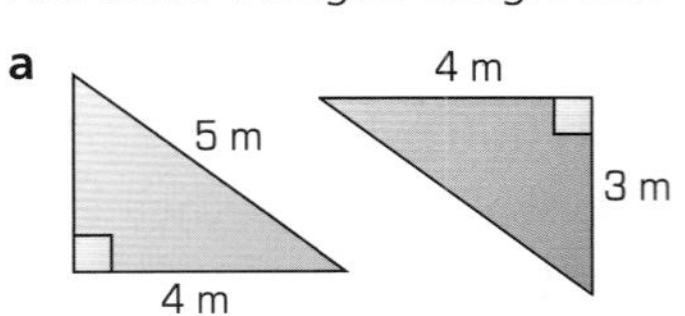

b

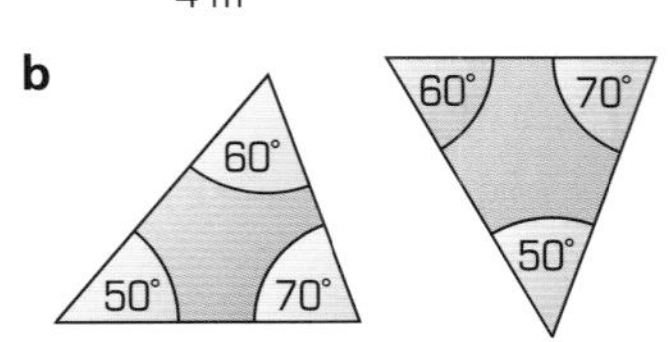

Give reasons for your answers.

2 These triangles are similar.
Find the unknown sides and angles.

a

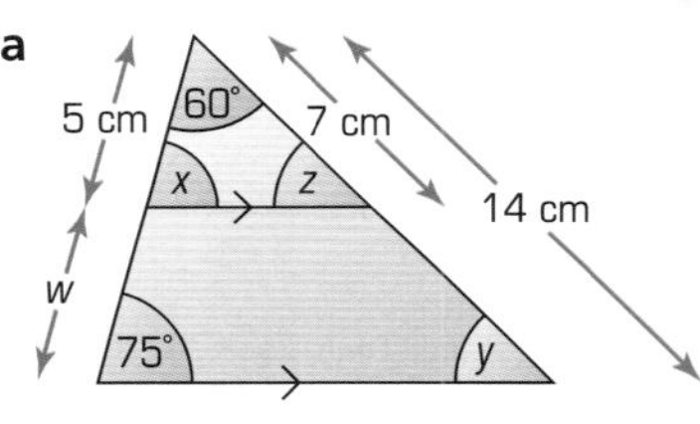

b

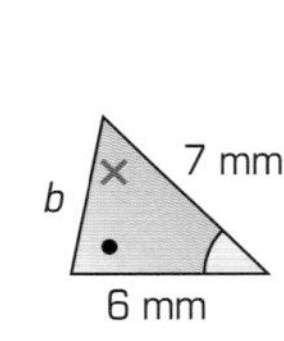

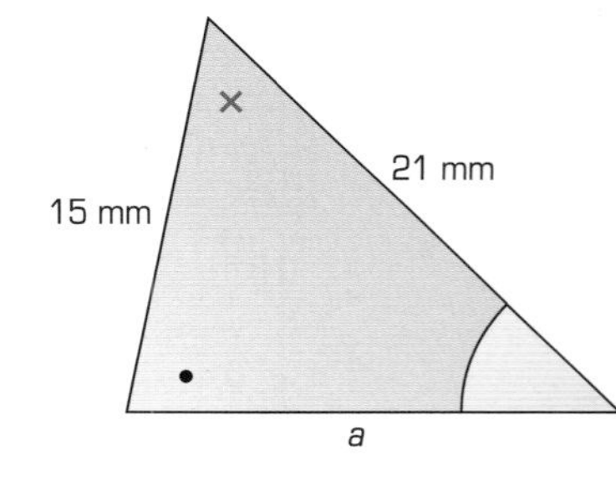

3

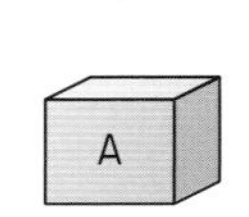

Volume = 9 cm^3

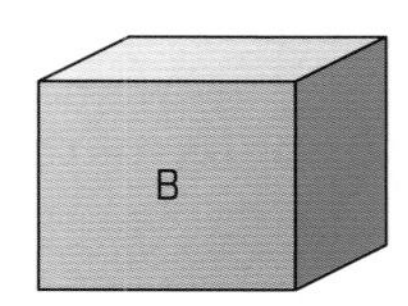

Volume = 72 cm^3

These cuboids are similar.
Find the ratio of:
a their surface areas
b their lengths.

1 Problem solving and revision

This unit will show you how to:

- Generate fuller solutions to problems.
- Explore connections in mathematics.
- Represent problems and synthesise information.
- Justify generalisations, arguments or solutions.
- Solve substantial problems by breaking them into simpler tasks.
- Use trial and improvement.
- Understand and use proportionality.
- Recognise limitations on accuracy of measurements.
- Estimate calculations by rounding numbers to one significant figure.
- Give reasons for choice of presentation.
- Pose extra constraints.
- Recognise that measurements given to the nearest whole unit may be inaccurate by up to one half of the unit in either direction.
- Solve problems using properties of angles and polygons.
- Understand upper and lower bounds.

You will use four stages in an exam question to maximise your marks:

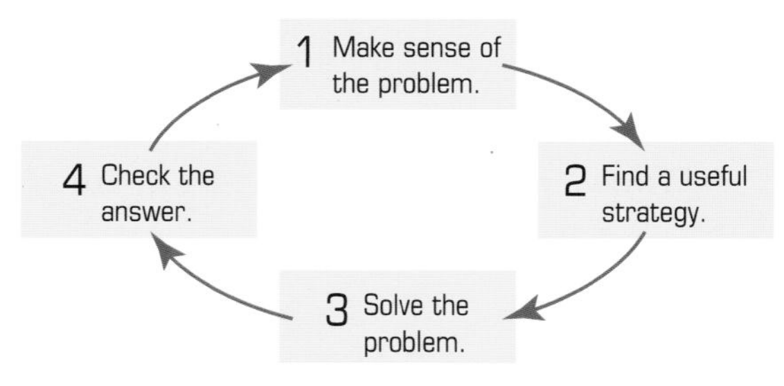

Before you start

You should know how to ...

- Calculate with integers, fractions, decimals and percentages.
- Use proportional reasoning and change.
- Make and justify estimates and approximations.
- Use algebra to represent unknown values.
- Know and use the index laws.
- Solve a pair of simultaneous linear equations.
- Change the subject of formulae.
- Write an expression for the general term of a sequence.
- Use angle facts and properties of polygons.
- Visualise 2-D representations of 3-D shapes.
- Transform 2-D shapes.
- Construct 2-D shapes.
- Use Pythagoras' theorem.
- Use units of measurement.
- Use compound measures.
- Know and use formulae for perimeter, area and volume.
- Understand upper and lower bounds.
- Find summary values for data.
- Interpret graphs and diagrams.
- Use probabilities including relative frequency.

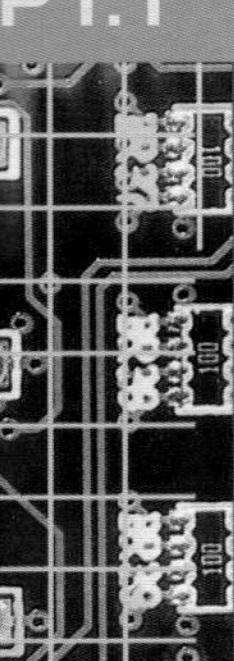

P1.1 Making sense of the problem

This spread will show you how to:

- Generate fuller solutions to problems.
- Explore connections in mathematics.
- Represent problems and synthesise information.

KEYWORDS

Problem Strategy
Accuracy
Unitary method

Problems in mathematics may involve lots of words and few numbers.

You should make sure that you:

- Understand what the question is asking you to do
- Extract all the relevant information
- Give your answer in appropriate units.

example

Kwame goes to a currency exchange bureau.
The exchange rate for pounds sterling to European euros is quoted as £1 to €1.44.
The exchange rate for pounds sterling to US dollars is quoted as £1 to $1.67.

How many euros would Kwame get for $10?

- You want to find the number of euros.
- Your answer will be in €.

You could:

- Convert $10 to £.
- Then convert £ to €.

Or use a unitary method:

- Convert $1 to €.
- Then ×10 to scale up.

Kwame opts for the first method:

Pounds → Dollars: × 1.67; Dollars → Pounds: ÷ 1.67

$\Rightarrow \quad \$10 = £(10 \div 1.67)$
$= £5.988\,023\,952$

Never round a number within a problem, only at the end.

Pounds → Euros: × 1.44; Euros → Pounds: ÷ 1.44

$\Rightarrow \quad £5.988\,023\,952 \times 1.44$
$= €8.622\,754\,491$
$= €8.62$

Round your answer to an appropriate degree of accuracy.
For most currencies, this will be two decimal places.

Check that your answer is sensible.

There are fewer € to the £ than there are $ to the £.
$\Rightarrow$ You would expect the number of € to be less than the number of $.

Exercise P1.1

1 Calculate each of these quantities to 2 decimal places.

a 12% of \$45 **b** $0.56 \times$ £34.21

c $\frac{3}{8}$ of €85.91 **d** 17.5% of \$563.98

e The \$ to £ exchange rate if \$6.36 = £4.20

f The annual percentage increase if a house value rises from £160 000 to £195 000 in one year.

2 **Puzzle**

Harry and Wendy's bank accounts when multiplied together make $^{-}24$.

When added they make $^{-}5$. Harry has more money than Wendy.

The figures on their accounts are both integers (in pounds £).

How much money does each person have?

3 Two rectangles have a combined area of 100 cm^2.

Their dimensions are shown in the diagram. Work out the value of n in cm.

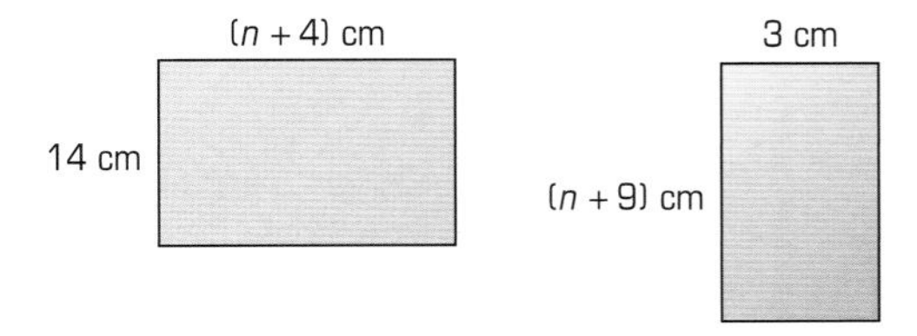

4 The probability of drawing a yellow ball out of a bag containing yellow, red and green balls is $\frac{5}{16}$.

There are 46 red balls and the probability of drawing a red ball is $\frac{24}{104}$. Work out:

a the number of yellow balls

b the probability of drawing a green ball.

5 Peter's teacher asked him to expand the expression $(x-4)^2$.

Peter said the answer was $x^2 + 16$.

Show that Peter is wrong.

Write, in its simplest terms, the answer that Peter should have got.

6 Show, without drawing the graph, that only three of the given points lie on the line $y = 3x - 4$:

(1.2, $^{-}0.4$) (8.3, 19.9)

(12.2, 32.6) ($^{-}3.5$, $^{-}14.5$)

Sketch the graphs of $y = 3x - 4$ and $y = {}^{-}3x$ on the same axes and describe anything you notice.

7 The mean value of these cards is $3x - 2$.

a Work out the expression on the missing card.

b Show that there is another set of three cards, one of which is $3x$, which also has a mean value of $3x - 2$.

8 **Investigation**

The units digits of 3^4 and 3^5 are 1 and 3 respectively.

Is it possible to predict the units digit of 3^n?

Extend this investigation to consider m^n.

9 The volume of any pyramid is always $\frac{1}{3}$ of the volume of a prism with the same base and vertical height.

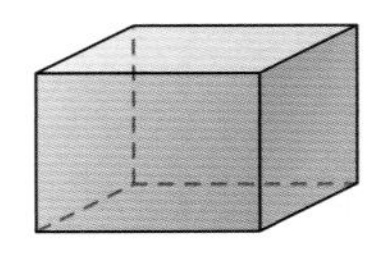

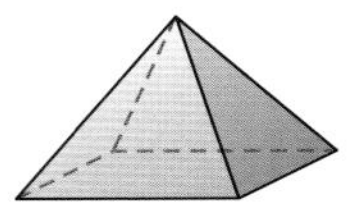

a Write a formula for the volume of a square-based pyramid with base b and vertical height h.

b Using your own variables for the dimensions, sketch and label a tetrahedron and write a general formula for its volume.

P1.2 Answering the question

This spread will show you how to:

- Generate fuller solutions to problems.
- Justify generalisations, arguments or solutions.

KEYWORDS
Explain
Justify
Probability

You will often be asked to explain your answer.
You should know what your answer means as well as being able to work it out.

example

Mary's height, h, is given in metres by the expression $h^2 + 4 = 5.62$
What is Mary's height to the nearest centimetre?

$h^2 + 4 = 5.62$
$h^2 = 5.62 - 4$
$= 1.62$

Remember:
The square root of a number can be positive **or** negative.

$\Rightarrow \quad h = \pm\sqrt{1.62} \quad$ so $\quad h = {}^-1.272\,79\ldots \quad$ or $\quad h = 1.272\,79\ldots$

Mary's height cannot be a negative value, so it must be 1.27 m to the nearest cm.

You may be asked to **justify** an answer that is already given.

example

There are 18 coloured balls in a bag, including red ones.
Michaela chooses a ball at random.

Explain why the probability that Michaela does **not** choose a red ball cannot be greater than 0.95.

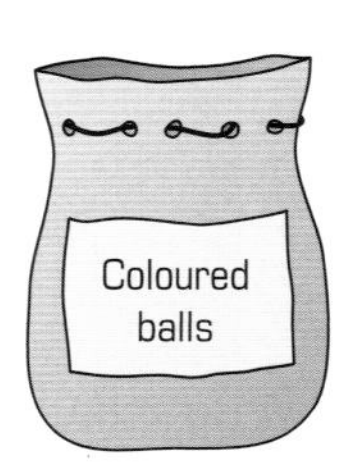

This question involves a double negative. Read it carefully to ensure that you understand it.

There is at least one red ball out of 18.
$\Rightarrow$ The minimum probability of choosing a red ball is $\frac{1}{18}$.
$\Rightarrow$ The maximum probability of not choosing a red ball is $\frac{17}{18}$.

$\frac{17}{18} = 0.944\,444\,44\ldots$ (or $0.9\dot{4}$)
$0.9\dot{4} < 0.95$

So the probability that Michaela does **not** choose a red ball cannot be greater than 0.95.

Exercise P1.2

1 Here is a box.

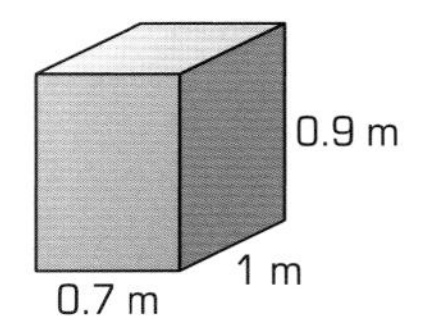

Ashley has a roll of material measuring 0.9 m by 4 m. Is it sufficient to cover the box completely?

2 Decide which box of 'Aunty May's Super Sweets' are the best value.

Does the answer surprise you?
Explain why.

3 The expressions in the box are attempts to fully factorise the expression $8m^2 - 16mt$.

$m^2(8m - 16t)$ $4m^2(2 - 8t)$

$2m(4m - 8t)$

$8m(m - 2t)$ $8m(m + 2t)$

a Decide which one is correct.
Justify your answer.

b Which of the others would multiply out to give $8m^2 - 16mt$?

4 Sketch the curve for the function $y = x^2$. Explain how you can use this to sketch curves of $y = x^2 + a$, where a is an integer.

5 The graph shows a straight line $y = {}^-2x + 1$.

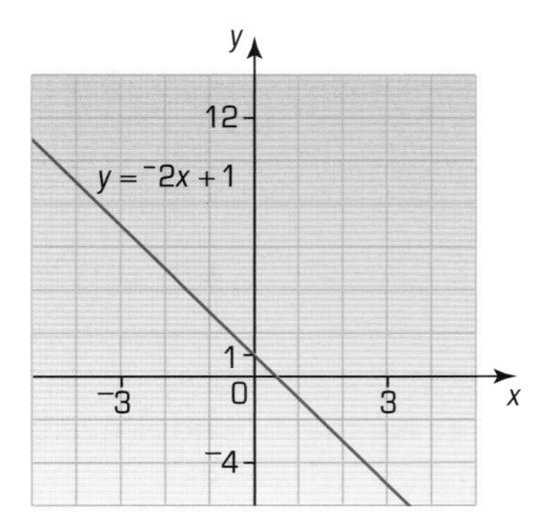

a Does the point $({}^-63, 127)$ lie on the line? Explain how you know.

b Write the coordinates of the point that lies on both the lines $y = {}^-2x + 1$ and $y = 2x + 1$.

c Explain how you know that $y = x + 3$ and $y = 2x + 4$ will intersect.
Work out the point of intersection.

6 The table below shows the probability of taking a counter out of a bag.

	Orange	Green	Red	Blue
Probability	0.58	0.3	0.1	?

a What is the probability that a blue counter will be taken out?

b Explain how you know that there must be at least 50 counters in the bag.

7 A 4 cm piece of string is cut from a length T cm long. The remaining length is then cut into five equal pieces. If the length of each piece is less than 9 cm:

a Write an inequality for the length of each piece involving T.

b What is the maximum value T can have? Explain your answer.

Choosing a strategy

This spread will show you how to:

- Solve substantial problems by breaking them into simpler tasks.
- Use trial and improvement where a more efficient method is not obvious.

KEYWORDS
Task
Strategy
Trial and improvement

You can often solve a complicated problem by breaking it into simpler tasks.

example

The hall floor at Roxford School needs varnishing. The diagram shows a floor plan.

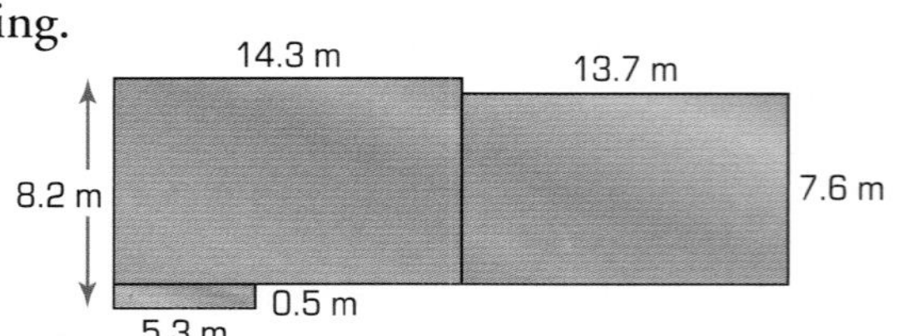

The varnish costs £5.63 per tin.
Each tin covers an area of 22 m^2.
Two coats of varnish are needed.

Allowing for 5% waste, calculate the cost of varnishing the hall floor.

Break the problem into smaller tasks:

1 Calculate the area to be varnished:

$14.3 \times (8.2 - 0.5) = 110.11\ m^2$
$5.3 \times 0.5 = 2.65\ m^2$
$13.7 \times 7.6 = 104.12\ m^2$ +
Total area $= 216.88\ m^2$

Two coats required
$\Rightarrow \quad 216.88\ m^2 \times 2 = 433.76\ m^2$

2 Calculate the number of tins required:

$433.76\ m^2 \div 22\ m^2 = 19.716\,363\,64$ tins
Allow for 5% wastage: $19.716\,363\,64 \times 1.05 = 20.702\,181\,82$ tins

Round up: 21 tins required.

3 Calculate the cost of varnish:

$21 \times £5.63 = £118.23$

The cost of varnishing the hall floor is £118.23.

You can often use trial and improvement as a valid strategy to solve problems.

example

$2.\square + 2.\triangle = 4.9$
$2.\square \times 2.\triangle = 5.98$
Find the single digits $\square$ and $\triangle$.

Compile a table of possible values, where $2.\square + 2.\triangle = 4.9$

□	△	2.□ × 2.△	Comment
1	8	5.88	Too small
4	5	6	Too large
2	7	5.94	Too small
3	6	5.98	Correct

$\square = 3$ and $\triangle = 6$ (or vice versa).

Be systematic in choosing your next numbers.

Exercise P1.3

1 Two integers in the range $^{-}100$ to $^{+}100$ have a product of $^{-}1600$ and a sum of 60. Use trial and improvement to calculate the two integers.

2 The diagram shows the net of a skip.

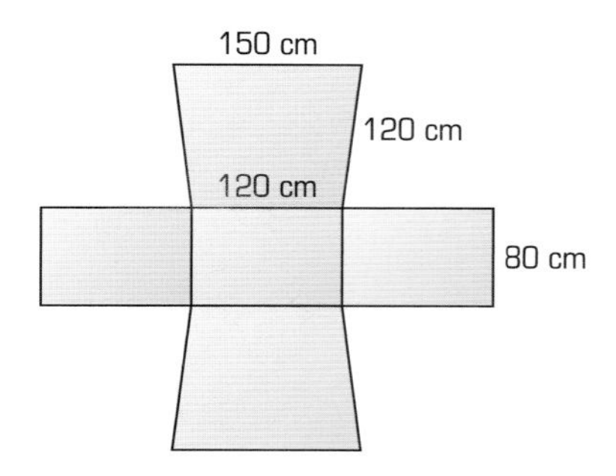

Calculate the cost of painting the **outside only** of all five sides.
The skip needs three coats of paint.
The cost of paint: £2.30 per tin.
Coverage: 2.2 m per tin.

3 The counters shown are put into a bag.

What extra counters would you need to put into the bag to make p(choose an 8) = $\frac{2}{11}$?
Is there more than one solution to this problem? Explain your answer.

4 Turf is available from Mortonhall Garden Centre in 10 m lengths 1 m wide.
Each 10 m length costs £15.60.
Calculate the cost of turfing this circular lawn. Show clearly how you arrived at your answer.

Radius = 6.1 m

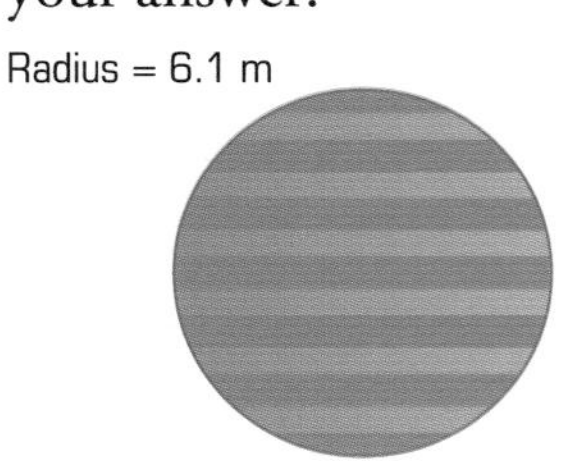

5 In each of these equations, use trial and improvement to work out the value of x correct to 1 decimal place.

a $x^4 = 27.9841$

b $x^2 + x + 1 = 9.9916$

c $(x - 6)^3 - 5 = 6.32$

6 The diagram shows a boat with two triangular sails. Calculate the length of binding needed to go around the perimeter of both sails.

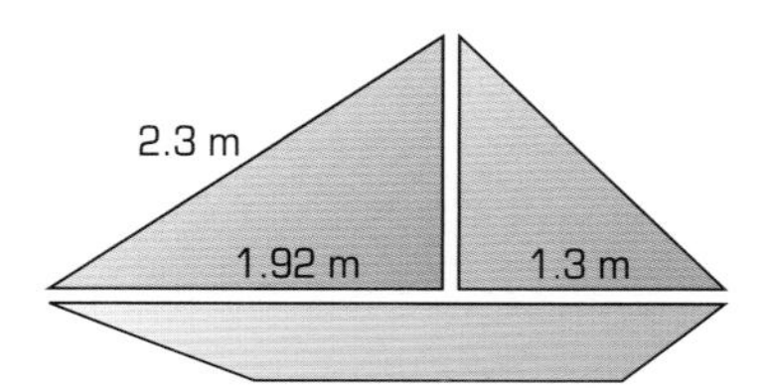

7 Two astronauts orbit the Earth.
The radius of their orbits is 2.3×10^7 km and 5.8×10^8 km from the Earth respectively.
Assuming that the Earth is a perfect sphere, calculate the minimum and maximum distances apart that they could be.

8 On his annual visit to France, a space traveller sees these exchange rates:

\$1.65 = £1 £0.90 = €1.20

On his planet the exchange rate for his currency (Marses) is 3.4 M = \$1.

a How many euros will the space traveller receive in exchange for 672 M?

b Can he afford three nights in a hotel that costs €52 per night?

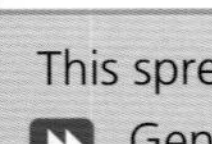

P1.4 Proportional reasoning

This spread will show you how to:

- Generate fuller solutions to problems.
- Understand and use proportionality.

KEYWORDS

Proportion, Scale factor, Ratio, Unitary method

Proportional reasoning involves solving a problem by using a scale factor. You can often scale quantities mentally.

example

Six litres of red paint and five litres of white paint make Candy Pink.
Grace has 12.5 litres of white paint.
How much red paint does she need to make Candy Pink?

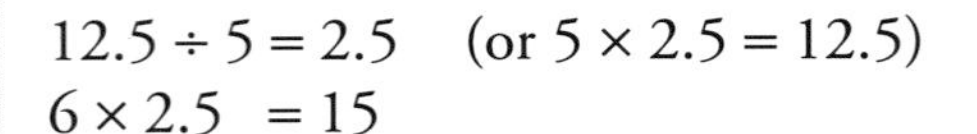
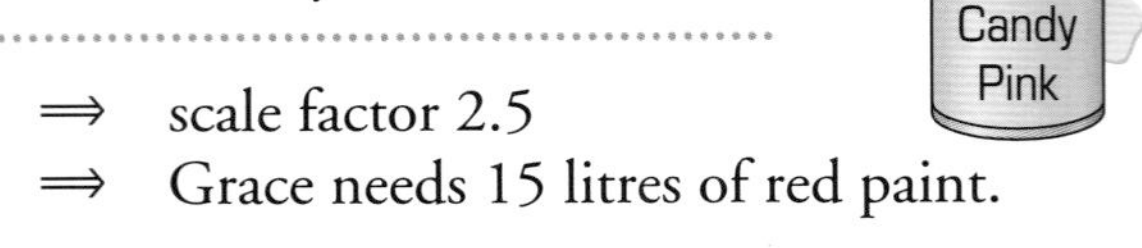

$12.5 \div 5 = 2.5$ (or $5 \times 2.5 = 12.5$) $\Rightarrow$ scale factor 2.5

$6 \times 2.5 = 15$ $\Rightarrow$ Grace needs 15 litres of red paint.

The **unitary method** involves finding the value of a single unit of the quantity, and then scaling accordingly.

example

The price of 64 pots of ink is £78.72.
How much would 37 pots of ink cost?

÷64, ×37:

64 pots = £78.72

1 pot = £1.23

37 pots = £45.51

37 pots of ink will cost £45.51.

In problems involving two or more quantities, it helps to know the **proportion** of each quantity.

example

A wallpaper pattern consists of areas of red, yellow and blue in the ratio 2 : 4 : 9.
A roll of this wallpaper is 0.9 m wide and 8.7 m long.
What area is coloured yellow?

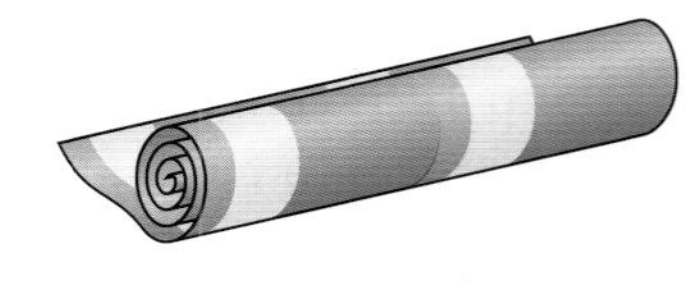

Add the components: $2 + 4 + 9 = 15$

Then the proportion of yellow is $\frac{4}{15}$.

Area of the roll $= 0.9 \text{ m} \times 8.7 \text{ m} = 7.83 \text{ m}^2$

Area that is yellow $= \frac{4}{15} \times 7.83 \text{ m}^2 = 2.088 \text{ m}^2$

$\frac{4}{15} = 0.266\,666\ldots$
Using $\frac{4}{15}$ is better than 0.2667 or 26.7% because it is more accurate.

Exercise P1.4

1 The table shows the sales of anti-horsefly powder sold by a company over a year:

Month	Sales (kg)
January	1230
February	2450
March	2563
April	2789
May	4623
June	4568
July	7654
August	6543
September	6788
October	4689
November	4283
December	2987

a Look at the sales trend and comment on anything you notice.
b What is the proportion of sales in July compared with the whole year?
c What is the proportion of sales in the summer months (June, July, August) compared with the winter months (December, January, February)?

2 Mary's car uses 0.32 litres of fuel for every 2.5 miles it travels.
Use this information to calculate the amount of fuel she would use on a 40-mile trip.

3 The list gives the food required to make a meal for six people.

530 g of chips
360 g of cabbage
200 g of carrots
? g of mushrooms
1.2 kg of beef

The same meal for 14 people requires 300 g of mushrooms.
Work out the missing figure on the list.

4 **Puzzle**
Harry travels at an average speed of 102 km per hour on the motorway.
The car uses, on average, 6 ml of fuel for every 60 m travelled.
How many litres of fuel does the car use travelling for 1 hour at this speed?

5 Explain the relationship between each of these pairs of numbers.

a	1	2	3	4	5
b	6	12	18	24	30

a	1	2	3	4	5
b	1	4	9	16	25

a	1	2	3	4	5
b	1	0.5	0.333...	0.25	0.2

6 **Investigation**
Match each table 1 to 5 with one of the equations A to E.

A $xy = {}^{-}120$ **B** $y = 3x^2$
C $y = x + 3$ **D** $x^2y = 120$
E $y = 3x$

Table 1

x	y
⁻3	0
⁻1	2
1	4
4	7
10	13

Table 2

x	y
⁻5	75
⁻3	27
0	0
2	12
9	243

Table 3

x	y
⁻2	30
⁻1	120
1	120
2	30
$\sqrt{12}$	10

Table 4

x	y
⁻4	30
⁻2	60
1	⁻120
3	⁻40
5	⁻60

Table 5

x	y
⁻7	⁻21
⁻2	⁻6
0	0
3	9
4	12

In which table does x vary directly with y?

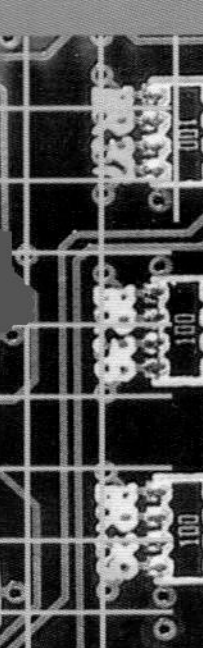

P1.5 Geometrical reasoning

This spread will show you how to:

- Generate fuller solutions to problems.
- Represent problems and synthesise information in geometric form.

KEYWORDS
Angles on a straight line
Properties
Strategy

You can use some simple but effective methods to solve geometry problems.

- Sketch a diagram.
- Write down what you know (properties, angles, lengths).
- Give letters to unknown angles or lengths.
- Devise a strategy.
- Show your working out, giving explanations.

You may need to find other angles or lengths in order to reach the one you require.

Try to list the sequence of calculations mentally, without actually working them out. Working backwards may help.

example

Calculate the size of the angle x.

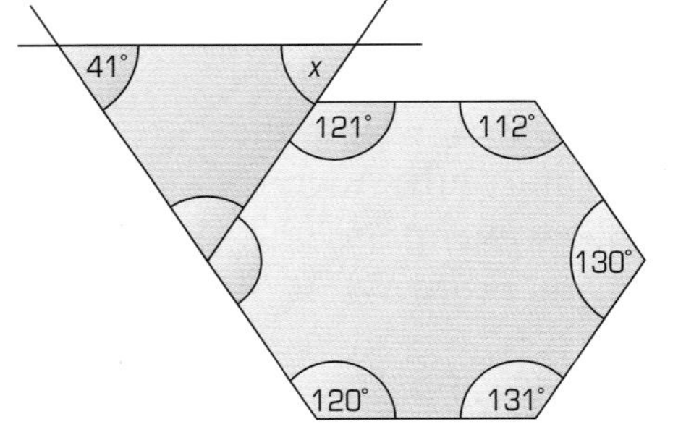

Sketch a diagram, labelling the angles a and b:

Strategy:
- To find x you need to know b
- To find b you need to know a

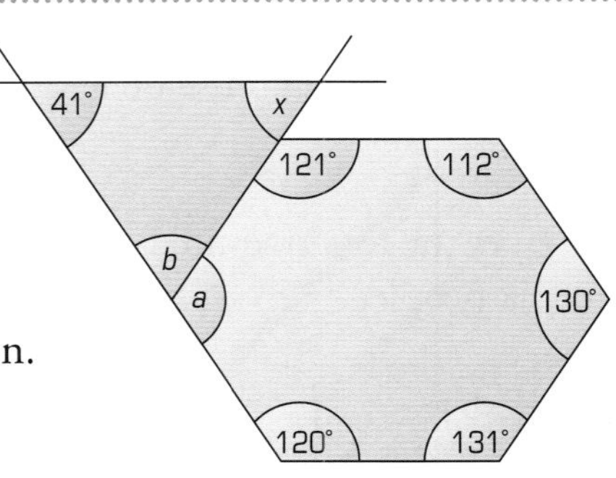

First find the sum of the interior angles of a hexagon.

Sum of angles $= (6 - 2) \times 180°$
$= 4 \times 180° = 720°$

$a = 720° - (121° + 112° + 130° + 131° + 120°)$
$= 106°$

$b = 180° - 106° = 74°$ (angles on a straight line)

$x = 180° - (41° + 74°) = 65°$ (angles in a triangle)

$\Longrightarrow \quad x = 65°$

Remember:
The sum of the interior angles of an n-sided polygon $= (n - 2) \times 180°$

Exercise P1.5

1 Find the angles marked with letters in these diagrams.

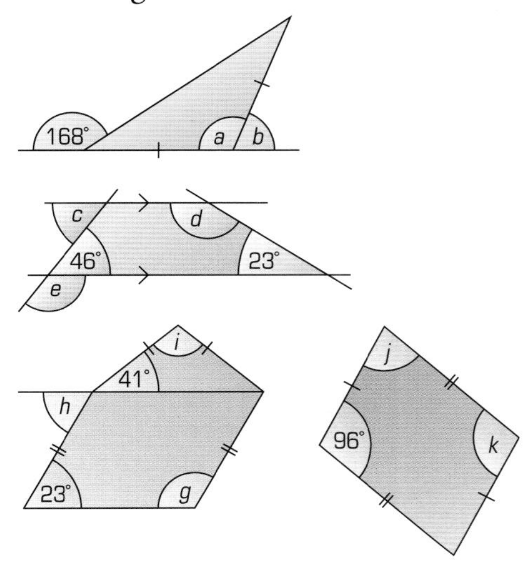

2 Work out the length of side AD in the diagram.

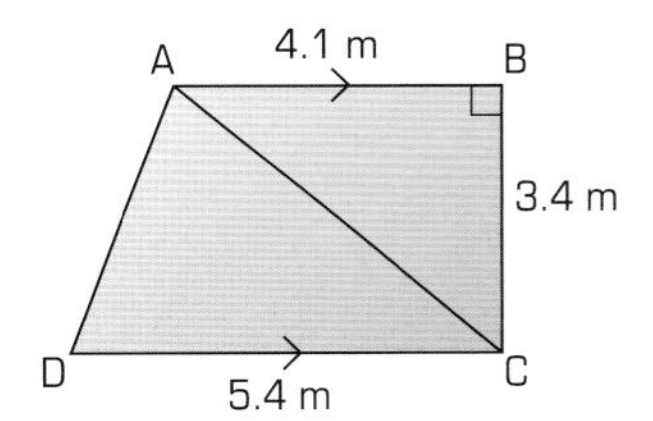

3 **Investigation**
How many different triangles can you make by joining any three vertices of a regular pentagon?

4 Any polygon can be split into triangles starting from one vertex.

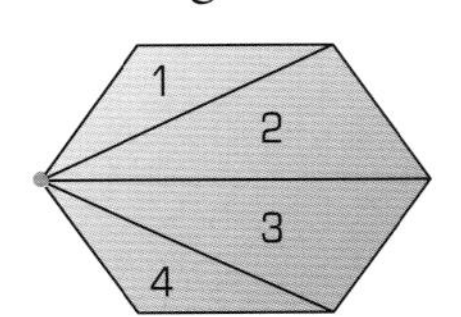

A hexagon can be split into four triangles using this method.

Explain how you can use this to work out the interior angle sum of any polygon.

5 This rectangle is cut along the dotted line to produce two triangles.

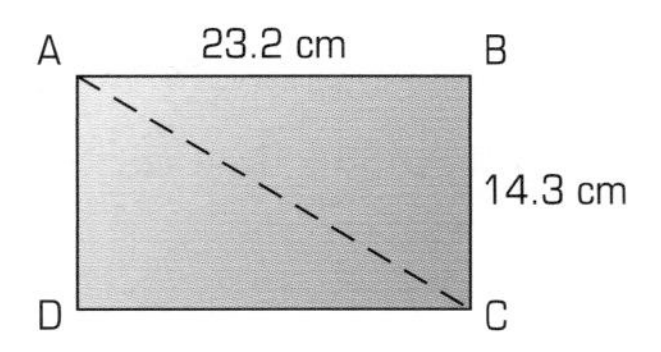

a Work out the perimeter of triangle ABC (to the nearest mm).

b If the length AC is n centimetres, write an expression in n for the length of AB.

6 One of the interior angles of a regular polygon is 160°.
How many sides does the polygon have?

7 The circle and square have areas in the ratio 1 : 2.5 respectively.

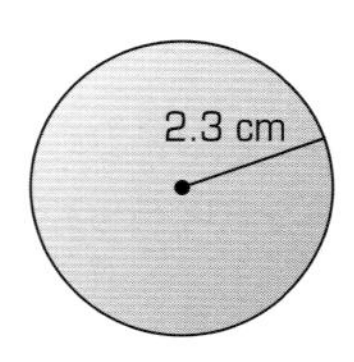

a Work out the side length of the square to the nearest centimetre.

b Give one possible set of dimensions of a rectangle whose area would be a half that of the circle.

8 Show that the perimeter of the sector of the circle shown is $12\pi + 32$.

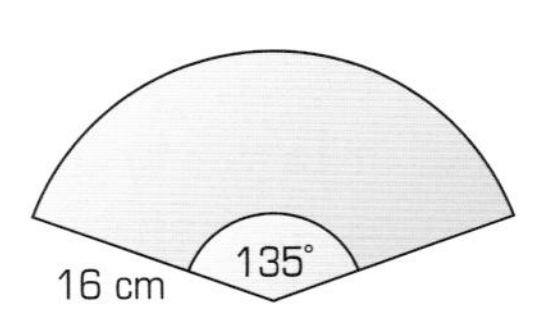

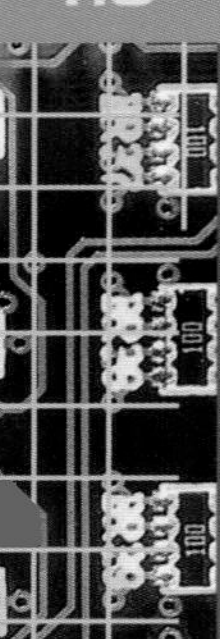

P1.6 Appropriate accuracy

This spread will show you how to:

- Recognise limitations on accuracy of data and measurements.
- Estimate calculations by rounding.
- Recognise that measurements may be inaccurate.
- Understand upper and lower bounds.

KEYWORDS

Accuracy
Approximate
Upper bound
Lower bound

A measurement is only as accurate as the instrument used to measure it.

You can use this tape measure to measure lengths to the nearest centimetre.

The length measured is 48 cm (to nearest cm).

You can use this ruler to measure lengths to the nearest millimetre.

The length measured is 6.5 cm (to nearest mm).

- **You should give answers to problems to an appropriate degree of accuracy.**

The accuracy depends on the accuracy of the measurements given.

example

The radius of a circle is 4.3 cm, measured to the nearest mm.
Calculate the area of the circle, giving your answer to an appropriate degree of accuracy.

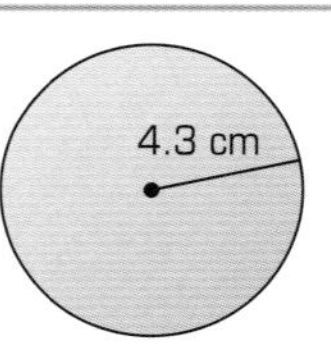

Area of a circle $= \pi r^2$
$= \pi \times 4.3 \text{ cm} \times 4.3 \text{ cm}$
$= 58.088\,048\,16 \text{ cm}^2$

Area of circle $= 58.09 \text{ cm}^2$ (to the nearest mm^2)

Radius is given to the nearest mm, so area should be to the nearest mm^2.
$1 \text{ cm}^2 = 10 \text{ mm} \times 10 \text{ mm}$
$= 100 \text{ mm}^2 \Rightarrow 1 \text{ mm}^2 = 0.01 \text{ cm}^2$

In the example, the answer is only as accurate as the radius quoted.
However, you can work out the largest and smallest possible values for the area.

The smallest possible value, or **lower bound**, of r is 4.25 cm.
$\Rightarrow$ Area (minimum) $= \pi \times 4.25 \text{ cm} \times 4.25 \text{ cm}$
$= 56.75 \text{ cm}^2$ (to the nearest mm^2)

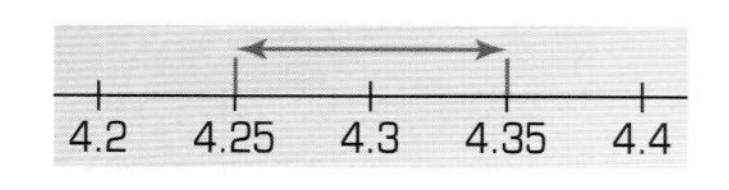

The largest possible value, or **upper bound**, of r is 4.35 cm.
$\Rightarrow$ Area (maximum) $= \pi \times 4.35 \text{ cm} \times 4.35 \text{ cm}$
$= 59.45 \text{ cm}^2$ (to the nearest mm^2)

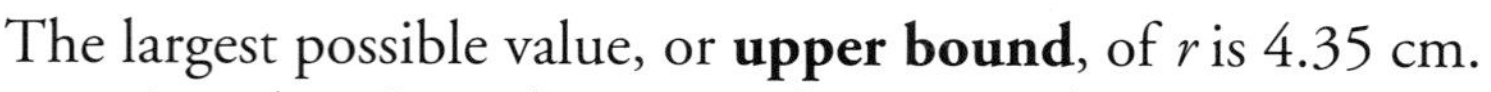

Check your answers by rounding numbers to 1 sf.
Then an approximate estimate for area $\cong 3 \times 4 \text{ cm} \times 4 \text{ cm} = 48 \text{ cm}^2$
This is fairly close to 58 cm^2.

Exercise P1.6

1 Simplify each of these algebraic expressions.

a $4x - 5xy - 6yx - 7$

b $cm + mc + 2c - 5m - 6mc$

c $7n^2 - 2n - 6n - n^2$

d $6n(n^2 - 2n) - 6n^3 - 4n^2$

2 Calculate the missing side on cuboid A if cuboids A and B have volumes in the ratio 1 : 2.3 respectively.

A  B 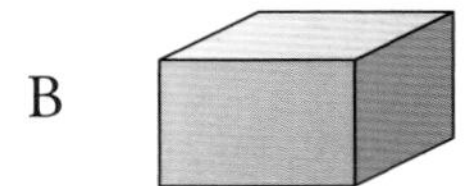

A $L = 3.2$ cm, $H = ?$, $W = 3.4$ cm

B $L = 4.5$ cm, $H = 3.2$ cm, $W = 3.8$ cm

3 The lengths of six pieces of string are given. The accuracy of the measurement is given in brackets.

a 55 mm (to the nearest mm)

b 42 m (to the nearest m)

c 23.4 cm (to 1 decimal place)

d 0.41 m (to the nearest cm)

e 97.4 cm (to the nearest mm)

f 1.255 m (to the nearest mm)

Work out the minimum and maximum lengths each of these pieces of string could actually be.

4 Use a mental method to decide which of these results is closest to the correct answer to $1.17 \times 2.95 \times 3$.

10.3542 1.03547
103.5 0.103545

Hint: Round the numbers, then estimate the answer.

5 Write each of these expressions without brackets and check your answer by using substitution.

a $(a + b)^2$ **b** $(a - b)^2$

c $a(b + c)^2$ **d** $(a + b)^2 - b^2$

6 **a** Make a pencil-and-paper estimate of the number of marbles with a radius of 0.4 cm that can fit into a cylindrical jar with a base of radius 3.51 cm and a height of 14.2 cm.

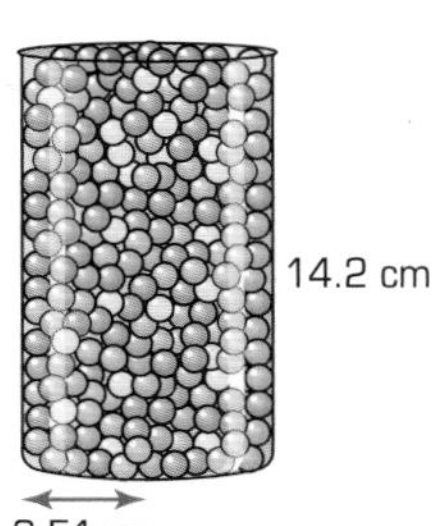

b Calculate the volume of water that could be poured into the empty jar.

Hint: Volume of a sphere $= \frac{4}{3}\pi r^3$

7 **Puzzle**

Lord Havitall orders a dinner table for his Manor House. He is quoted a price of £423 per m^2 for the surface area of the table. He wants it to be circular and to be able to seat 20 people.

Mrs Havitall would like a square table. By how much would the costs differ?

(Assume that each person needs 0.9 m of space around the table.)

8 **Investigation**

Consider these numbers:

3.45	4.67	2.34	4.67	8.46	3.48

Does doubling each of the numbers in the table double the mean?

Investigate the effect on the mean and range of doing other things to the numbers (for example squaring, adding or subtracting a constant).

Are your results true for any set of numbers?

P1 Summary

You should know how to ...

1 Generate fuller solutions to problems.

2 Recognise limitations on accuracy of data and measurements.

3 Solve problems using properties of angles, of parallel and intersecting lines, and of triangles and other polygons.

4 Use proportional reasoning to solve a problem. Understand and use proportionality and calculate the result of any proportional change using multiplicative methods.

Check out

1 Ann walked 450 m in 8 minutes 32 seconds.

a What was her average speed?

b Investigate the times it would take her to travel the same distance by other means, for example by bicycle, jogging, or by car. Use sensible estimates for the average speed in each of these cases, giving justification where necessary.

2 A company is commissioned to manufacture 125 000 metal rods of length 1.62 cm ± 3%.

a Explain what ±3% means.

b Express the possible lengths of the metal rods as an inequality: □ $\leqslant x \leqslant$ □

The cost of manufacturing a metal rod is 0.42 pence per cm.

c Show the effect that the ±3% tolerance could have on manufacturing costs. Use calculations in your explanation.

3 Use diagrams to show that the sum of the interior angles of any polygon is always a multiple of 180°.

You could design this as a poster for younger students.

Hint: Use your knowledge of triangles.

4 **a** Copy and complete the table of values for the relationship $y = \frac{3}{x}$.

x	1	3	5	7	9	11
y						

b Use your table to create similar tables for:

i $y = \frac{6}{x}$ **ii** $y = \frac{3}{2x}$ **iii** $y = \frac{1}{x}$

c Describe a real-life situation where two variables might be related in this way.

5 Expressions and formulae

This unit will show you how to:

- Square a linear expression, expand the product of two linear expressions of the form $x \pm n$ and simplify the corresponding quadratic expression.
- Establish identities such as $a^2 - b^2 = (a + b)(a - b)$.
- Construct and solve linear equations with integer coefficients.
- Add simple algebraic fractions.
- Use formulae from mathematics and other subjects.
- Substitute numbers into expressions and formulae.
- Derive and use more complex formulae, and change the subject of a formula.
- Generate fuller solutions to problems.
- Distinguish between practical demonstration and proof.
- Justify generalisations, arguments or solutions.
- Identify exceptional cases or counter-examples.

You can demonstrate a statement by showing it to be true in a particular case.

Before you start

You should know how to ...	Check in
1 Multiply a single term over a bracket.	**1** Which expansion gives a different simplified expression to the other two? $3(2x + 4) - 2(x + 5)$ $(14x + 1) - (10x - 1)$ $6(2x - 4) + 3(7 - 2x)$
2 Solve a linear equation.	**2** Solve these equations: **a** $3(2x + 4) = 4(3x - 2)$ **b** $10 - 6y = 4$ **c** $\frac{z + 7}{4} = \frac{3 - 2z}{6}$
3 Add and subtract numerical fractions.	**3** Find the lengths x and y. Show your working.

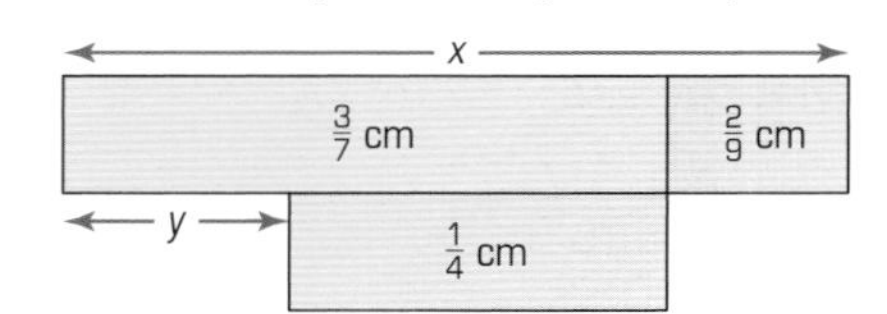

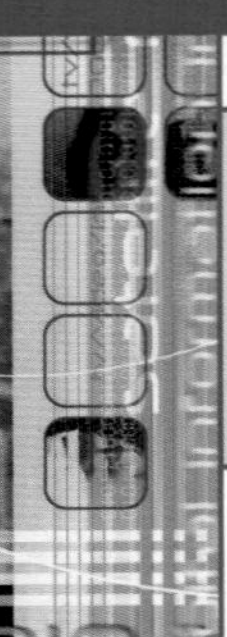

A5.1 Proof

This spread will show you how to:

- Distinguish between a practical demonstration and a proof.
- Justify generalisations, arguments or solutions.
- Identify exceptional cases or counter-examples.

KEYWORDS
Prove
Counter-example
Demonstrate
Generalise
Statement

Here is a mathematical statement.

'The sum of any three consecutive integers is always a multiple of 3.'

You can **demonstrate** that the statement is true by showing examples.

$3 + 4 + 5 = 12$ a multiple of 3

$31 + 32 + 33 = 96$ a multiple of 3

$117 + 118 + 119 = 354$ a multiple of 3

The statement is true in these three cases.

You **prove** that it is true by generalising to all possible cases.

- Take three consecutive integers: n, $n + 1$, $n + 2$
- Add them together: $n + (n + 1) + (n + 2)$
  $= 3n + 3$
  $= 3(n + 1)$
- Justify the statement: '$3(n + 1)$ is a multiple of 3 because it is $3 \times$ an integer.
  Therefore the sum of any three consecutive numbers is always a multiple of 3.'

- **A counter-example can be used to disprove a statement.**

example

Disprove this statement: 'All prime numbers are odd'.

2 is a prime number, so the statement is false.

You may need to prove statements from different areas of maths, such as geometry and algebra. It helps if you know how to generalise these special number sequences.

- Consecutive numbers — These differ by 1 each time, starting from 1.
  1, 2, 3, 4, ... — nth term $T(n) = n$ — Write n, $n + 1$, $n + 2$, $n + 3$, ...
- Even numbers — These differ by 2 each time, starting from 2.
  2, 4, 6, 8, 10, 12, ... — nth term $T(n) = 2n$ — Write $2n$, $2n + 2$, $2n + 4$, ...
- Odd numbers — These differ by 2 each time, starting from 1.
  1, 3, 5, 7, 9, 11, ... — nth term $T(n) = 2n - 1$ — Write $2n - 1$, $2n + 1$, $2n + 3$, ...

example

Prove that the sum of any two odd numbers is always even.

First demonstrate it:

$3 + 5 = 8$ even

$37 + 53 = 90$ even

$111 + 13 = 124$ even

Then prove it:

$(2n - 1) + (2m - 1) = 2n + 2m - 2$
$= 2(n + m - 1)$

This is even because it is a multiple of 2.

Use m for a **different** odd number.

Exercise A5.1

1 'The sum of any five consecutive integers is always a multiple of five.'

a Demonstrate whether or not this statement is true by trying out some different sets of five consecutive integers.

b Prove, using algebra, that this statement is always true.

2 For each statement given below:

a Try out some numbers of your own to demonstrate whether or not the statement is true.

b Prove, using algebra, the statements that you have decided are true.

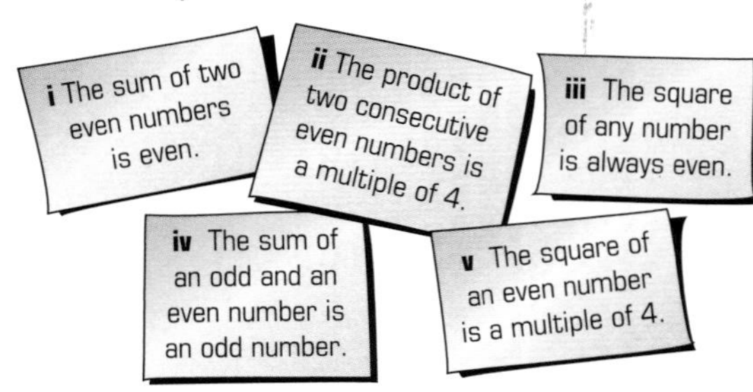

3 Find a counter-example for each of these statements:

a When you add one to a number and square it, the answer is always even.

b When you treble a prime number, the answer is always odd.

c The product of three consecutive numbers is always a multiple of four.

4 In this pyramid, the number in each brick is found by adding the two numbers beneath it.

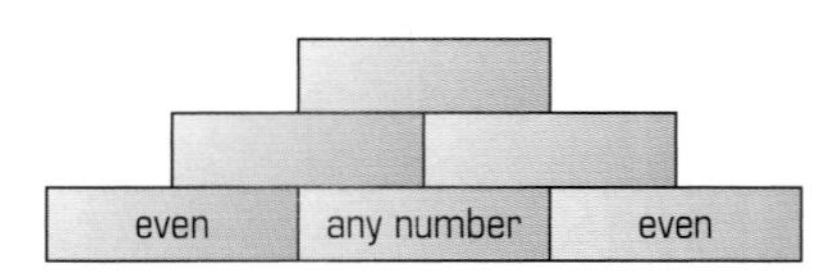

Prove that the top number will always be a multiple of 2.

5 **a**
- Take a two-digit number: 57
- Reverse it: 75
- Find the difference: $75 - 57 = 18$
  18 is a multiple of 9.

Investigate the statement: 'The difference between any two-digit number and the same number with its digits reversed is always a multiple of 9.'

> **Hint:** Express the original number as $10t + u$, for example, 57 is $10 \times 5 + 7$.

b **i** Prove that a two-digit number with equal tens and units digits is always divisible by 11.

ii Prove that a three-digit number with hundreds and units summing to give the tens digit is always a multiple of 11.

6 Here is a triangle.

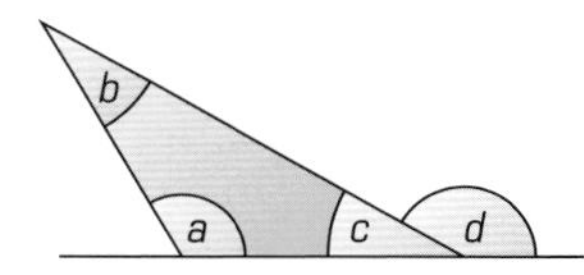

a Write c in terms of d.

b Write c in terms of a and b.

c Prove that $d = a + b$ using your answers to parts **a** and **b**.

7 Prove that the angle in a semicircle (angle B) is always 90°.

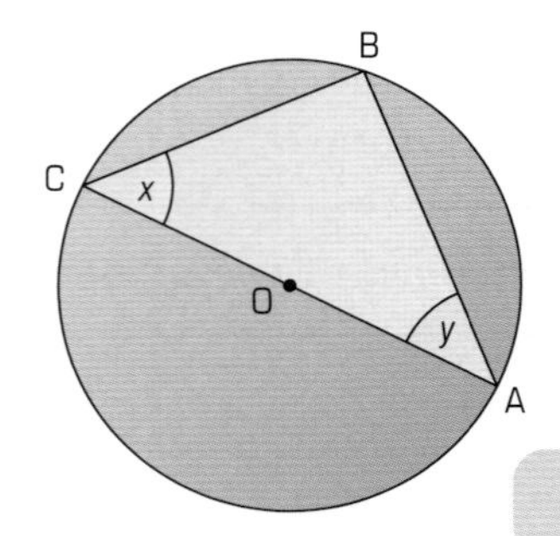

> **Hint:** Join B to O.

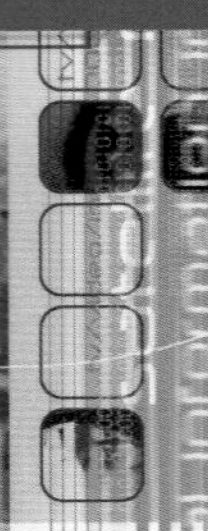

A5.2 Further expanding

This spread will show you how to:

- Expand the product of two linear expressions.
- Simplify the corresponding quadratic expression.
- Establish identities such as $a^2 - b^2 = (a + b)(a - b)$.

KEYWORDS

Identity, Expression, Expand, Factorise, Difference of two squares, Identically equal to (≡)

You can expand double brackets containing more than one letter.

$(2x + 3y)(4x - 2y)$

$$= (2x \times 4x) + (2x \times (^-2y)) + (4x \times 3y) + (3y \times (^-2y))$$
$$= 8x^2 - 4xy + 12xy - 6y^2$$
$$= 8x^2 + 8xy - 6y^2$$

Remember: multiply each term in one bracket by each term in the other.

Remember to collect like terms.

Be careful with negatives – use an extra bracket.

Sometimes the terms in x cancel each other out.

example

Expand and simplify these expressions.

a $(x + 2)(x - 2)$ **b** $(x + 4)(x - 4)$ **c** $(x + 11)(x - 11)$

a $(x + 2)(x - 2)$
$= x^2 + 2x - 2x + 4$
$= x^2 - 4$

b $(x + 4)(x - 4)$
$= x^2 + 4x - 4x + 16$
$= x^2 - 16$

c $(x + 11)(x - 11)$
$= x^2 + 11x - 11x + 121$
$= x^2 - 121$

► In general, $(a + b)(a - b) \equiv a^2 - b^2$
This identity is called the **difference of two squares**.

Remember:
≡ means 'is identically equal to'.
$2x + x \equiv 3x$ is an **identity** – it is true for all values of x.
$2x + 1 = 3$ is an **equation** because it is only true for $x = 1$.

The difference of two squares can help you to expand or factorise some expressions very quickly.

example

a Expand the expression $(x + 8)(x - 8)$.
b Factorise the expression $x^2 - 196$.

a $(x + 8)(x - 8)$ is in the form $(a + b)(a - b)$, so $(x + 8)(x - 8) = x^2 - 64$
b $x^2 - 196$ is in the form $a^2 - b^2$, so $x^2 - 196 = (x + 14)(x - 14)$

$64 = 8^2$

$14 = \sqrt{196}$

You can use the difference of two squares in some calculations.

example

Evaluate $101^2 - 99^2$.

$101^2 - 99^2$ is in the form $a^2 - b^2$, so $101^2 - 99^2 = (101 + 99)(101 - 99)$
$= 200 \times 2 = 400$

Exercise A5.2

1 Carefully expand and simplify these expressions. You should find the solution in the right-hand box.
For the odd one out, find its solution.

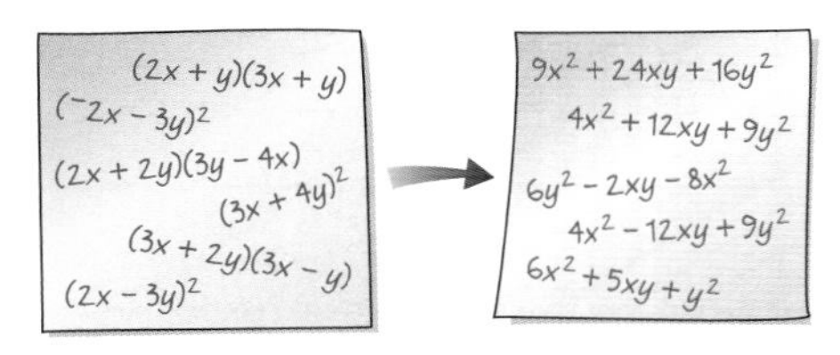

2 Expand and simplify:

a $(2x + 4)(^-2x + 1)$

b $(3y + 2)^2$

c $(3z + 4)(2z - 3)$

d $(3x + 4)(2x + 2) + (3x + 1)^2$

e $(2x + 4y)^2 - (3x - 5y)^2$

3 Write an expression for each of the following quantities.
Expand and simplify your expressions.

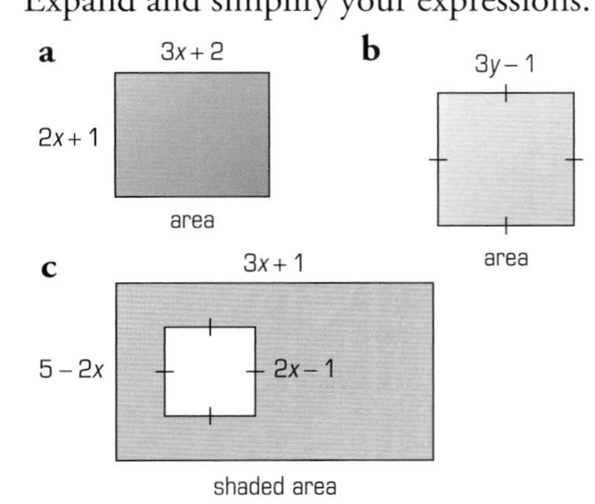

4 Copy and complete this table of the differences of two squares.

Factorised form	Expanded form
$(x + 1)(x - 1)$	
$(x + 12)(x - 12)$	
	$x^2 - 100$
$(2x + 1)(2x - 1)$	
$(3x + 2)(3x - 2)$	
	$16x^2 - 9$

5 Factorise these expressions, using the difference of two squares.

a $x^2 - 25$ **b** $x^2 - 36$

c $x^2 - 81$ **d** $x^2 - 225$

e $x^2 - \frac{1}{4}$ **f** $x^2 - y^2$

g $4x^2 - 25$ **h** $9x^2 - 36$

i $4x^2 - 9y^2$ **j** $m^2 - n^2$

k $25m^2 - 100n^2$ **l** $9m^2 - 16y^2$

6 Explain why $16y^2$ is always a square number, but $y^2 + 16$ is not a square number.

7 **a** Multiply out these expressions and show they are all equivalent to $x^2 - y^2$.

i $2y(x - y) + (x - y)^2$

ii $(x - y)(x + y)$

iii $x(x - y) + y(x - y)$

b The area of this shape can be obtained by dividing it in different ways.

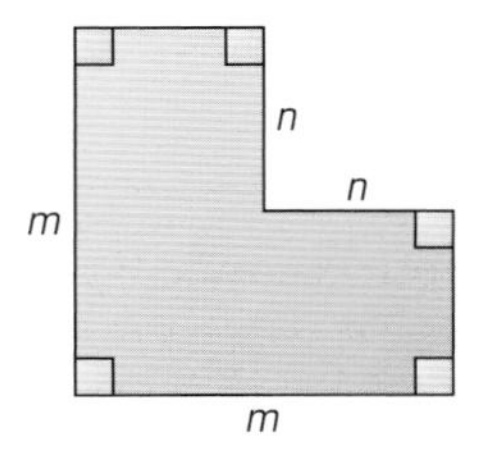

Write three different expressions for the area and prove that they are equivalent.

8 Without a calculator, evaluate:

a $100\,002^2 - 99\,998^2$

b $1\,600\,003^2 - 1\,399\,997^2$

c $0.3^2 - 0.1^2$

d $3.501^2 - 3.499^2$

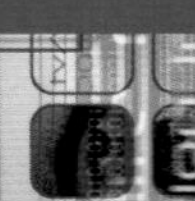

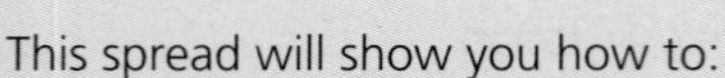

A5.3 Solving equations with brackets

This spread will show you how to:
- Construct and solve linear equations with integer coefficients.
- Expand and simplify the product of two linear expressions.

KEYWORDS
Expand   Equation
Collect like terms

You can use your skills in expanding brackets to solve equations.

example

Solve the equation $(x + 4)(x - 3) = (x + 2)^2$

- Expand the brackets: $(x + 4)(x - 3) = x^2 + 4x - 3x - 12$ $(x + 2)(x + 2) = x^2 + 2x + 2x + 4$
- Collect like terms: $= x^2 + x - 12$ $= x^2 + 4x + 4$
- Equate both sides: $x^2 + x - 12 = x^2 + 4x + 4$
- Solve the equation: $x - 12 = 4x + 4$

$\Rightarrow$ $^{-}12 = 3x + 4$

$\Rightarrow$ $^{-}16 = 3x$

$\Rightarrow$ $^{-}\frac{16}{3} = x$ or $x = {}^{-}5\frac{1}{3}$

It simplifies to a two-sided linear equation.

Some problems may require you to construct the equation yourself.

example

The areas of the rectangle and the square are equal.
Find the dimensions of each shape (all lengths given in cm).

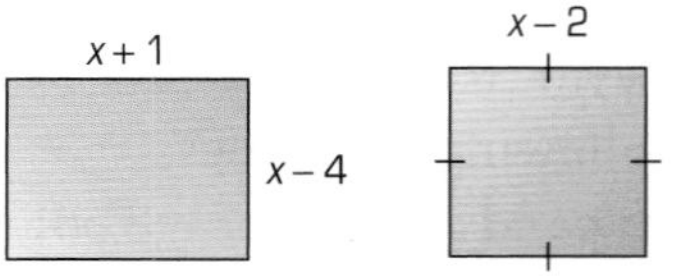

$$(x + 1)(x - 4) = (x - 2)^2$$
$$x^2 - 4x + x - 4 = x^2 - 2x - 2x + 4$$
$$^{-}3x - 4 = {}^{-}4x + 4$$
$$x - 4 = 4$$
$$x = 8$$

For the square, $(x - 2) = 8 - 2 = 6$
So its dimensions are 6 cm by 6 cm.

For the rectangle, length $= (x + 1) = 8 + 1 = 9$, width $= (x - 4) = 8 - 4 = 4$
So its dimensions are 9 cm by 4 cm.

You may need to use your knowledge from different areas of mathematics.

example

In the graph, curve 1 has equation $y = (x + 1)(x + 3)$.
Curve 2 has equation $y = (x - 2)(x - 5)$.
Find the coordinates of the point A.

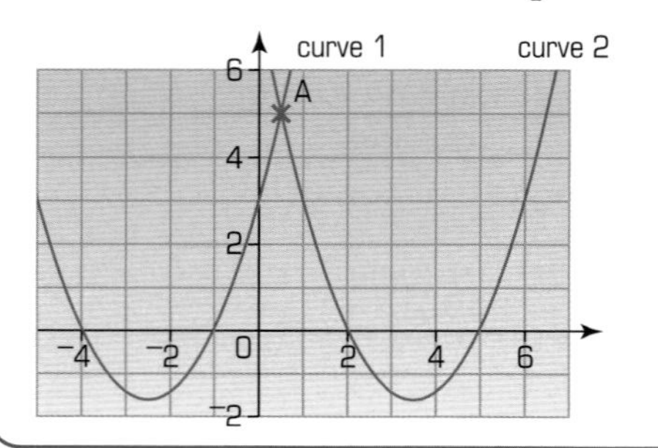

Point A is on both curves, so at A:

$$(x + 1)(x + 3) = (x - 2)(x - 5)$$
$$x^2 + 4x + 3 = x^2 - 7x + 10$$
$$4x + 3 = {}^{-}7x + 10$$
$$11x = 7$$
$$x = \frac{7}{11}$$

Find y by substituting in either equation:

$$y = (x + 1)(x + 3) = (\tfrac{7}{11} + 1)(\tfrac{7}{11} + 3)$$
$$= 5\tfrac{115}{121}$$

Use fractions rather than decimals because they are exact.

So the coordinates of A are $(\frac{7}{11}, 5\frac{115}{121})$.

Exercise A5.3

1 Solve these equations by expanding the brackets.

a $(x+3)(x+6) = (x+5)^2$
b $(y+4)(y-3) = (y+3)(y-1)$
c $(z+3)^2 = (z-2)^2$
d $(2m+1)^2 = (m+1)(4m-1)$
e $(n-3)(n+3) = n(n-18)$
f $(p+2)^2 + (2p-1)^2 = 5p(p+1)$

2 In each case, form an equation and solve it to find the value of the unknown.

a

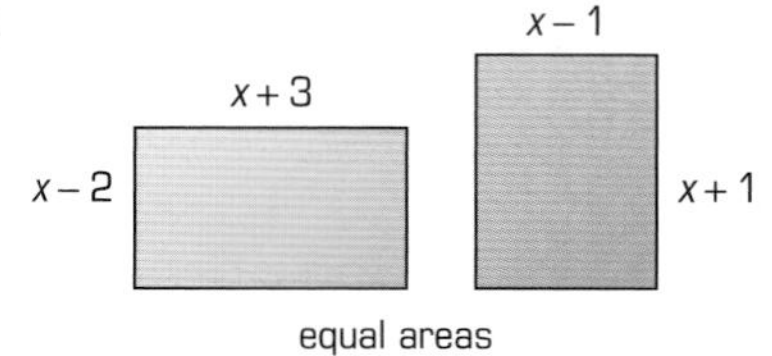

b

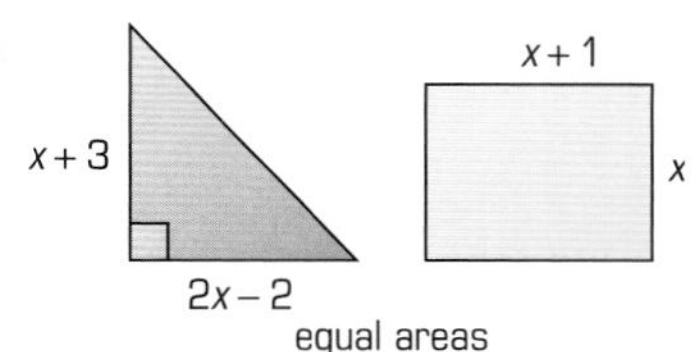

c

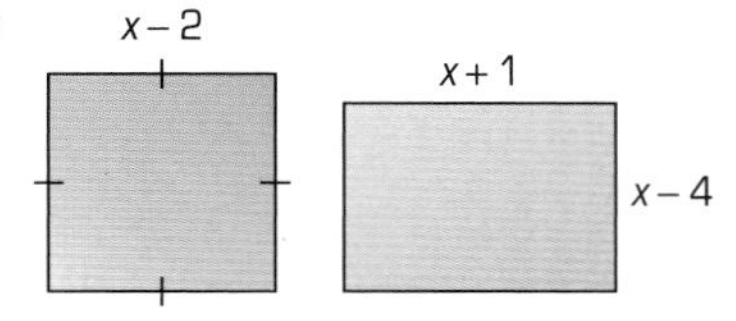

Area of the square is one less than the area of the rectangle.

3 Use Pythagoras' theorem to find the value of y in each case.

a **b**

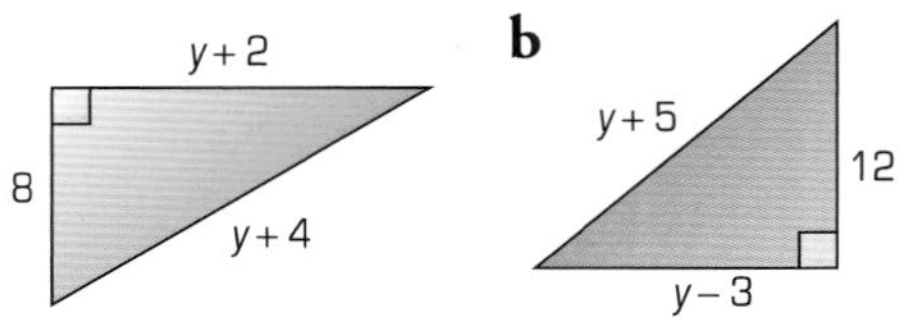

4 The cylinder is hollow.
The half sphere is solid.
The surface areas of these solids are equal.
What is the radius of the sphere?

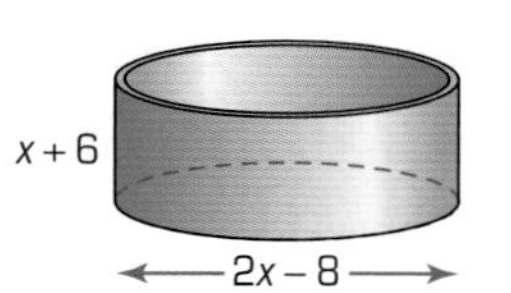

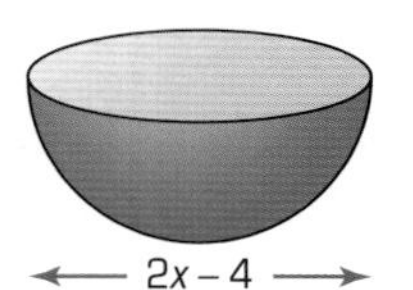

Useful formulae:
Curved surface area of cylinder = $2\pi rh$
Surface area of sphere = $4\pi r^2$

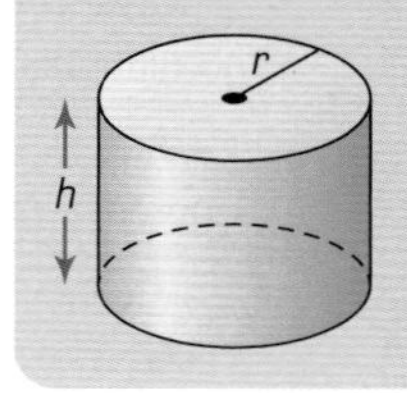

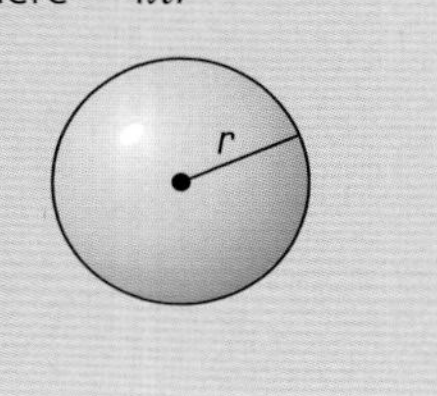

5 Solve:
$(x+2)(x+4)(x+8) = x(x^2 + 14x + 48)$

6 This rectangle is 7 cm wide.
Its diagonal is 1 cm more than its length.
Find the area of the rectangle.

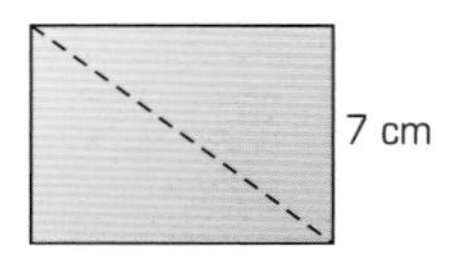

7 Prove that the difference between the mean of the squares of five consecutive numbers and the square of the mean of these same five numbers is 2.

Hint: Remember to demonstrate this to yourself with numbers first to be sure you understand it.

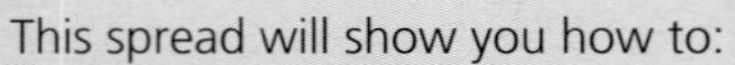
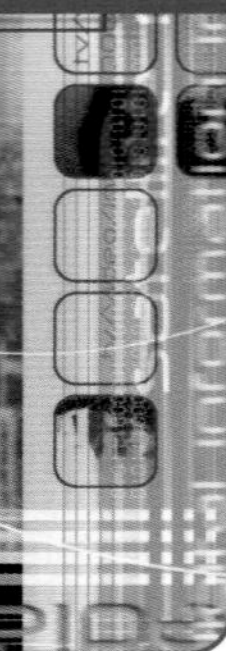

A5.4 Algebraic fractions

This spread will show you how to:

- Add simple algebraic fractions.
- Simplify or transform algebraic expressions.

KEYWORDS
Numerator
Algebraic fraction
Denominator
Lowest common denominator
Cancel

Algebraic fractions work in the same way as numerical fractions.

Numerical example	Steps in working	Algebraic example
$\frac{3}{7}+\frac{2}{9}$		$\frac{3x}{7}+\frac{2x}{9}$
$=\frac{27}{63}+\frac{14}{63}$	Make the denominators equal	$=\frac{27x}{63}+\frac{14x}{63}$
$=\frac{27+14}{63}$	Add the numerators	$=\frac{27x+14x}{63}$
$=\frac{41}{63}$	Collect like terms	$=\frac{41x}{63}$ or $\frac{41}{63}x$

63 is the lowest common denominator of 7 and 9.

These are the same – it does not matter if you $\times 41$ then $\div 63$, or $\times \frac{41}{63}$.

Subtraction works in the same way as addition.

example

Write an expression for the difference in the heights of these houses.

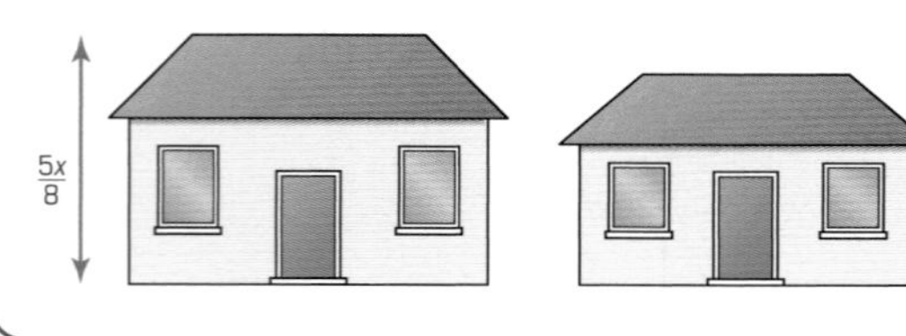

Difference in height $=\frac{5x}{8}-\frac{y}{4}$

$=\frac{5x}{8}-\frac{2y}{8}$

$=\frac{5x-2y}{8}$

You cannot simplify this further because $5x$ and $^{-}2y$ are unlike terms.

If you know how to multiply and divide numerical fractions, then algebraic fractions are easy.

Numerical example	Steps in working	Algebraic example
$\frac{5}{8}\div\frac{5}{12}$		$\frac{2}{5y}\div\frac{4x}{7}$
$\frac{5}{8}\times\frac{12}{5}$	Convert to a × and turn the second fraction upside-down	$\frac{2}{5y}\times\frac{7}{4x}$
$\frac{5}{8}\times\frac{12}{5}=\frac{1}{2}\times\frac{3}{1}$	Cancel common factors	$\frac{2}{5y}\times\frac{7}{4x}=\frac{1}{5y}\times\frac{7}{2x}$
$\frac{1\times 3}{2\times 1}=\frac{3}{2}$, or $1\frac{1}{2}$	Multiply numerators and multiply denominators	$\frac{1\times 7}{5y\times 2x}=\frac{7}{10xy}$

Exercise A5.4

1 If you need to, remind yourself how to deal with numerical fractions first. Check your solutions using the $a^{b}/_{c}$ fraction button on your calculator.

a $\frac{2}{9} + \frac{1}{8}$ **b** $\frac{3}{5} - \frac{1}{2}$

c $2\frac{1}{2} + 1\frac{1}{3}$ **d** $3\frac{1}{5} - 2\frac{1}{8}$

e $\frac{2}{3} + \frac{1}{7}$ **f** $\frac{2}{9} \div \frac{3}{4}$

g $\frac{5}{8} \times \frac{2}{15}$ **h** $1\frac{1}{3} \div \frac{8}{9}$

2 Simplify:

a $\frac{x}{5} + \frac{x}{4}$ **b** $\frac{x}{7} - \frac{y}{2}$

c $\frac{4}{x} + \frac{5}{x}$ **d** $\frac{10}{x} - \frac{2}{x}$

e $\frac{2}{x} + \frac{3}{y}$ **f** $\frac{4}{p} - \frac{3}{q}$

g $\frac{2}{x} + \frac{5}{x^2}$ **h** $\frac{3}{m} + \frac{2}{4m}$

i $\frac{2}{5}x - \frac{1}{3}x$

j Half of b – one-fifth of b

3 Simplify, remembering to cancel common factors before multiplying:

a $\frac{x}{5} \times \frac{x}{4}$ **b** $\frac{y}{10} \div \frac{3}{y}$

c $\frac{2}{3}b \times \frac{1}{2}b$ **d** $(\frac{2}{3}x)^2$

e $\frac{4m}{5} \div \frac{1}{2}$ **f** $\frac{5x}{3} \div \frac{x}{2}$

g $\frac{xy}{3} \times \frac{9}{x^2}$ **h** $\frac{3ab}{c} \div \frac{a^2}{c}$

4 Add these fractions by finding the lowest common denominator in each case.

a $\frac{2x}{5} + \frac{x}{10}$ **b** $\frac{3}{2x} + \frac{4}{x}$

c $\frac{5}{x} + \frac{6}{x^2}$ **d** $\frac{11}{a} + \frac{12}{ab}$

5 True or false?

$$\frac{a}{b} + \frac{c}{d} = \frac{adbc}{bd}$$

6 Correct each of these calculations.

a $\frac{1}{x} + \frac{1}{y} = \frac{2}{x+y}$

b $\frac{3x}{5} \times \frac{2x}{3} = \frac{{}^1\cancel{3x}^1}{5} \times \frac{2\cancel{x}^1}{\cancel{3}_1} = \frac{2}{5}$

c $\frac{a}{3} \div \frac{5}{a} = \frac{a}{3} \times \frac{a}{5} = \frac{2a}{15}$

7 a Write a simplified formula for the perimeter of this rectangle.

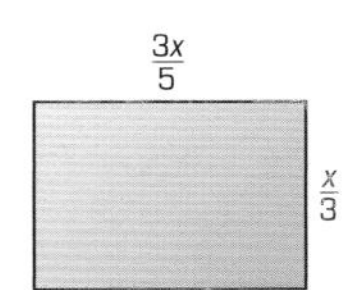

b If the perimeter of the rectangle is 70 cm, form an equation and solve it to find x and, hence, the dimensions of the rectangle.

8 Form an equation and solve it to find the value of x in each case.

a **b**

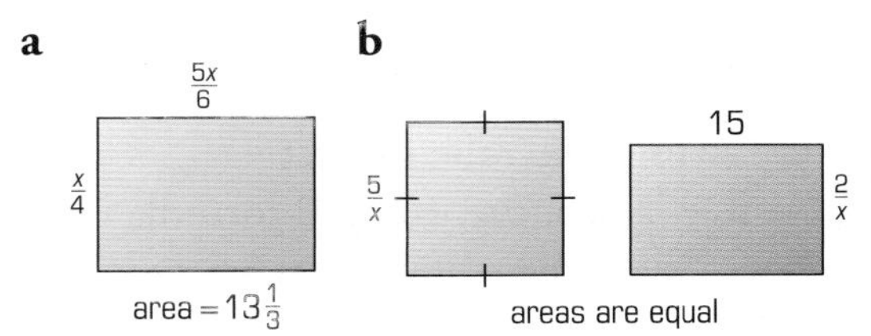

9 Show that the sum of $\frac{1}{4}y$ and $\frac{1}{8}y$ is the same as the product of $\frac{1}{2}y$ and $\frac{y+2}{4}$.

10 Challenge

Prove that $\left(\frac{8}{x} + \frac{4}{x^2}\right) \div \frac{2}{x^2}$ is equal to $2(2x + 1)$.

A5.5 Using and rearranging formulae

This spread will show you how to:

- Use formulae from mathematics and other subjects.
- Substitute numbers into expressions and formulae.
- Change the subject of a formula.

KEYWORDS

Formula
Substitute
Rearrange
Subject
Inverse operation

A **formula** describes the relationship between variables.

To convert temperature in degrees Celsius (°C) to degrees Fahrenheit (°F), you can use the formula:

0° 10° 20° 30° 40° 50° 60° 70° 80° 90° 100° °C

40° 60° 80° 100° 120° 140° 160° 180° 200° °F 32° 212°

$F = \frac{9C}{5} + 32$

You can convert from Celsius to Fahrenheit by substituting values into the formula.

For example, to convert $^-12$ °C into °F:

$F = \frac{9 \times {}^-12}{5} + 32$

$= \frac{^-108}{5} + 32$

$= 10.4 \quad \Longrightarrow \quad {}^-12$ °C is equal to 10.4 °F.

It is not easy to use this formula to evaluate a Celsius temperature given a Fahrenheit temperature.
You can change the subject of a formula by rearranging it.

example

a **i** Rearrange the formula
$v = u + at$
to make t the subject.

ii Use your rearranged formula to find the value of t when $v = 5$, $u = 2$ and $a = {}^-1$.

b **i** Rearrange the formula
$V = \pi r^2 h$
to make r the subject.

ii Find the value of r when $V = 140$ and $h = 5$.
Give your answer to 3 significant figures.

---

a **i** $v = u + at$ subtract u

$v - u = at$ divide by a

$\frac{v - u}{a} = t$

ii $t = \frac{5 - 2}{^-1} = {}^-3$

b **i** $V = \pi r^2 h$

$\frac{V}{\pi h} = r^2$ divide by πh

$\sqrt{\frac{V}{\pi h}} = r$ square root

ii $r = \sqrt{\left(\frac{140}{\pi \times 5}\right)} = 2.99$ (to 3 sf)

▶ You rearrange a formula by using inverse operations.

Remember to perform any operations on both sides of the formula.

Exercise A5.5

1 Hero's formula gives the area of any triangle with sides a, b and c:
$A = \sqrt{(s(s-a)(s-b)(s-c))}$
where $s = \frac{1}{2}(a + b + c)$.

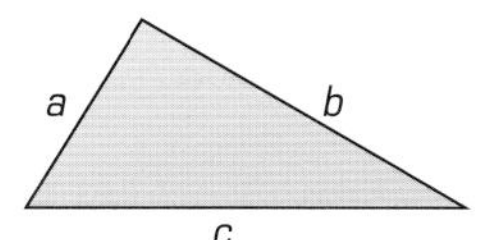

Use the formula to find the area of a triangle with sides 3 cm, 5 cm and 6 cm.

2 The volume of a pyramid is found using the formula:
$V = \frac{1}{3}lbh$.

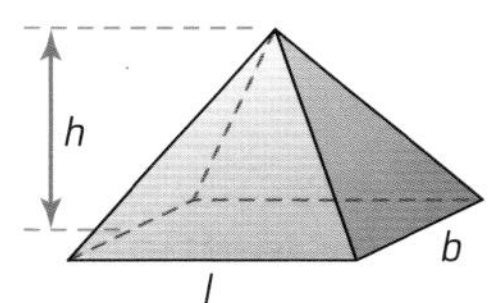

Adapt this formula to find the volume of half a square-based pyramid whose height is equivalent to its base length.

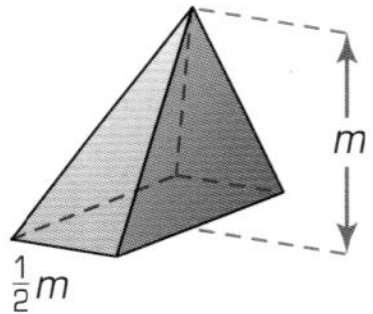

3 Which solid has the largest volume?

a $V = \pi r^2 h$
Cylinder, height 5 cm, radius 2 cm

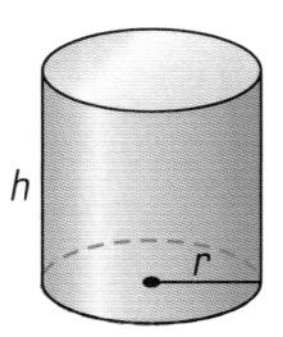

b $V = \frac{4\pi r^3}{3}$
Sphere, radius 2 cm

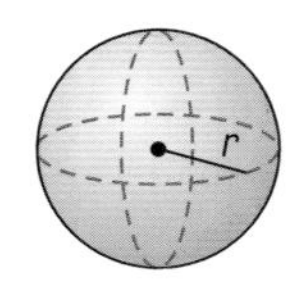

c $V = \frac{\pi^2(r_1 + r_2)(r_1 - r_2)^2}{4}$
Torus,
outer radius 3 cm,
inner radius 1 cm

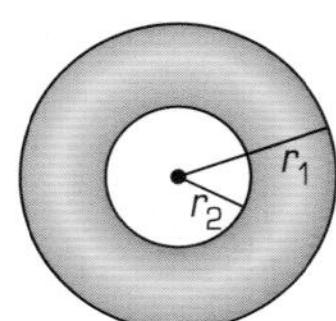

4 The volume of a sphere is found using the formula:
$V = \frac{4}{3}\pi r^3$

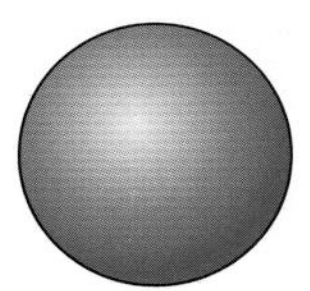

a Make r the subject of this formula.
b Hence, find the radius of a sphere with volume 500π cm^3.

5 This formula gives the time, T, that it takes a pendulum of length, l, to make one complete swing.

$$T = 2\pi\sqrt{\frac{l}{g}}$$

a Make **i** l and **ii** g the subject of this formula.
b Using your solutions from part **a**, find
i l when $g = 9.8$ and $T = 20$
ii g when $l = 200$ and $T = 30$.

6 Here are two formulae:

$$W = a + b + \frac{5\sqrt{(a^2 + b^2)}}{3}$$

$$X = \frac{1}{2}ab + \frac{(a^2 + b^2)}{9}$$

Let $a = 2.3$ and $b = 0.8$.
Which is larger: W or X?

7 The distance, d, between two coordinates (x_1, y_1) and (x_2, y_2) can be found using the formula:

$$d = \sqrt{(x_2 - x_1)^2 + (y_2 - y_1)^2}$$

Use the formula to decide the type of triangle joining the points (2, 3), (5, 7) and (6, 0).

A5.6 Rearranging harder formulae

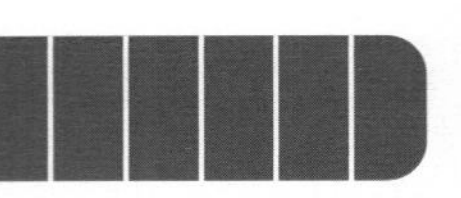

This spread will show you how to:
- Derive and use more complex formulae, and change the subject of a formula.

KEYWORDS
Formula Subject
Inverse operations
Derive

When you rearrange a difficult formula, you need to take care over inverse operations. It helps to read the formula, starting from the intended subject.

example

a Make x the subject of the formula
$P = R - wx$

b Make x the subject of the formula
$k = \frac{s}{x} + m$

a $P = R - wx$
Read the formula: 'multiply x by w, then **subtract from** R'.
Make the formula easier by **adding** wx:
$P + wx = R$
$wx = R - P$ subtract P
$x = \frac{R - P}{w}$ divide by w

b $k = \frac{s}{x} + m$
Read the formula: 'x **divided into** s, then add m'.
$k - m = \frac{s}{x}$ subtract m
Make the formula easier by **multiplying** by x:
$x(k - m) = s$
$x = \frac{s}{k - m}$ divide by $(k - m)$

Some problems may require you to **derive** a formula before rearranging it.

example

The diagram shows a quarter circle inside a square.

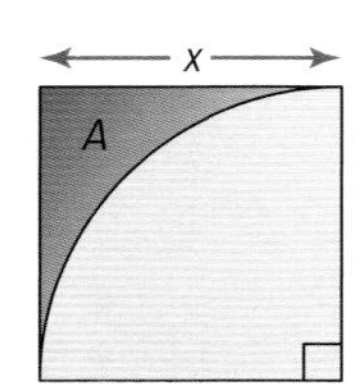

a Write a formula for A, the area shaded purple in the diagram.

b Rearrange the formula to make x the subject.

a A = area of square − area of quarter circle
$A = x^2 - \frac{\pi x^2}{4}$ factorise
$A = x^2(1 - \frac{\pi}{4})$

Using Area of a circle = πr^2

b
$A = x^2(1 - \frac{\pi}{4})$
$\frac{A}{1 - \frac{\pi}{4}} = x^2$ $\div(1 - \frac{\pi}{4})$
$\sqrt{\left(\frac{A}{1 - \frac{\pi}{4}}\right)} = x$ square root

Exercise A5.6

1 By reading the layers, rearrange each formula to make x the subject:

a $y = mx + c$ **b** $k = x^2 - p$

c $w = \frac{x + y}{z}$ **d** $m = \sqrt{x}$

e $l = \frac{z(x - k)}{w}$ **f** $w = kx^2$

g $m = \frac{x^3 - p}{q}$ **h** $z = r(\frac{mx + c}{p} - q)$

2 Match up the formulae that are rearrangements of one another:

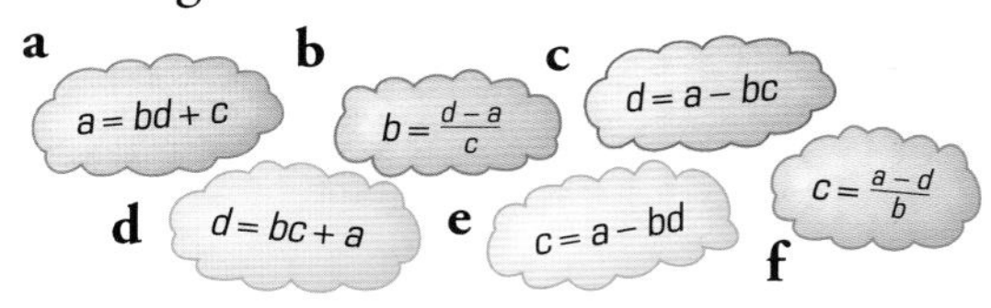

3 Rearrange these formulae to make y the subject. Take care as they involve negative and fractional terms.

a $m = p - y$ **b** $m = \frac{p}{y}$

c $z = pz - y$ **d** $t = w - my$

e $p = k(t - y)$ **f** $p = \frac{k}{y} + q$

g $z = \frac{t}{y} - r$ **h** $m = k - \frac{p}{y}$

4 Use your knowledge of fractions to rearrange these formulae to make p the subject.

a $m = \frac{1}{2}(p - e)$ **b** $k = m + \frac{1}{3}p$

c $t = \frac{4}{3}p - r$ **d** $k = m - \frac{1}{2}p$

e $z = \frac{1}{5}p^2$ **f** $w = \frac{1}{m}(p - q)$

5 This physics formula shows the relationship between the focal length (f) of a lens and the distances of the object (u) and the image (v) from the lens.

$$\frac{1}{v} + \frac{1}{u} = \frac{1}{f}$$

a By adding $\frac{1}{v}$ and $\frac{1}{u}$, derive an equivalent formula for $\frac{1}{f}$.

b How would you now make f the subject of this formula?

c Using a similar procedure, make u the subject.

6 Show that $m = \frac{1}{k}(w - \frac{p}{x})$ can be rearranged to give $x = \frac{p}{w - mk}$.

7 Factorisation can be useful when rearranging more difficult formulae. For example, $ax + b = cx + d$. It is difficult to make x the subject since it appears on both sides of the formula. Follow these steps.

$ax + b = cx + d$ → Collect x-terms on one side → $ax - cx = d - b$

$ax - cx = d - b$ → Factorise to pull all x-terms together → $x(a - c) = d - b$

$x(a - c) = d - b$ → Read and reverse the layers as before → $x = \frac{d - b}{a - c}$

Make x the subject of:

a $px - q = m - rx$

b $k(p - x) = q(x + w)$

8

x

y

An ellipse is a 'squashed' circle.
You can find its area using the formula $A = \pi xy$.

a Use this information to write a formula for the shaded area in this elliptical shape.

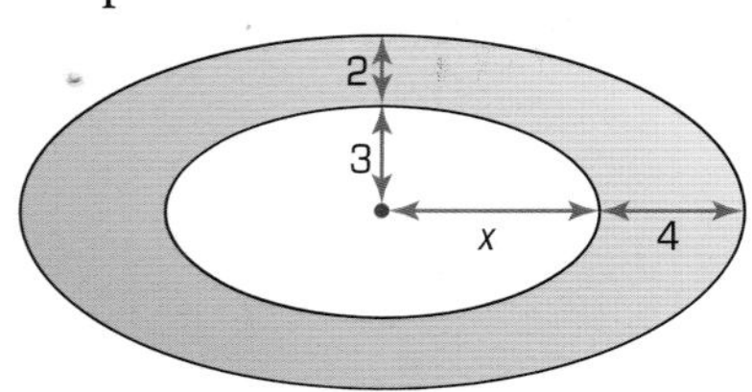

b Rearrange your formula to make x the subject.

A5 Summary

You should know how to ...

1 Square a linear expression and expand the product of two linear expressions of the form $x \pm n$.

Check out

1 **a** Expand these expressions.

i $(2x - 3y)(5x + 4y)$

ii $(6a - 2b)^2$

b Factorise these expressions.

i $x^2 - 100$ **ii** $x^2 - 25$

iii $4x^2 - 9$ **iv** $a^2 - b^2$

c Using your answer to, **b** simplify $\frac{a^2 - b^2}{2a + 2b}$.

2 Establish identities.

2 **a** The areas of the square and rectangle are the same.

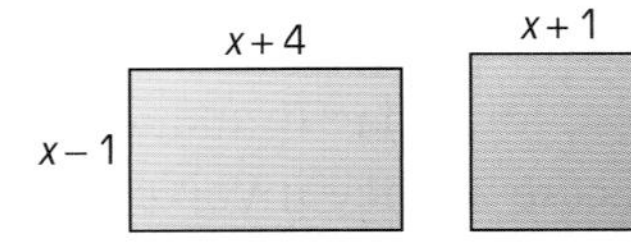

Write this mathematically.

For what value of x would this happen?

b Show that the sum of these two fractions is $\frac{ab - c}{b^2}$.

3 Change the subject of a formula.

3 Make e the subject of these formulae.

a $\frac{me + z}{p} = k$

b $we^2p + k = m$

c $\frac{1}{2}(e + f) = k$

d $\frac{w - e}{m} = R$

4 Generate fuller solutions to problems.

4 **a** Find a set of five consecutive integers for which the average of their squares is two more than the square of their average.

b Prove that your result in part **a** is the case for any five consecutive integers.

3 Statistical reports

This unit will show you how to:

- Design a survey or experiment to capture the necessary data from one or more sources.
- Determine the sample size and degree of accuracy needed.
- Design data collection sheets.
- Identify possible sources of bias.
- Identify what extra information may be required to pursue a further line of enquiry.
- Find the median and quartiles for data sets.
- Select, construct and modify suitable graphical representation to progress an enquiry.
- Identify key features present in the data.
- Analyse data to find patterns and exceptions.
- Examine critically the results of a statistical enquiry, and justify choice of statistical representation in written presentations.
- Recognise limitations on accuracy of data.
- Give reasons for choice of presentation.
- Justify generalisations, arguments or solutions.
- Pose extra constraints and investigate whether particular cases can be generalised further.

Statistics play an important role in sport.

Before you start

You should know how to ...

1 Draw a stem-and-leaf diagram.

Check in

1 Steve measured the arm span, in cm, of the boys in a class.

159 156 168 162 171 170 175 164 161
158 175 179 160 178 181 155 162 163
162 170 180 174 167 183 162 178 159

Draw a stem-and-leaf diagram to represent these data.

D3.1 Planning an investigation

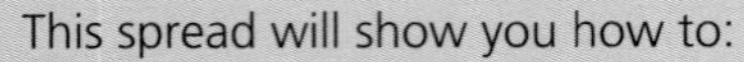
This spread will show you how to:

- Design a survey to capture the necessary data.
- Design, trial and if necessary refine data collection sheets.

KEYWORDS

Bias | Hypothesis
Data | Pilot survey

Lauren wants to compare how long it takes people to write their name forwards and backwards.
She decides on two **hypotheses**:

Remember:
A hypothesis is an idea based on limited evidence.

- Left-handed people take longer to write their name forwards than right-handed people.
- Left-handed people take longer to write their name backwards than right-handed people.

Lauren carries out a **pilot survey**, to help plan her investigation.
She asks seven friends to write their name forwards and backwards ...

First Lauren finds that the data is too similar.

Lauren needs to time to the nearest tenth of a second.

The room is noisy for some of the time.

She needs to ensure that the conditions remain the same.

There is only one left-handed person.

She needs a sampling strategy to reduce **bias**.

Some names are easier to write than others!

She needs to take the number of letters into account.

Lauren collects these results:

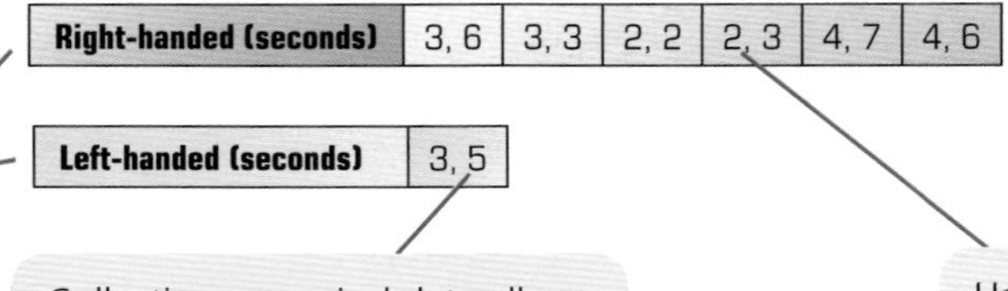

Right-handed (seconds)	3, 6	3, 3	2, 2	2, 3	4, 7	4, 6

Left-handed (seconds)	3, 5

Key
3, 6 means:
forward time 3 s
backwards time 6 s

Having two groups allows Lauren to make **comparisons**.

Collecting numerical data allows Lauren to calculate **averages** and measures of **spread**.

Having two pieces of data from each person allows Lauren to draw a **scatter graph**.

- When you plan an investigation, your plan should:
  - Give reasons for choosing the investigation
  - Outline any practical problems and how bias can be minimised
  - Allow collection of reliable data relevant to the problem
  - Allow a range of relevant calculations and graphs.

Exercise D3.1

In this unit you will carry out your own statistical investigation.
You are going to survey how long it takes people to write their names.

1 Decide on your specific aim or hypothesis.
Write down your reasons for choosing it.

2 Decide whom you are going to collect data from, specifying for example:
a their ages
b boys and/or girls

Your sources of data should be relevant to your hypothesis.

3 Decide how you are going to collect your data and the degree of accuracy needed.

4 Design a data collection sheet to record the data you are going to collect.

5 Write down what measures you will be able to calculate from the data you collect.
Explain what you hope these measures will tell you.

6 Write down which graphs you will be able to draw to display the data you collect.
Write about how these graphs will be effective in highlighting key points.

7 Carry out a pilot survey.
- Collect data appropriate to your aim from about seven people (for example, if your survey is Year 9 girls then the people in the pilot should also be Year 9 girls).
- Look at your survey results:
  - Do you need to make changes to your data collection sheet?
  - Did you find any unexpected problems?
  - Did you collect all the data you needed to do your chosen calculations and graphs?
  - Think about any improvements you could make to your survey.

Gathering data

This spread will show you how to:

- Discuss how data relate to a problem.
- Identify possible sources of bias and plan how to minimise it.
- Determine the sample size and degree of accuracy needed.

KEYWORDS

Sample	Bias
Random	Census
Stratified	Population
Systematic	Representative

Lauren restricts her investigation to the **population** of Year 9 students in her school.

The whole population takes part in a census.

There are only 13 left-handed students in Year 9 so Lauren decides to ask all of them. This is a **census**.

Lauren now has to decide how to choose her representative sample of right-handed students.

- She starts by asking for copies of the registers of all Year 9 classes.
- After crossing off all left-handed students Lauren numbers the other students 1–174.
- Lauren decides to survey 15% of the year group.
  There are 187 students in total.
  She samples 15 right-handed students: $\frac{(15 + 13)}{187 \times 100} = 14.97\%$.

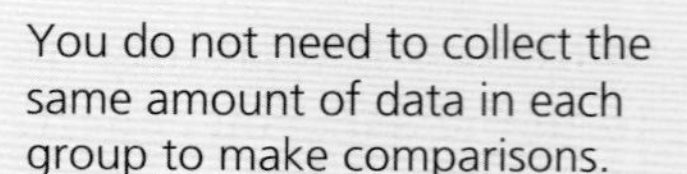

You do not need to collect the same amount of data in each group to make comparisons.

She could use the random number generator on a calculator.

This is **simple random** sampling.

She could choose a random starting point (say the third person), then choose every eleventh.

This is **systematic** sampling.

She could divide Year 9 into form groups and randomly choose people in proportion from each group.

This is **stratified** sampling.

- Lauren decides to use a simple random sample for right-handed students.
  If anyone in her sample is absent it will be easy to choose a substitute free from personal bias.

Lauren times, using the same stopwatch, how long the students take to write their names forwards and backwards, to the nearest tenth of a second, while they are on their own in a quiet room.

Exercise D3.2

You should now be ready to choose a sample for the investigation you planned in D3.1.

1 You need to decide your population.

- You should know from your aim the population from which your sample will be chosen.

2 You need to decide your sample size.

- If you are going to compare two groups of people such as boys and girls or two different age groups, you will need enough data from each group to calculate reliable measures and draw valid graphs.

3 You need to decide your sampling method.

- Write down which method you are going to use.
- Write down why you decided to use this sampling method instead of other methods.
- Write down **in detail** exactly how your sample will be chosen.
- Write down what you will do if anyone in your chosen sample is unavailable when you actually collect the data. Consider:
  - Have you a large enough sample not to worry about one missing person or will you need to replace that person?
  - What will you do if several people are unavailable?

4 Collect your data.

Representing and analysing data

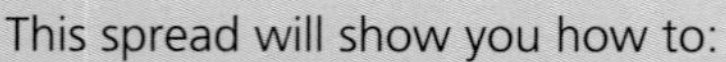

This spread will show you how to:

- Select, construct and modify suitable graphical representation to progress an enquiry.
- Identify key features present in the data.
- Find summary values that represent the raw data.
- Analyse data to find patterns and exceptions.

KEYWORDS

Scatter diagram

Correlation

Line of best fit

Lauren draws two data collection tables for her investigation.

Left-handed

Number of letters in name	7	4	5	5	5	4	9	8	6	8	6	7	6
Forwards time (seconds)	3.7	2.1	2.3	2.9	2.5	2.0	4.2	4.0	3.3	3.9	2.8	3.6	3.3
Backwards time (seconds)	6.1	2.4	2.6	3.6	3.0	2.5	6.8	6.2	4.8	5.9	5.8	5.6	4.2

Right-handed

Number of letters in name	7	6	7	6	5	8	7	5	5	5	7	7	9	5	4
Forwards time (seconds)	3.6	3.2	4.6	3.9	2.7	4.8	4.4	2.5	2.8	2.6	4.0	4.2	5.2	3.0	2.5
Backwards time (seconds)	4.2	3.8	6.3	5.2	4.3	6.2	5.8	3.8	4.0	3.5	5.9	6.8	7.3	3.7	3.2

From her tables Lauren can see that both left- and right-handed students take longer to write their name backwards than forwards.

Lauren draws two scatter diagrams to represent these data.

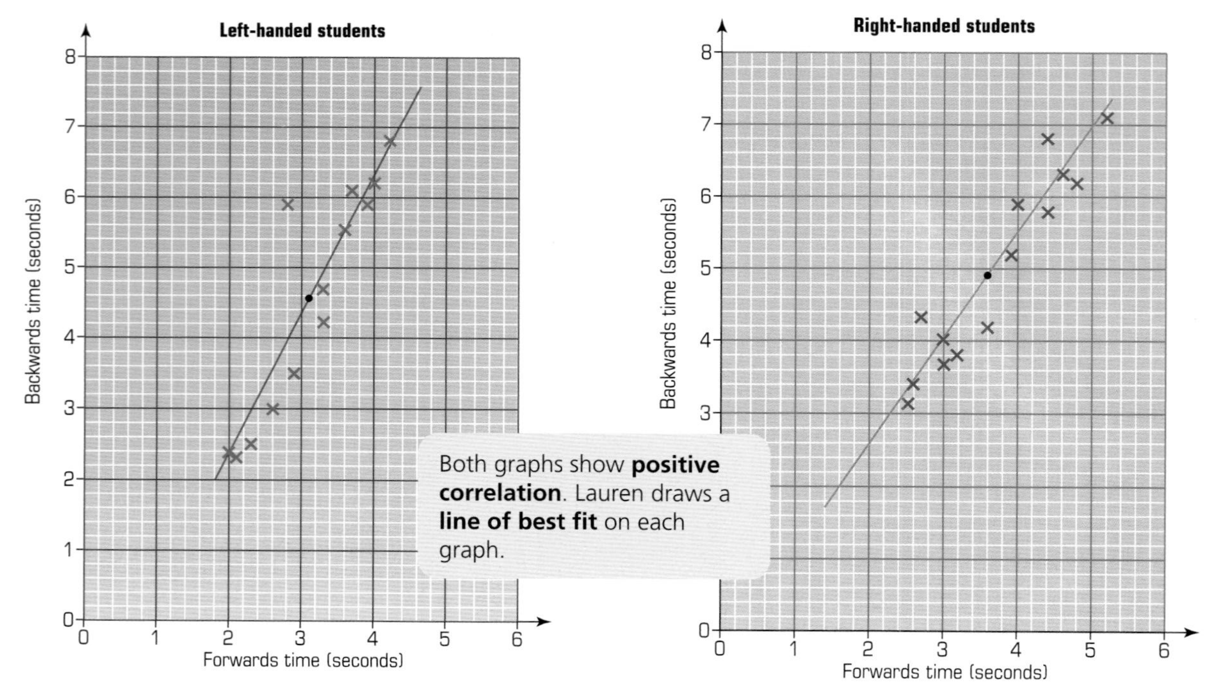

Lauren now needs to analyse the data further.
She must calculate averages and measures of spread for comparisons.

Exercise D3.3

You should now have collected your data.

1 You may want to sort your data into different groups.

2 Look at your data and write down if any of the data appears to be very different from the rest.

If there are any outliers, try to give a reason.
For example, data may have been recorded incorrectly.

A data value that is very different from the rest of the data is called an **anomaly** or an **outlier**.

3 Calculate appropriate measures of average and spread for your data.

- You may want to find more than one average, for example the mean and median.

Explain why these are the best measures for you to use.

- If you calculate more than one measure, write down what the different measures show.

4 Draw graphs to display your data.

- Your graphs should be appropriate for the data you collect.
- Ensure that your graphs are drawn on graph paper with a title, correct labels on axes, and a key.
- Explain what each graph shows.

Remember:

- You can only draw a scatter diagram if you have collected **bivariate** (paired) data.
- If you have collected a lot of data or you have grouped data you may want to draw a **cumulative frequency graph**.

This is an example of a cumulative frequency graph from Lucy's survey on page 76.

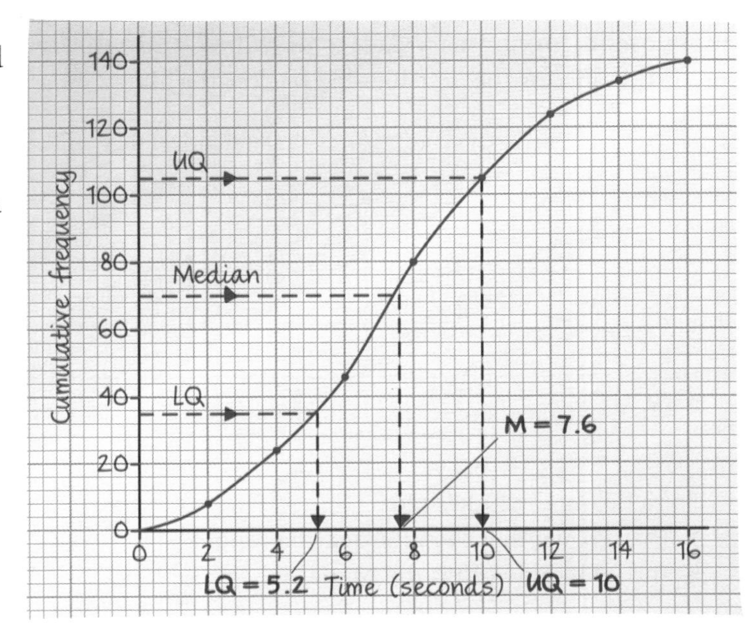

To look at how to draw a cumulative frequency graph, look at page 76.

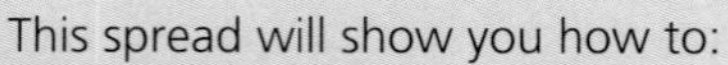

D3.4 Representing data

This spread will show you how to:

- Find the median and quartiles for data sets.
- Select, construct and modify suitable graphical representation to progress an enquiry.

KEYWORDS

Box-and-whisker plot
Stem-and-leaf diagram
Median
Quartiles

Lauren orders her data using back-to-back **stem-and-leaf diagrams**.
These will help her to find the medians, lower quartiles and upper quartiles.

Left-handed

Backwards		Forwards
8 2 1	6	
9 8 6	5	
8 2	4	0 2
6 0	3	3 3 6 7 9
6 5 4	2	0 1 3 5 8 9

Right-handed

Backwards		Forwards
3	7	
8 3 2	6	
9 8 2	5	2
3 2 0	4	0 2 4 6 8
8 8 7 5 2	3	0 2 6 9
2	2	5 5 6 7 8

Key:
2 | 4 | 0
4.2 s Backwards
4.0 s Forwards

For small data sets of size n, you can use these formulae:
Median = $\frac{1}{2}(n + 1)$th value
Lower quartile = $\frac{1}{4}(n + 1)$th value
Upper quartile = $\frac{3}{4}(n + 1)$th value

So if $n = 15$:
UQ = $\frac{3}{4}(15 + 1)$th value
= 12th value
Median = $\frac{1}{2}(15 + 1)$th value
= 8th value
LQ = $\frac{1}{4}(15 + 1)$th value
= 4th value

Remember to count from the **smallest** value each time.

Left-handed	Backwards (s)	Forwards (s)
Lowest time	2.4	2.0
Lower quartile	2.8	2.4
Median	4.8	3.3
Upper quartile	6.0	3.8
Highest time	6.8	4.2

Right-handed	Backwards (s)	Forwards (s)
Lowest time	3.2	2.5
Lower quartile	3.8	2.7
Median	4.3	3.6
Upper quartile	6.2	4.4
Highest time	7.3	5.2

Lauren uses these summary data to draw **box-and-whisker plots**.
Here is the one for the 'left-handed backwards' data.

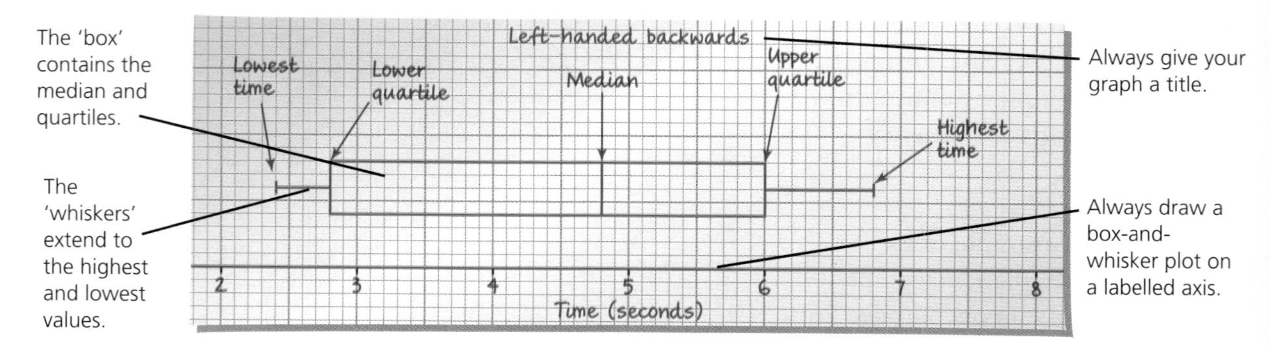

- For a box-and-whisker plot:
  - The box width is the interquartile range
  - The whisker width is the range.

Exercise D3.4

In this exercise you will draw a box-and-whisker plot for your data.

1 Find this information from your data:

a lowest value
b lower quartile
c median
d upper quartile
e highest value.

Write the values in a table.

> **Hints:**
> To find these values you should first order your data.
>
> ▶ If your data are not grouped, then you can either list them or draw a stem-and-leaf diagram.
>
> ▶ If your data are grouped and you do not have the raw data, then you need to draw a cumulative frequency graph.

See page 76 to remind yourself how to draw a cumulative frequency graph.

2 Draw the box-and-whisker plots for your data.

> **Hints:**
>
> ▶ If you are drawing more than one box-and-whisker plot and you want to use them to compare different groups, then all the plots need to be drawn to the same scale.
>
> ▶ It is best to draw all the graphs on the same page using just one axis.
>
> ▶ Remember to label the axes, and give a title to each box-and-whisker plot.
>
> ▶ You do not need to label each value as LQ, Median, UQ ....

D3.5 The shape of a distribution

This spread will show you how to:
- Identify key features present in the data.
- Analyse data to find patterns and exceptions, look for cause and effect and try to explain anomalies.

KEYWORDS
Distribution
Skew
Interpret
Box-and-whisker plot

Lauren draws four box-and-whisker plots on the same diagram. This enables her to compare the shape of the **distributions**.

A distribution is a set of data.

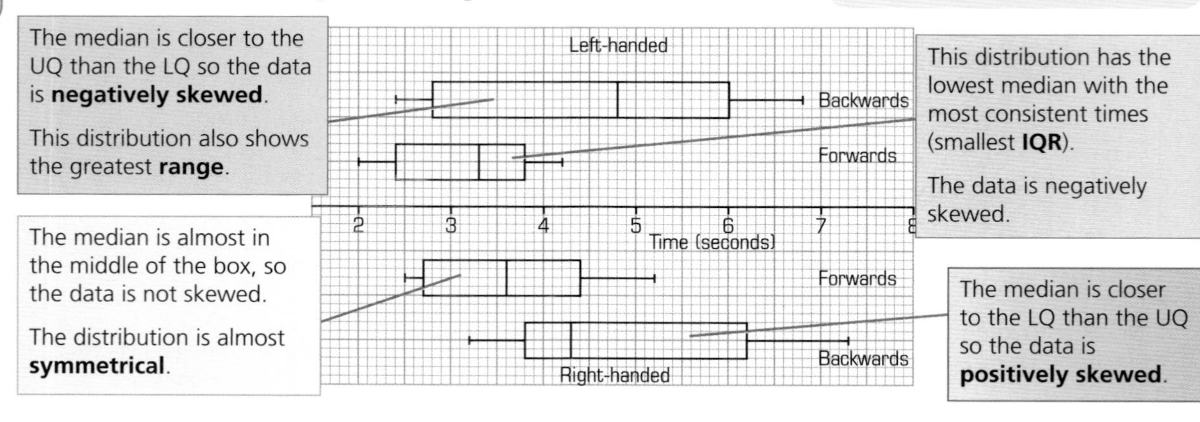

The median is closer to the UQ than the LQ so the data is **negatively skewed**.

This distribution also shows the greatest **range**.

This distribution has the lowest median with the most consistent times (smallest **IQR**).

The data is negatively skewed.

The median is almost in the middle of the box, so the data is not skewed.

The distribution is almost **symmetrical**.

The median is closer to the LQ than the UQ so the data is **positively skewed**.

- **Skewness** describes the shape of a distribution.

- A negatively skewed distribution has the median closer to the UQ.

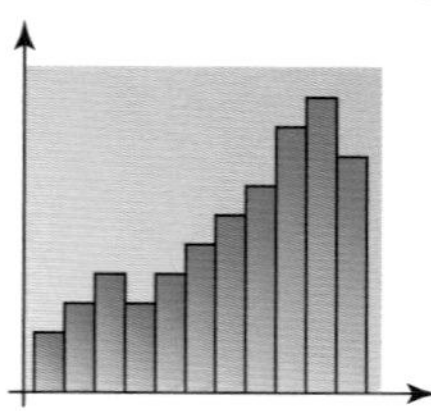

- A positively skewed distribution has the median closer to the LQ.

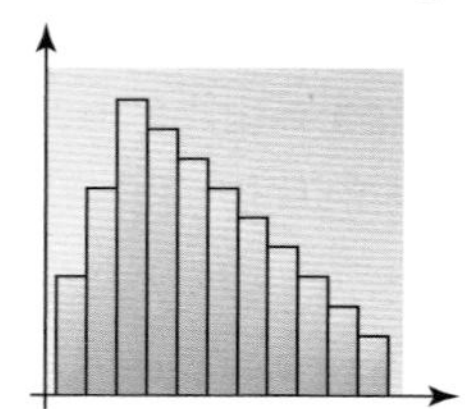

- A symmetrical distribution has the median in the middle of the box.

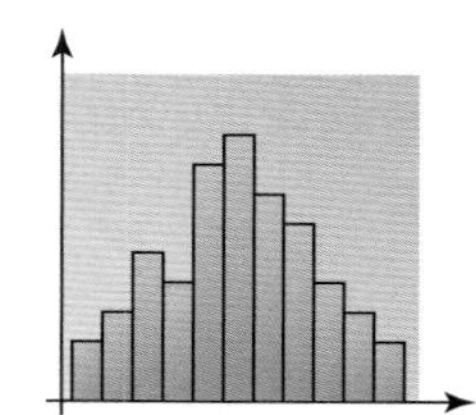

Lauren also calculates the mean time for each group.

Left-handed	Mean time (s)
Forwards	3.1
Backwards	4.6

Right-handed	Mean time (s)
Forwards	3.6
Backwards	4.9

The means show that:
- The right-handed sample took longer than the left-handed sample
- Both samples took longer to write their name backwards than forwards.

- When making comparisons you should summarise and interpret graphs and calculations.
- You should comment on differences and similarities in averages and spread.

Exercise D3.5

In this exercise you will interpret your measures and graphs and make comparisons.

1 Write down your initial observations and your interpretations throughout your investigation next to the measures and graphs.

> **Hints:**
> - If you are comparing graphs, ideally they should be on the same page.
> - An example of an observation is 'The scatter graph shows positive correlation'.
> - An example of an interpretation is 'As the time taken to write your name forwards increases, the time taken to write your name backwards also increases'.

2 Draw a summary table for all the measures of average and spread that you have found.

3 Comment on differences and similarities between the data sets.

> **Hints:**
> - Your comments should focus on differences or similarities of average values and measures of spread.
> - Remember that comments on individual values are relevant to your sample only and do not give an indication of the overall picture.
> - Relate your comments to your original aims.

Remember to highlight any anomalies or unusual results that look out of place.
If you find any anomalies can you suggest a possible cause?

D3.6 Statistical reports

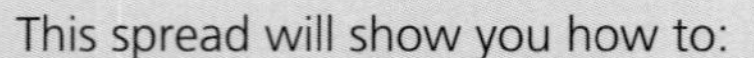

This spread will show you how to:

- Examine critically the results of a statistical enquiry.
- Recognise limitations of any assumptions and their effect on conclusions drawn.
- Identify what extra information may be required to pursue a further line of enquiry.
- Identify possible sources of bias and plan how to minimise it.

KEYWORDS
Limitations
Outlier
Distribution

Lauren looks critically at her results.

		Median	Mean
Left–handed (LH)	Forwards	3.3	3.1
	Backwards	4.8	4.6
Right–handed (RH)	Forwards	3.6	3.6
	Backwards	4.3	4.9

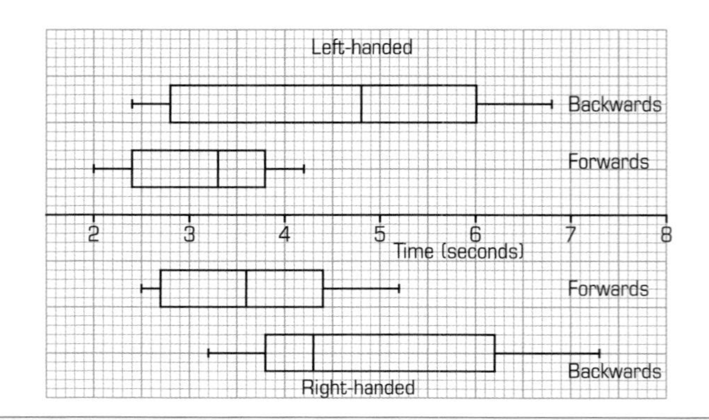

She discusses her findings in a report.

RH backwards has a much lower median than mean.
The positive skew shown in the box-and-whisker plot has affected the mean, making it very high.

RH forwards has an almost symmetrical distribution, and the mean and median are the same.

The LH data both show negative skew.

Lauren then remarks on the initial data.

In the LH data, the number of letters in a name (name size) is quite different to the name size in the RH data.
I took a census for the LH data, but I took a random sample for the RH data.
I cannot alter the distribution in the LH group, but I could have chosen a group of RH students with a distribution similar to the LH students using stratified sampling.

To see if this would affect her results, Lauren could group the times according to name size, and then calculate the mean for each group.

In a statistical report you should:

- Comment on patterns and give plausible reasons for exceptions (outliers)
- Relate the data back to the original problem
- Recognise possible limitations in your strategy and suggest improvements
- Comment constructively on the practical consequences of the work.

Exercise D3.6

In this exercise you will write your final report for your data handling investigation.

1 Look at your initial hypothesis.
Either:
- **a** Accept it – the evidence of your survey supports your hypothesis
- **b** Reject it – the evidence of your survey does not support your hypothesis
- **c** State that you cannot reach a conclusion – your graphs and measures do not provide strong evidence either way.

> **Hints:**
> - You should give reasons, based on your graphs and calculations, for your decision.
> - Do not ignore any anomalies, but try to give reasons for them.

2 Look critically at your data.
Comment on:
- **a** the similarities in your different sets of data
- **b** how you could have improved any aspect of the data collection
- **c** whether there is any evidence of bias in the data collection.

3 Comment on any aspect of your investigation that you could change.
In particular, write down:
- **a** any assumptions you made when you began the investigation, and how they affected your results
- **b** any improvements you would make if you were to do the investigation again, and how these improvements might affect the results.

4 Write down any practical consequences of your investigation.

Summary

You should know how to ...

1 Identify possible sources of bias and plan how to minimise it.

2 Examine critically the results of a statistical enquiry, and justify choice of statistical representation in written presentations.

3 Recognise limitations on accuracy of data and measurements.

Check out

1 Merlin surveyed the speed of cars that drove down the High Street of his town between 8 am and 8.30 am on two consecutive Monday mornings.

One Monday was during school term time, the other Monday was during school half term.

Identify possible sources of bias in Merlin's survey and write a plan to minimise any bias.

2 Merlin grouped his data in the table shown.

Speed, miles per hour	Under 30	30–40	Over 40
Week X – frequency	23	22	15
Week Y – frequency	56	18	16

Describe, with reasons, what diagrams and statistical measures you would use to analyse this data.

3 Explain why the results of Merlin's survey may not be accurate.

4 Applying geometrical reasoning

This unit will show you how to:

- Solve problems using properties of angles, of parallel and intersecting lines, and of triangles and other polygons.
- Understand and apply Pythagoras' theorem.
- Visualise and use 2-D representations of 3-D objects.
- Analyse 3-D shapes through 2-D projections, including plans and elevations.
- Begin to use sine, cosine and tangent in right-angled triangles to solve problems in 2-D.
- Calculate lengths, areas and volumes in right prisms, including cylinders.
- Recognise limitations on the accuracy of measurements.
- Generate fuller solutions to problems.
- Justify generalisations, arguments or solutions.
- Pose extra constraints and investigate whether particular cases can be generalised further.

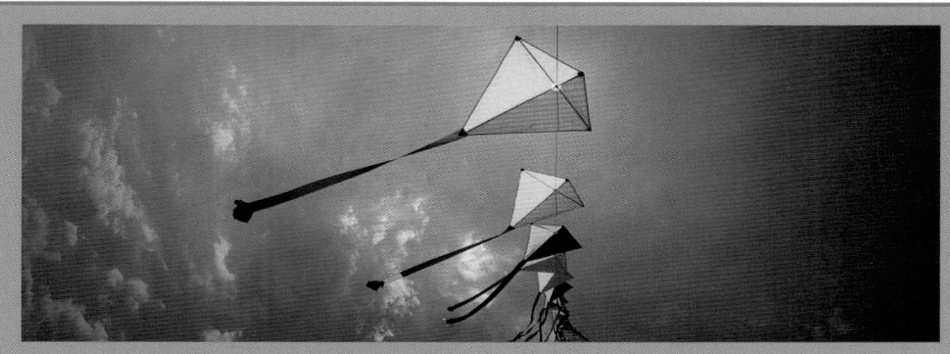

You can use geometrical reasoning to design your own kite.

Before you start

You should know how to ...	Check in
1 Identify angles in parallel and intersecting lines.	1 Label the angles equal to x in each diagram. **a** 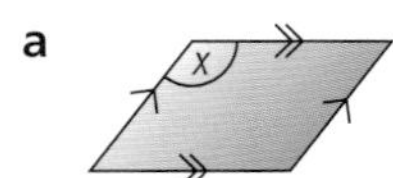 **b** 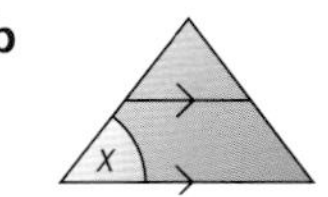
2 Apply Pythagoras' theorem to finding a missing length.	2 Find the unknown length in each diagram. **a** 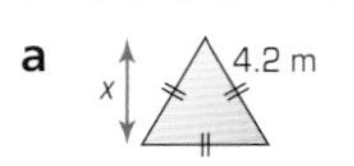 **b** 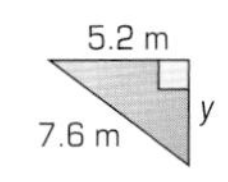
3 Round an answer to a number of significant figures.	3 Round each number to two significant figures. **a** 425.392 **b** 0.4823 **c** 4.285 73 **d** 0.000 483 **e** 599.38 **f** 248.36
4 Calculate the area of a shape, including a circle.	4 Find the areas of these shapes. **a** 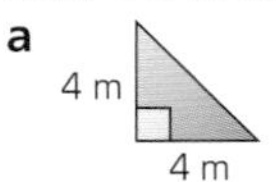 **b** 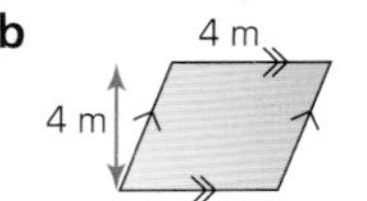 **c** 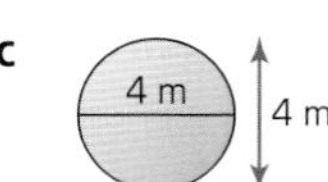

Properties of angles

This spread will show you how to:

- Solve problems using properties of angles.

KEYWORDS

Quadrilateral, Corresponding, Symmetry, Bisect, Alternate

Sam is designing a kite.
Here is its basic shape.

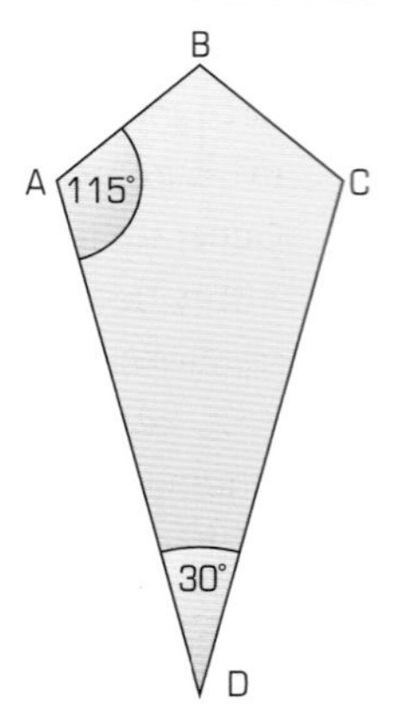

Sam wants to find angles ABC and BCD:

$\angle BCD = \angle BAD = 115°$ (by symmetry)

$\angle ABC = 360° - (115° + 115° + 30°)$ (angles of a quadrilateral)

$= 100°$

Sam fixes stiff rods into the diagonals of the kite.

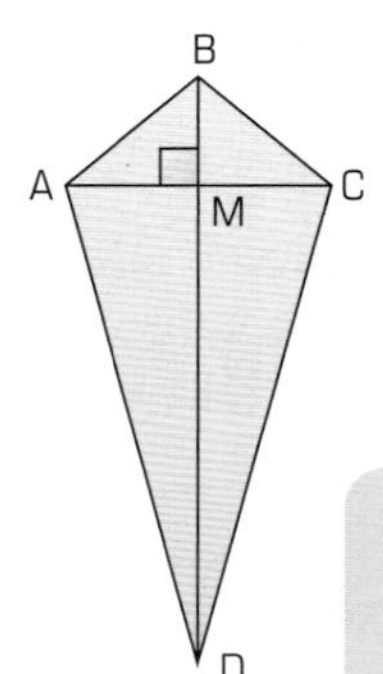

The line BD bisects $\angle ABC$ and $\angle ADC$.

$\angle ABM = 100° \div 2 = 50°$

$\angle ADM = 30° \div 2 = 15°$

So: $\angle BAM = 40°$ (angles of a triangle)

$\angle MAD = 75°$ (angles of a triangle)

Remember:
The diagonals of a kite intersect at 90°.

Sam designs a pattern to go on his kite.

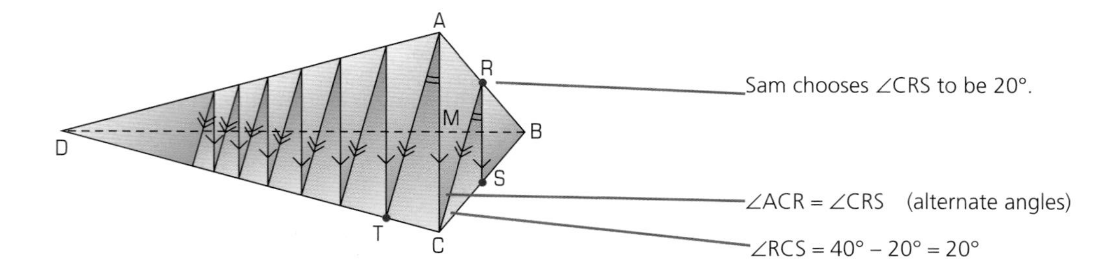

The blue angles in the 'tops' of the triangles to the left of the line AC correspond to $\angle CRS$.

Remember:
Corresponding angles are equal.

Exercise S4.1

1 This diagram shows two overlapping squares and a straight line.

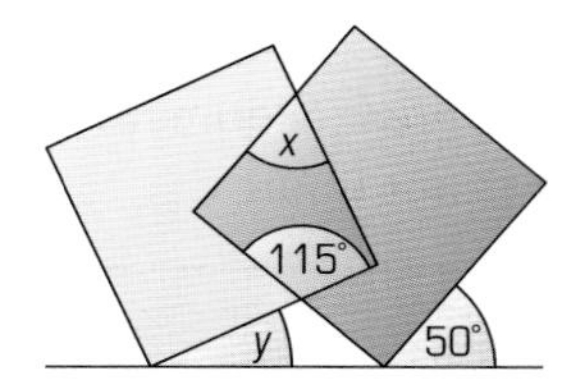

Calculate angles x and y.

2 ABCD is a kite.
Work out ∠BAD.

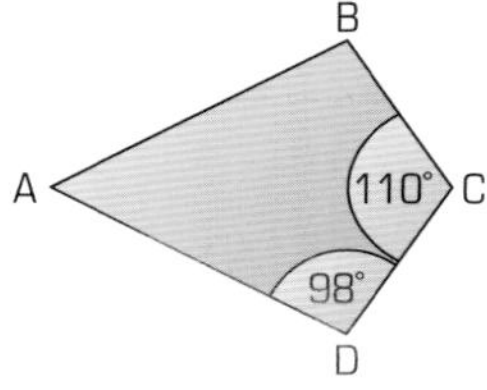

3 Find the angles marked with letters. Give reasons for your answers.

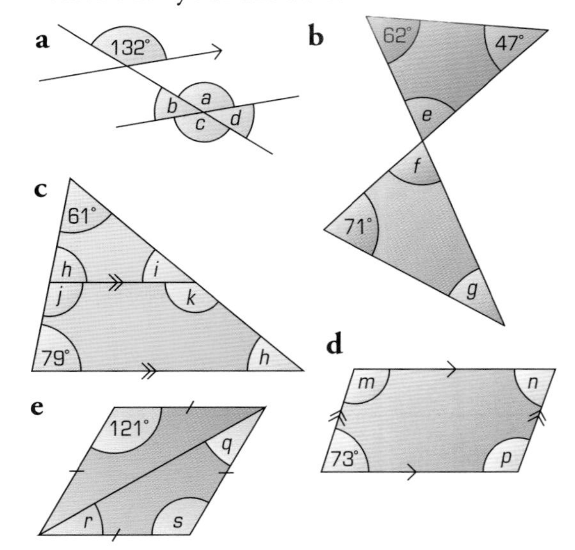

d

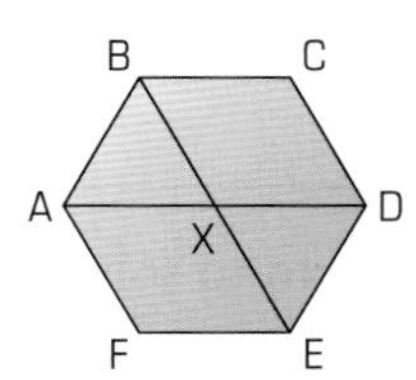

4 PQRS is a kite.

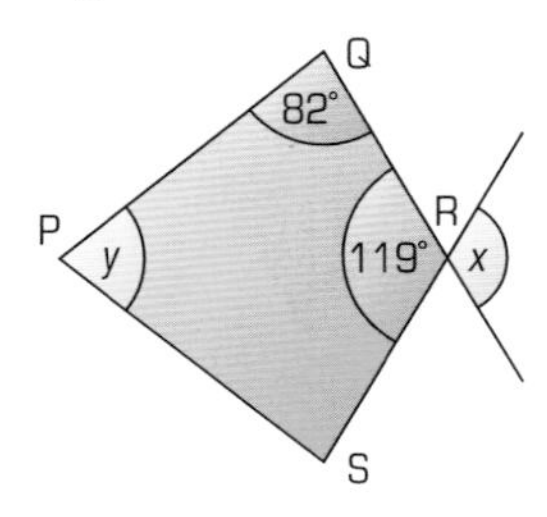

Find angles x and y.

5 Find the sum of the interior angles in these polygons.

a

b

6 There are two triangles connected together.

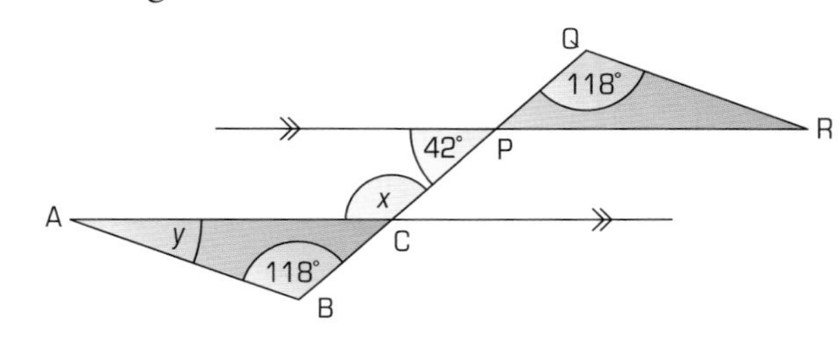

a Calculate x and y.
b Show that triangles ABC and PQR are similar.
c Show that AB and QR are parallel.

7 ABCDEF is a regular hexagon.

B C
A X D
F E

Find:
a ∠AFE **b** ∠EXD **c** ∠XBC

8 These two polygons are regular.

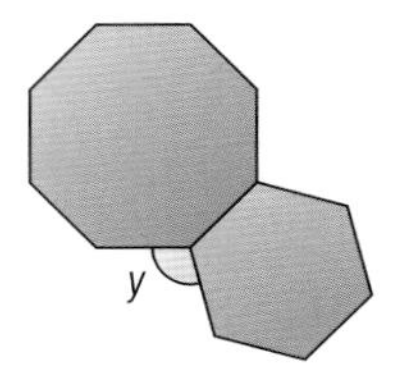

Find y.

9 Design a kite pattern like the one on page 204. You will need to work out the size of any unknown angles.

S4.2 Review of Pythagoras' theorem

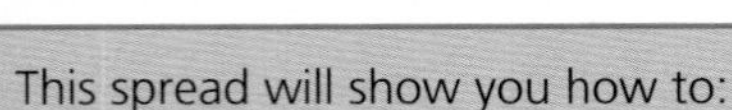

This spread will show you how to:

- Understand and apply Pythagoras' theorem.

KEYWORDS
Pythagoras' theorem
Hypotenuse Perimeter
Significant figures

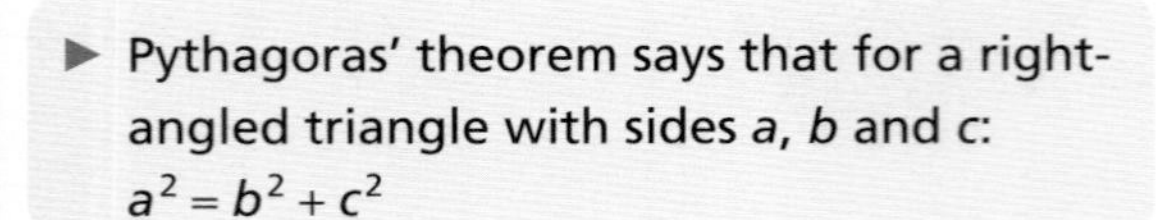

- Pythagoras' theorem says that for a right-angled triangle with sides *a*, *b* and *c*:
  $a^2 = b^2 + c^2$

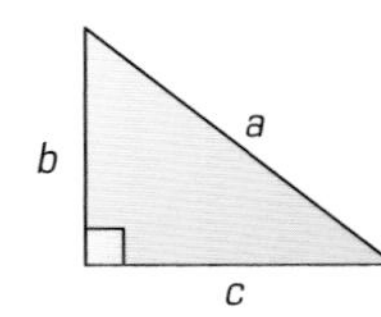

Remember:
Side *a* is called the **hypotenuse**.

If you know two sides of a right-angled triangle, you can find the third side.

example

Find the missing lengths in these triangles.
Give your answers to 3 significant figures.

a

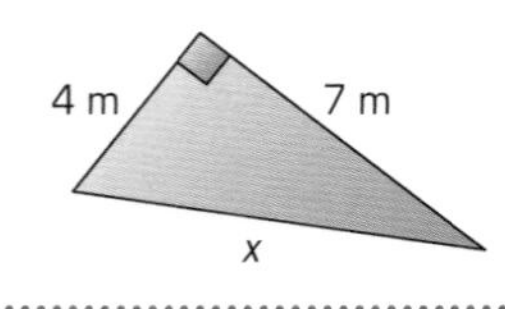

b

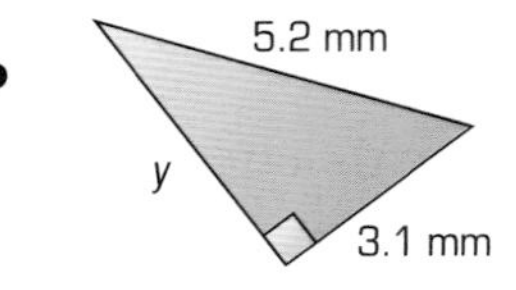

a

$a^2 = b^2 + c^2$
$x^2 = 4^2 + 7^2$
$x^2 = 65$
$x = \sqrt{65}$
$x = 8.06$ m to 3 sf

b

$a^2 = b^2 + c^2$
$5.2^2 = 3.1^2 + y^2$
$27.04 = 9.61 + y^2$
$y^2 = 27.04 - 9.61$
$= 17.43$
$y = 4.17$ mm to 3 sf

Hint:
Longest side **add**
Shorter side **subtract**

You can use Pythagoras' theorem to solve problems with more complicated shapes.

example

Tasneem is designing a kite with the dimensions shown.
Find the outside perimeter of the kite (to 1 dp).

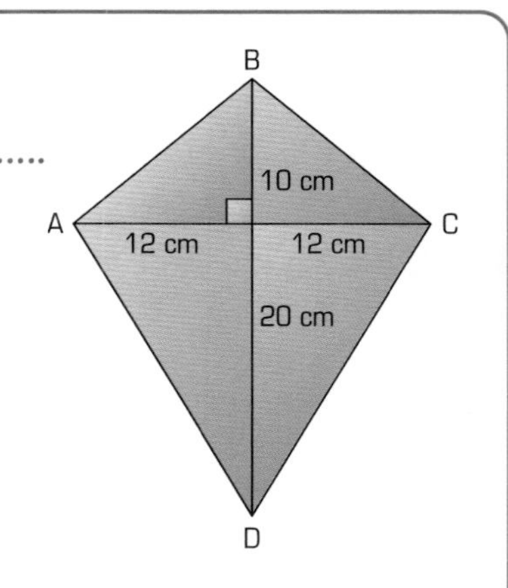

$AB^2 = 10^2 + 12^2 = 244$
$AB = \sqrt{244} = 15.620\,499\,35$
$AB = BC = 15.620\,499\,35$ cm
$DC^2 = 12^2 + 20^2 = 544$
$DC = \sqrt{544} = 23.323\,807\,58$ cm
$AD = DC = 23.323\,807\,58$ cm

Perimeter ABCD $= 15.620\,499\,35 \times 2 + 23.323\,807\,58 \times 2$
$= 77.888\,613\,86$

So the perimeter of Tasneem's kite is 77.9 cm to 1 dp.

Exercise S4.2

1 Find the unknown lengths in these triangles to an appropriate degree of accuracy.

a

2.1 m
x
3.4 m

b

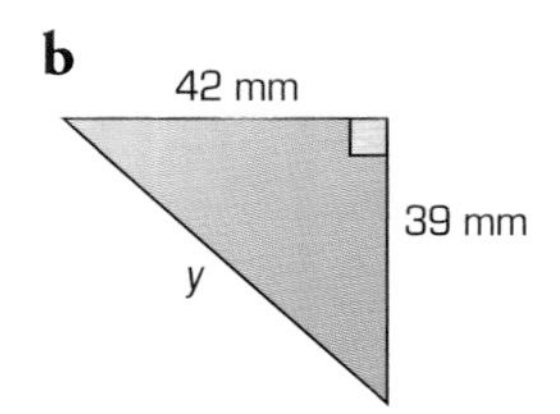

c

5.3 cm
p
2.1 cm

d

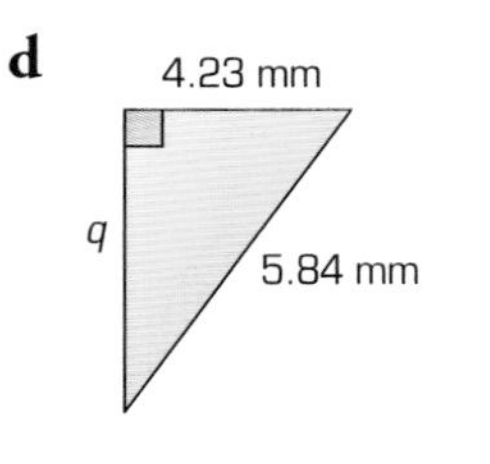

e

4.2 m
r

f

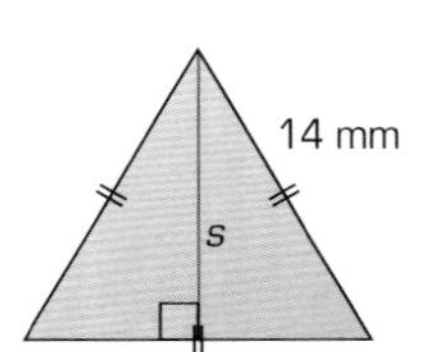

2 Find whole-number lengths for the missing sides in these right-angled triangles.
There may be more than one answer.

a

7
a
c

b

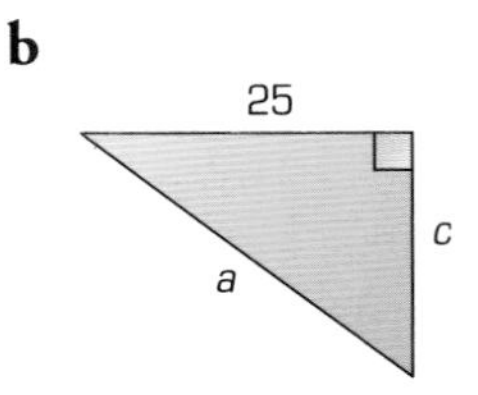

3 You walk due south for 5 km, then due west for 4 km. What distance are you from your starting point?
Give your answer to 1 decimal place.

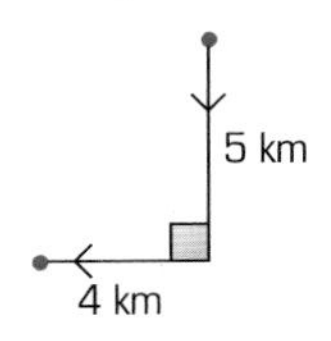

4 A 4.5 m ladder rests against a wall with its foot 1.6 m away from the wall.
How far up the wall will the ladder reach?
Give your answer to 1 dp.

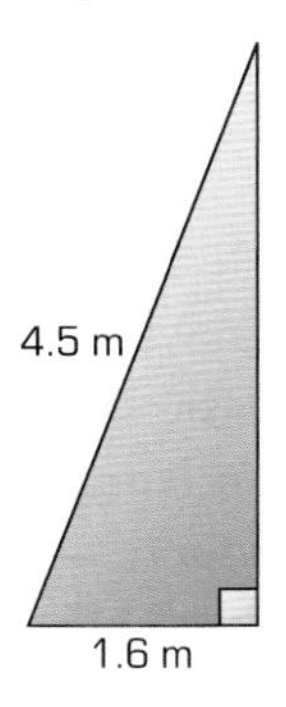

5 Each set of three numbers gives the lengths of sides of a triangle in cm.

Use Pythagoras' theorem to decide whether the triangles are right-angled or not.

a 5, 12, 13
b 12, 14, 16
c 10, 16, 22
d 16, 30, 34
e 26, 30, 64
f 6, 8, 11

6 Show that angle a is 90°.

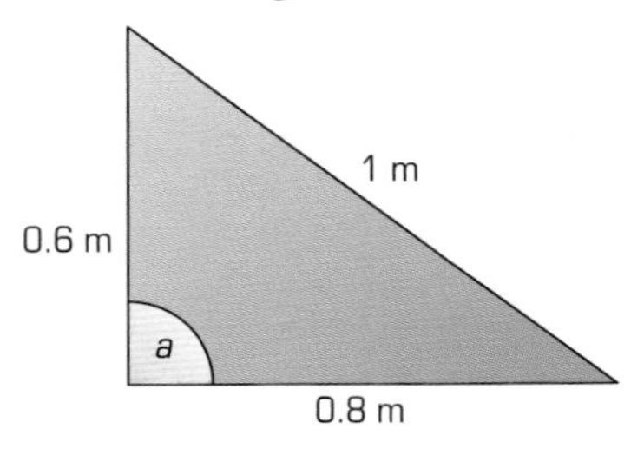

7 **Investigation**
If $a^2 = b^2 + c^2$, does:
a $a^3 = b^3 + c^3$
b $a^4 = b^4 + c^4$?

Explain your answers, giving examples.

S4.3 2-D representations of 3-D shapes

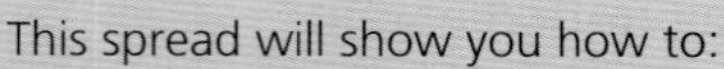

This spread will show you how to:

- Visualise and use 2-D representations of 3-D objects.
- Analyse 3-D shapes through 2-D projections, including plans and elevations.

KEYWORDS
Isometric
Elevation
Net
Plan
Projection
Cross-section
Plane
Pythagoras' theorem

Dean is making a model of a building.

He has drawn his basic shape on **isometric** paper.

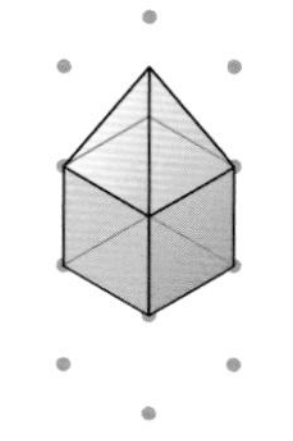

Dean can also represent his model by using **plans** and **elevations**.

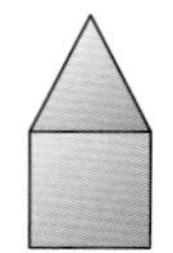

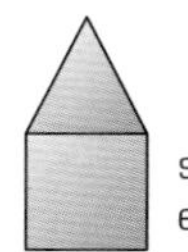

- Plans and elevations are **projections** of a 3-D object onto a 2-D surface or plane.

Note:
A common example of a projection is a shadow.

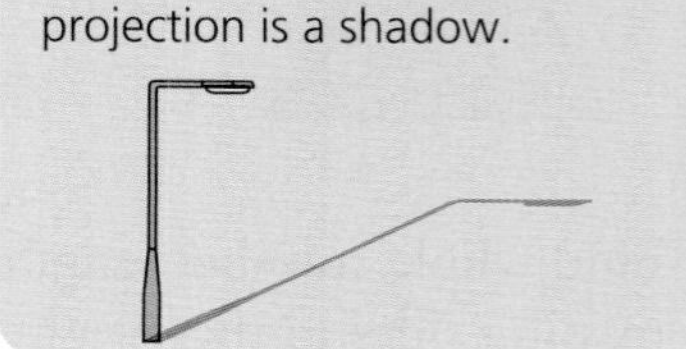

You can identify a 3-D shape from:

- its plan and elevations
- a **net**
- its **cross-sections**.

example

Here are the plan and elevations of a solid.

a Sketch the solid shape.
b Sketch the net of the solid shape.
c Triangle PQR is the front elevation of the shape.

Use Pythagoras' theorem to calculate the length PR.

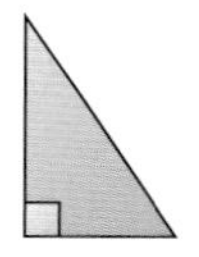

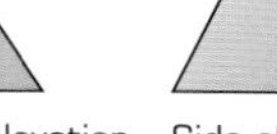

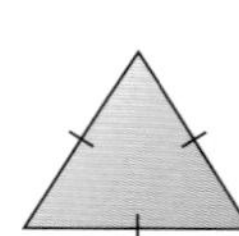

P, 4 cm, Q, 6 cm, R

a

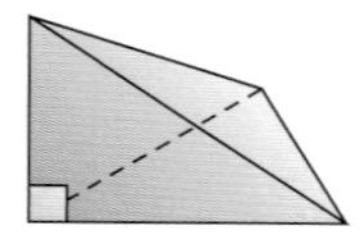

b

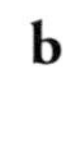

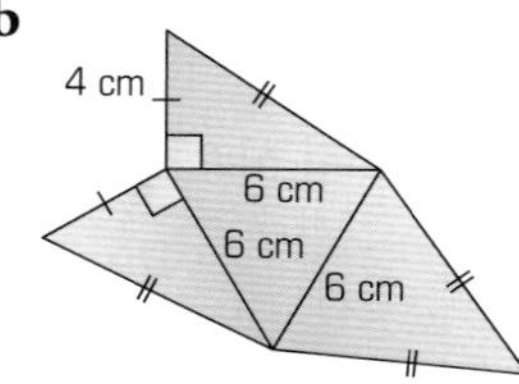

c

$$a^2 = b^2 + c^2$$
$$PR^2 = 4^2 + 6^2$$
$$= 52$$
$$PR = \sqrt{52} = 7.211$$

The length PR is 7.2 cm (to 1 dp)

Exercise S4.3

1 Draw 2-D representations of these shapes on isometric paper:

a a cuboid with sides 1 cm, 2 cm and 3 cm

b a triangular prism of length 3 cm and cross-section an equilateral triangle with sides 3 cm.

2 Sketch nets for the 3-D solids in question 1. Indicate measurements in your sketches.

3 Use the descriptions in the table to identify the 3-D solids.

	Front	Side	Plan
a	circle	circle	circle
b	triangle	triangle	square
c	circle	rectangle	rectangle
d	rectangle	rectangle	circle

4 Four solids are placed on an overhead projector. Below are the shadows projected on the screen.
Describe the possible solids for each shadow (there may be more than one answer).

a

b

c

d

5 You need to work with a partner.
Using five multilink cubes, make a solid (without your partner seeing it).

a Draw its plan, front and side elevations on isometric paper.

b Show the plans to your partner and ask them to make your solid with their cubes. Were they right?

Repeat with six cubes.

6 **Investigation**
Investigate the number of polyhedra which have a triangle as their front elevation.

7 Use Pythagoras' theorem to calculate the unknown lengths in these solids.

a

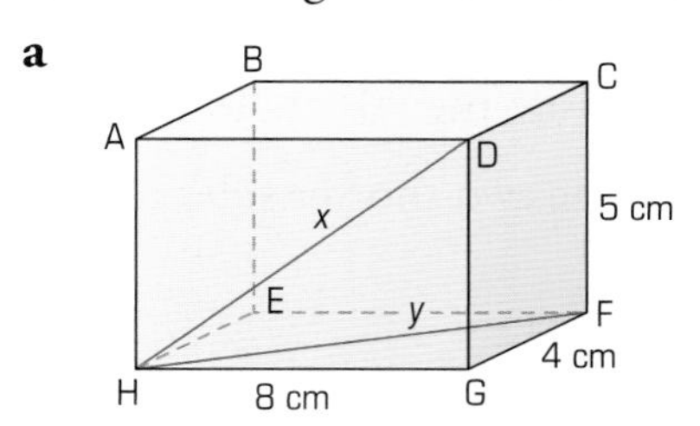

b

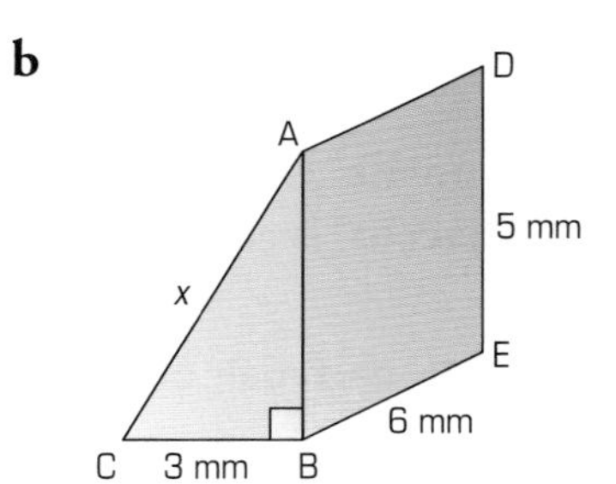

c

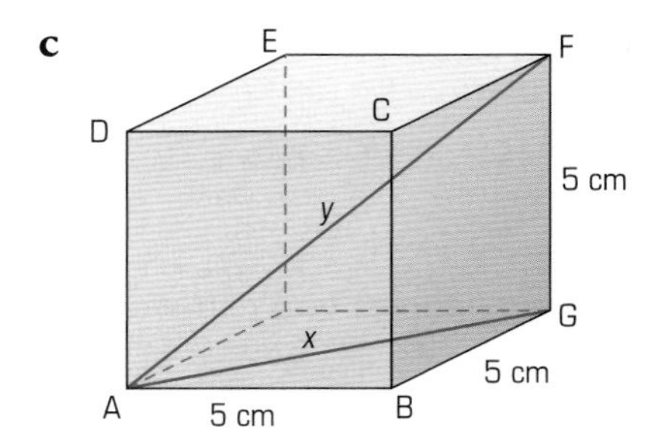

Hint: Consider the right-angled triangle:

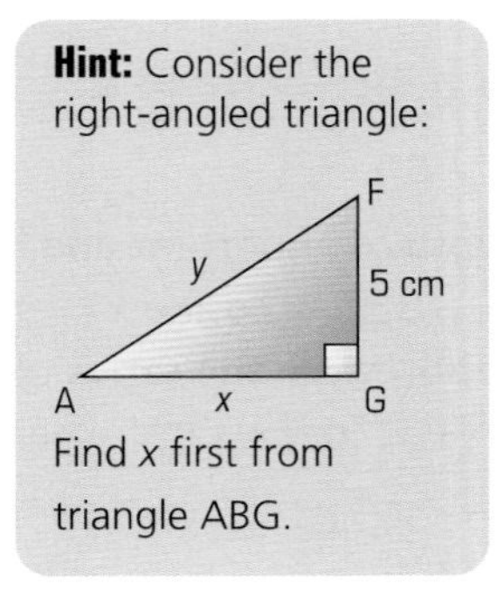

Find x first from triangle ABG.

8 Find the volume of each solid in question 7.

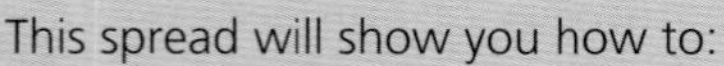

S4.4 Introducing trigonometry

This spread will show you how to:

- Begin to use sine, cosine and tangent in right-angled triangles to solve problems in two dimensions.

KEYWORDS

Hypotenuse, Opposite, Adjacent, Sine, Cosine, Tangent, Trigonometry

In a right-angled triangle, the **hypotenuse** is the longest side.
It is always opposite the 90° angle.

You can name the two shorter sides as well.

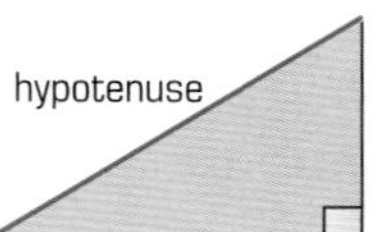

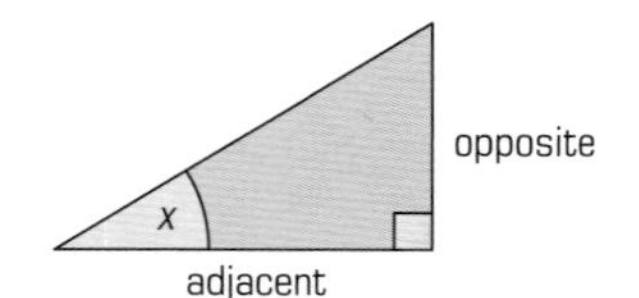

The side labelled '**opposite**' is opposite $\angle x$.
The side labelled '**adjacent**' is adjacent to $\angle x$.

The side labelled '**opposite**' is opposite $\angle y$.
The side labelled '**adjacent**' is adjacent to $\angle y$.

Hint: Adjacent means 'next to'.

- For any given angle, each pair of sides is always in the same ratio.

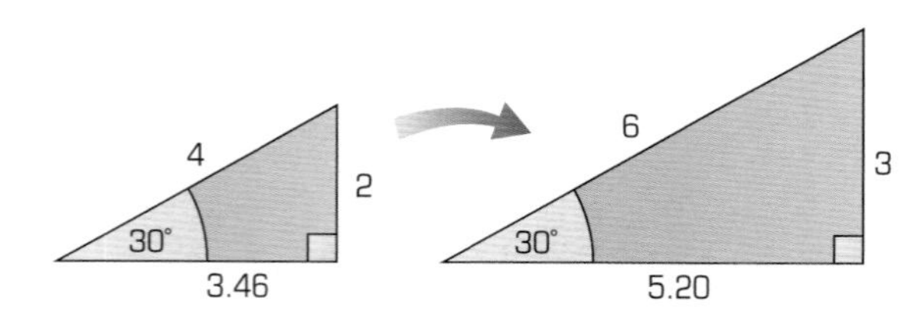

For each of these triangles you can work out three ratios:

$\frac{\text{opposite}}{\text{hypotenuse}} = \frac{2}{4} = 0.5$ $\qquad \frac{\text{opposite}}{\text{hypotenuse}} = \frac{3}{6} = 0.5$

$\frac{\text{adjacent}}{\text{hypotenuse}} = \frac{3.46}{4} = 0.87$ $\qquad \frac{\text{adjacent}}{\text{hypotenuse}} = \frac{5.20}{6} = 0.87$

$\frac{\text{opposite}}{\text{adjacent}} = \frac{2}{3.46} = 0.58$ $\qquad \frac{\text{opposite}}{\text{adjacent}} = \frac{3}{5.2} = 0.58$

Notice that the **ratios of lengths** are the same.
This is because the **angles** are the same in both triangles.

- These ratios are important and have special names:
  - sine $x = \frac{\text{opposite}}{\text{hypotenuse}}$ cosine $x = \frac{\text{adjacent}}{\text{hypotenuse}}$ tangent $x = \frac{\text{opposite}}{\text{adjacent}}$

  You can abbreviate the ratios to:
  - $\sin x = \frac{\text{opp}}{\text{hyp}}$ $\cos x = \frac{\text{adj}}{\text{hyp}}$ $\tan x = \frac{\text{opp}}{\text{adj}}$

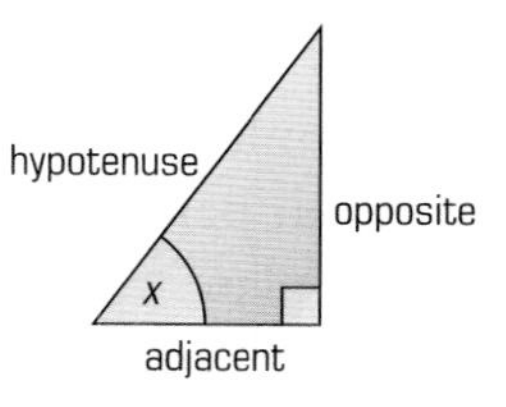

You can find the sine, cosine and tangent of any angle.
The branch of mathematics that deals with these ratios is called **trigonometry**.

Trigonometry literally means 'triangle measure'.

Exercise S4.4

1 Copy the diagrams and label the sides of the triangles hypotenuse, adjacent and opposite relative to the angle shown.

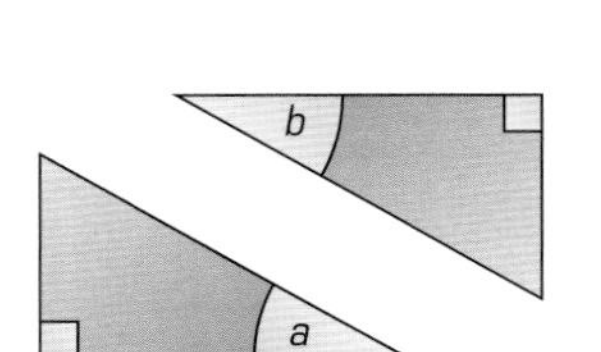

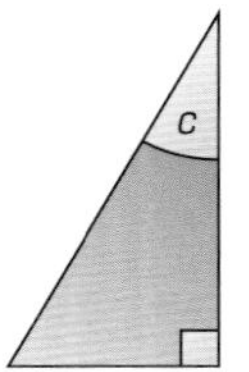

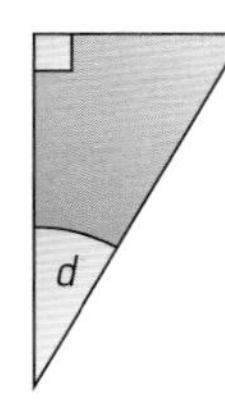

2 Using constructions, draw right-angled triangles with sides:

a 5 cm, 4 cm, 3 cm

b 7.5 cm, 6 cm, 4.5 cm

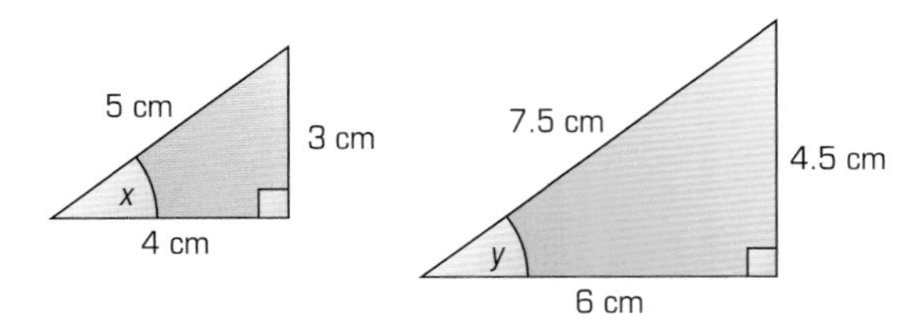

c Measure the size of angles x and y.

d Label each side opp, hyp, adj relative to x and y.

e Work out these ratios.

$\sin x = \frac{\text{opp}}{\text{hyp}}$ $\sin y = \frac{\text{opp}}{\text{hyp}}$

$\cos x = \frac{\text{adj}}{\text{hyp}}$ $\cos y = \frac{\text{adj}}{\text{hyp}}$

$\tan x = \frac{\text{opp}}{\text{adj}}$ $\tan y = \frac{\text{opp}}{\text{adj}}$

Write down anything you notice.

3 Write down the ratios:

a $\sin y$

b $\cos y$

c $\tan y$

for this triangle.

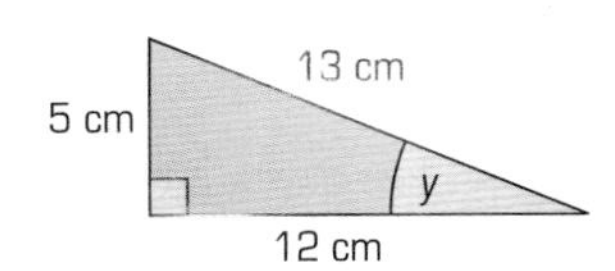

4 PQR is an equilateral triangle.

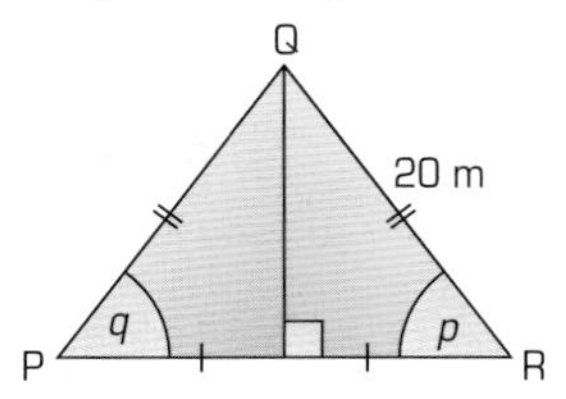

a Write down the ratios $\cos p$ and $\cos q$.

b How big are angles p and q?

c What is $\cos 60°$?

5 Here are two triangles:

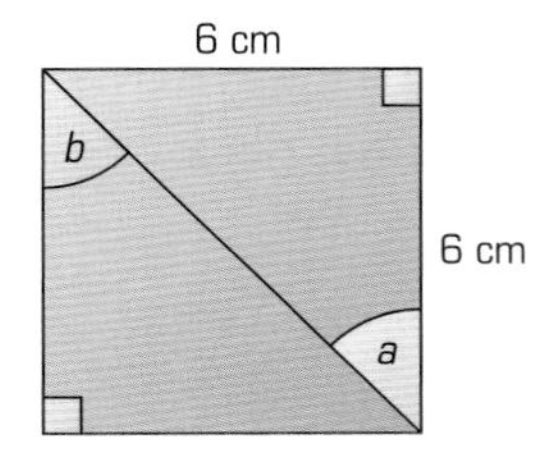

a Write down the ratios $\tan a$ and $\tan b$.

b Find angles a and b.

c What is $\tan 45°$?

6 You can find sin 40° on a calculator by pressing either [4] [0] [sin] or [sin] [4] [0]

Check which gives 0.642 78 on your calculator.

Use your calculator to find these ratios.

> **Hint:** Make sure your calculator is in degree mode.

a sin 30°	**b** sin 60°	**c** sin 90°
d sin 0°	**e** cos 30°	**f** cos 60°
g cos 90°	**h** cos 0°	**i** tan 0°
j tan 45°	**k** tan 30°	**l** tan 60°

7 If $\cos 60° = \frac{1}{2}$ what could x and y measure on this triangle?

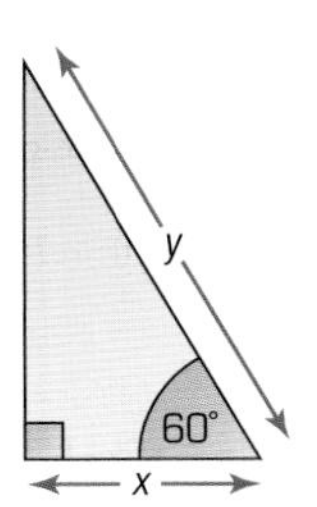

Using trigonometry

This spread will show you how to:

- Begin to use sine, cosine and tangent in right-angled triangles to solve problems in two dimensions.

KEYWORDS

Sine	Opposite
Cosine	Adjacent
Tangent	Hypotenuse

Remember:

- $\sin x = \frac{\text{opp}}{\text{hyp}}$ $\cos x = \frac{\text{adj}}{\text{hyp}}$ $\tan x = \frac{\text{opp}}{\text{adj}}$

SOH CAH TOA

'Six Old Horses Clumsy And Heavy Trod On Andrew.'

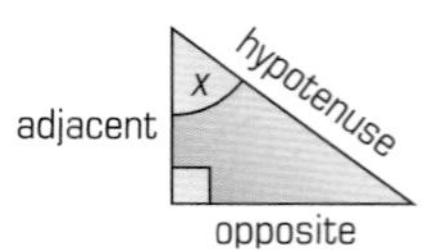

You can use sine, cosine and tangent to calculate lengths in triangles.

example

Find the length AB in the triangle ABC.
Give your answer to 1 decimal place.

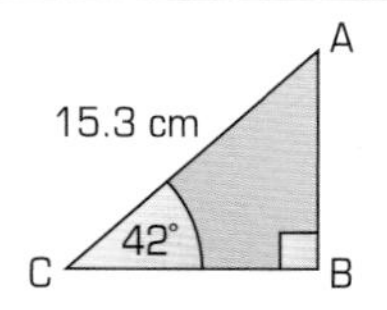

- Label the sides of the triangle.

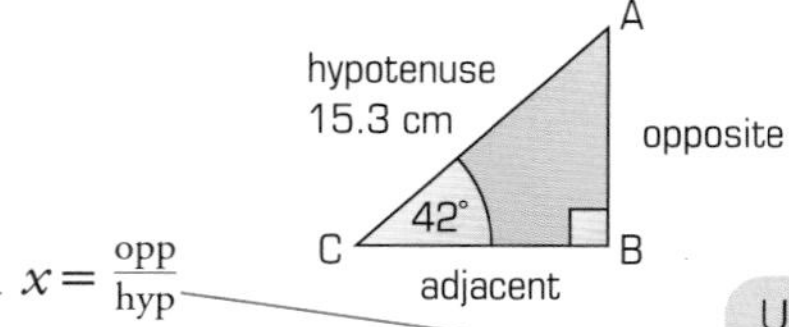

The opposite side is opposite to the angle that you are given.

- Decide which ratio to use. $\sin x = \frac{\text{opp}}{\text{hyp}}$
- Substitute the values given. $\sin 42° = \frac{\text{AB}}{15.3}$
- Multiply both sides by 15.3. $\text{AB} = 15.3 \times \sin 42°$
- Use your calculator. $\text{AB} = 10.2376 \ldots$

$\text{AB} = 10.2 \text{ cm}$ (to 1 dp)

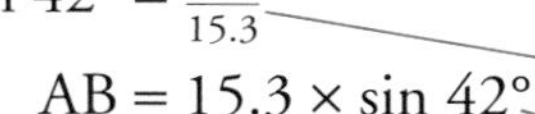

Use the ratio that contains:
- the length you are looking for
- and the length you know.

$15.3 \times \sin 42° = \frac{\text{AB}}{\cancel{15.3}} \times \cancel{15.3}$

You can also use sine, cosine and tangent to find angles in triangles.

example

Find angle P in the triangle PQR.
Give your answer to the nearest degree.

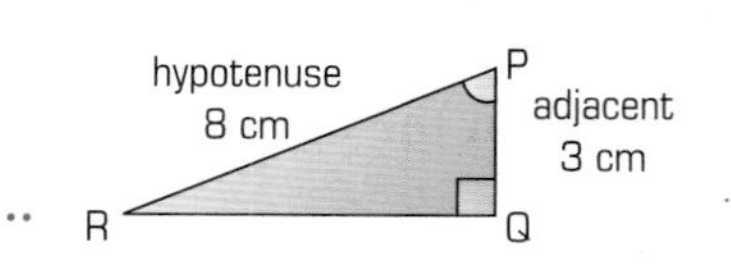

- Label the sides of the triangle.
- Decide which ratio to use. $\cos \text{P} = \frac{\text{adj}}{\text{hyp}}$
- Substitute the values given. $\cos \text{P} = \frac{3}{8} = 0.375$

$\text{P} = \cos^{-1}(0.375)$

- Use your calculator. $\angle \text{P} = 67.9756 \ldots°$

$= 68°$ (to nearest degree)

Use the ratio that contains **both** the lengths that you know.

Finding the angle depends on the keys on your calculator. $\boxed{\cos^{-1}}$ $\boxed{\sin^{-1}}$ $\boxed{\tan^{-1}}$ are used to find angles.
Here is one possibility: $\boxed{\text{shift}}$ $\boxed{\cos}$

Exercise S4.5

1. Use the sine ratio to find the unknown angles in these diagrams.

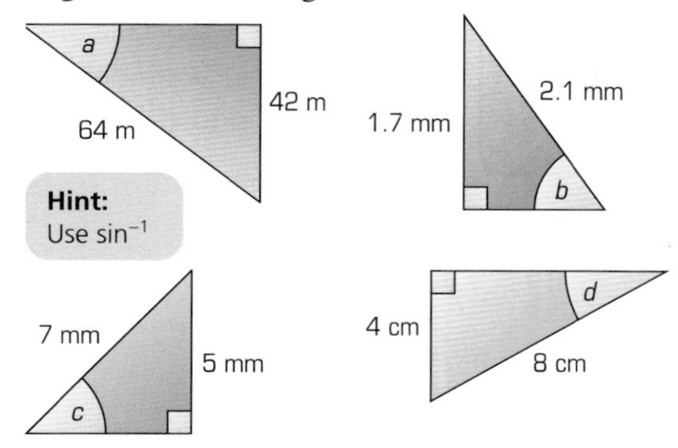

2. Use the cosine ratio to find the unknown angles in these diagrams.

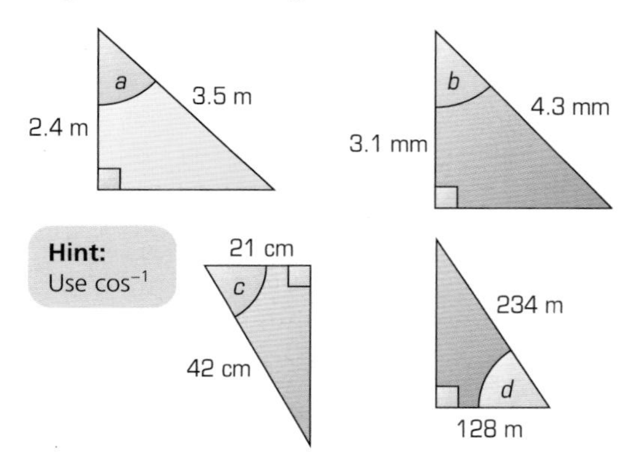

3. Use the tangent ratio to find the unknown angles in these diagrams.

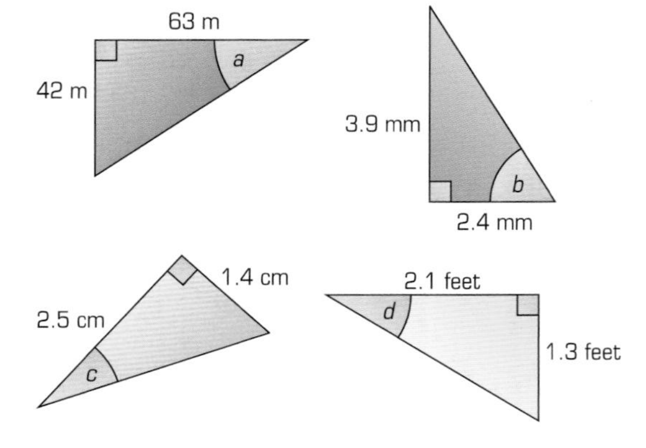

4. Use the appropriate ratio to find the unknown angles in these diagrams.

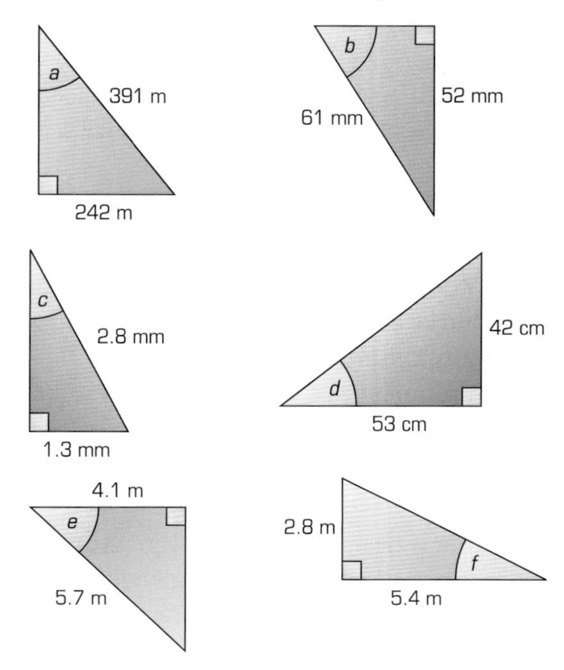

5. Find the angles between the diagonal and the shorter side of a rectangle with sides 5.7 mm and 10.3 mm.

6. Find the angle at the base of this isosceles triangle.

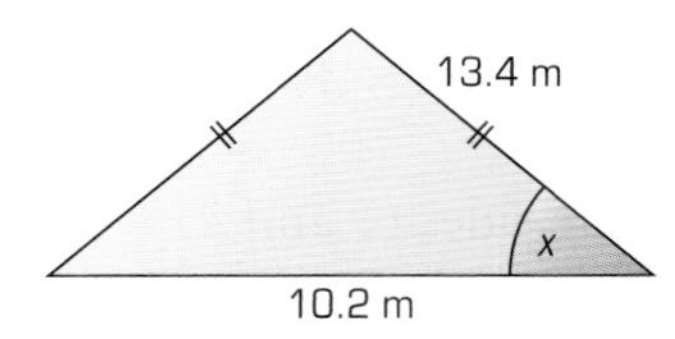

7. Find the angles between the sides at the base of an isosceles triangle with sides 20 mm, 20 mm and 15 mm.

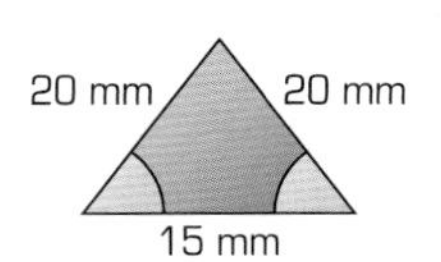

S4.6 Surface area and volume of a prism

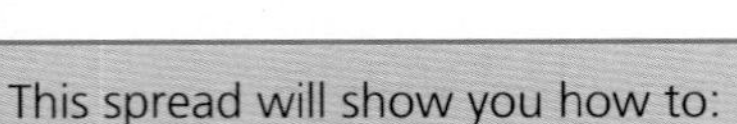

This spread will show you how to:

- Calculate lengths, areas and volumes in right prisms, including cylinders.
- Recognise limitations on the accuracy of measurements.

KEYWORDS

Volume, Surface area, Prism, Cylinder, Significant figures (sf)

Remember:

- For any prism:
  - Volume = area of cross-section × length
  - Surface area = total area of all its faces
- For any cylinder:
  - Volume $= \pi r^2 \times h$ or $\pi r^2 h$
  - Surface area $= 2\pi r^2 + 2\pi rh$

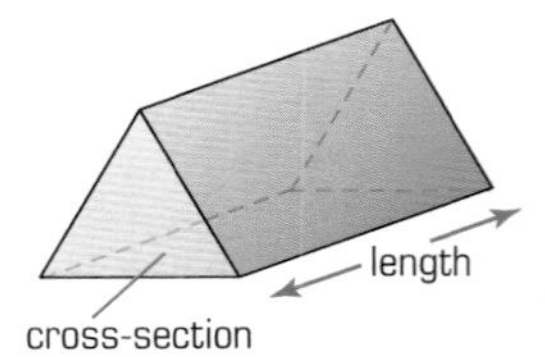

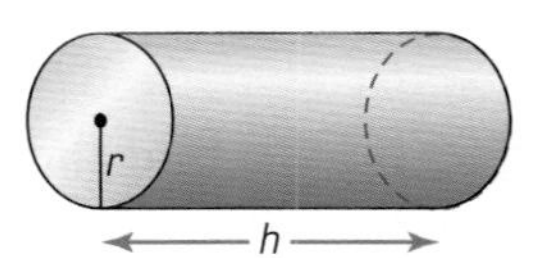

example

Maxine is designing a bread bin with the dimensions shown.
The cross-section is a rectangle with a quarter circle on top.

Calculate (to 3 sf):

a the volume of the bread bin

b the surface area of the bread bin.

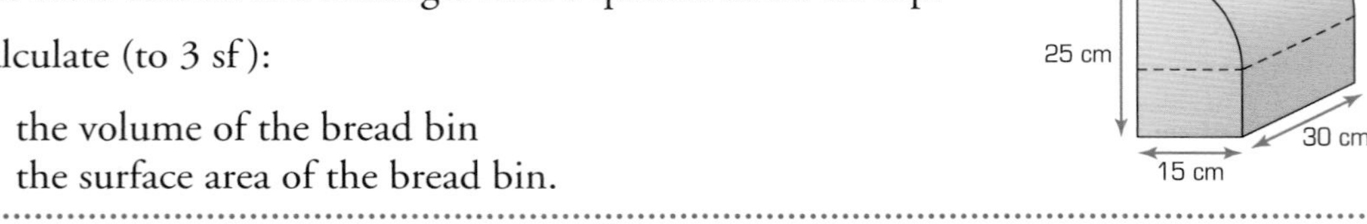

a Area A = 15 cm × 10 cm = 150 cm^2

Area B $= \frac{1}{4} \times \pi \times 15^2$

$= 176.714\,58$ cm^2

Total cross-section area = 150 + 176.714 58

= 326.714 58 cm^2

Volume = 326.714 58 × 30 = 9801.44 ... cm^3

= 9800 cm^3 (to 3 sf)

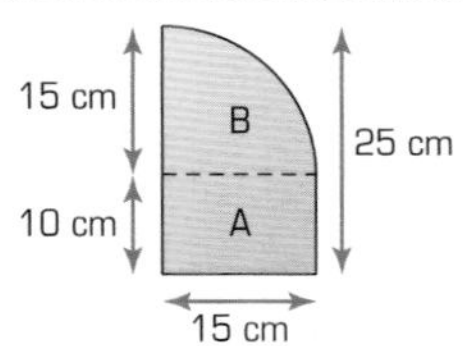

b Total area of the sides (1 and 2) = 326.714 58 × 2

= 653.429 16 cm^2

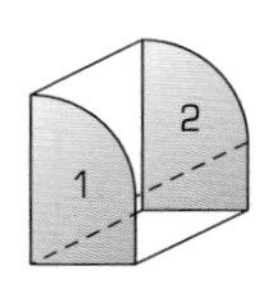

Area of the front, base and back (3, 4 and 5) = (10 × 30) + (15 × 30) + (25 × 30)

= 1500 cm^2

The curved top (6) is a quarter of the curved surface of a cylinder.

Area of curved surface of a cylinder $= 2\pi rh$

So area of curved top $= \frac{1}{4} \times 2 \times \pi \times 15 \times 30 = 706.86$ ...

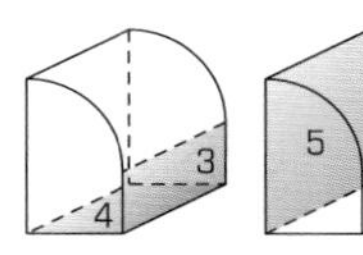

Total surface area (1, 2, 3, 4, 5 and 6) = 653.429 16 + 1500 + 706.86 ...

= 2860.2891 = 2860 cm^2 (to 3 sf)

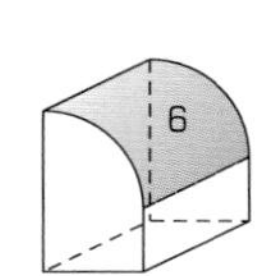

Exercise S4.6

1 Find the volume and the surface area of these solids.

a

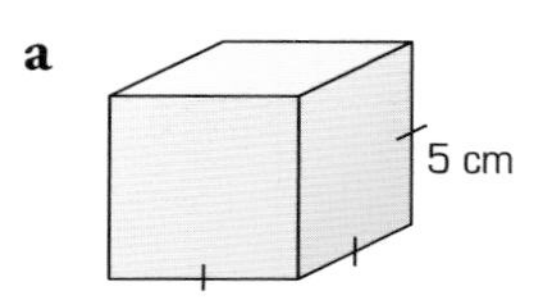

b

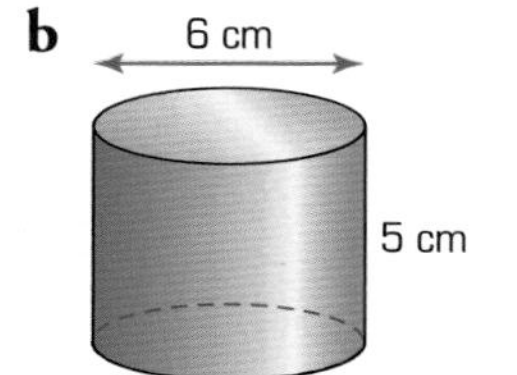

c

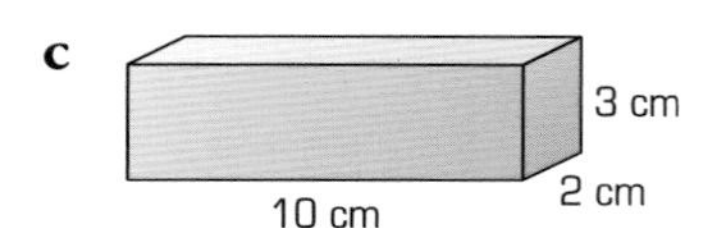

d

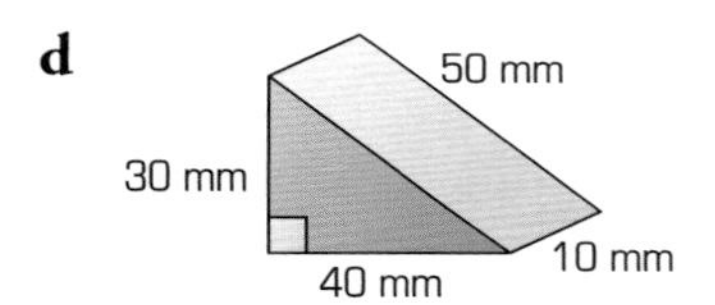

e

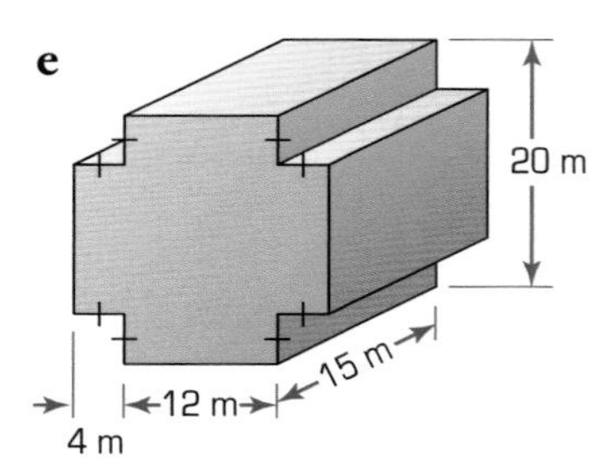

2 Here is a sketch of a swimming pool. Calculate the volume of water that the pool holds.

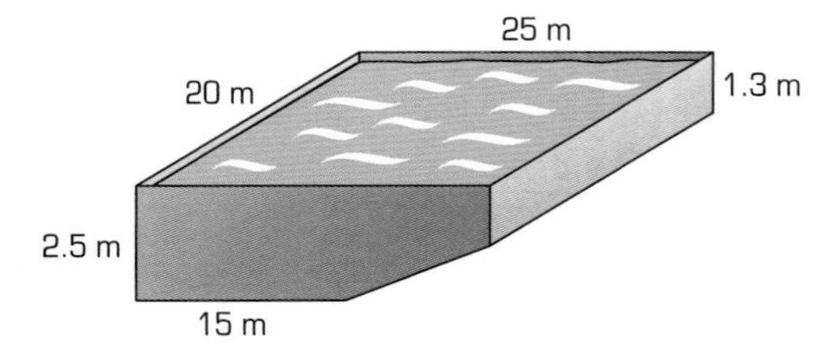

3 Find the surface area of this triangular prism (you will need to use Pythagoras' theorem to calculate the unknown length).

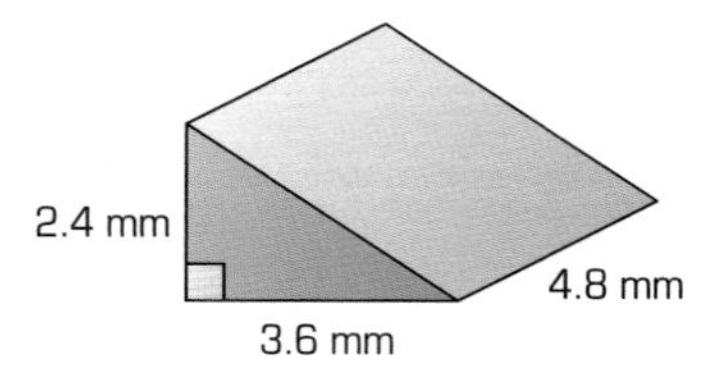

4 Find the volume of this triangular prism (you will need to use trigonometry to calculate the unknown lengths *a* and *b*).

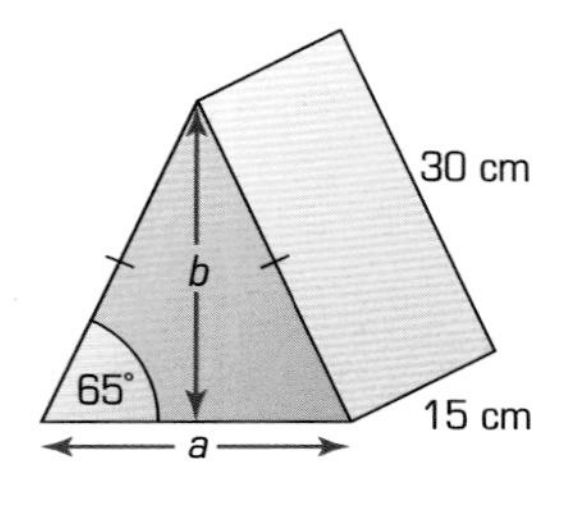

5 Problem solving

A cylinder has a diameter of 7.5 cm and a height of 10.5 cm. Find the diameter and height of some other cylinders with the same volume. Which of these cylinders has the smallest surface area?

You may find a spreadsheet useful here.

6 Large cylinders of sweets are made as shown here: They are packed neatly into individual boxes into which they just fit. What percentage of space in each box is wasted (i.e. only filled with air)?

If the dimensions of the cylinder and its box are doubled, is the wasted space doubled? Explain your answer.

Summary

You should know how to ...

1 Understand and apply Pythagoras' theorem.

2 Recognise limitations on the accuracy of measurements.

3 Generate fuller solutions to problems.

Check out

1 Find the unknown sides and angles.

a

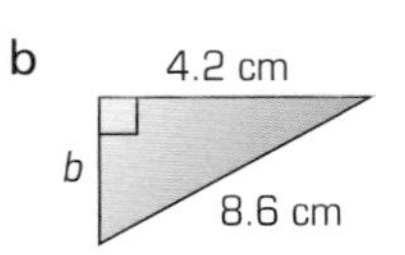

b

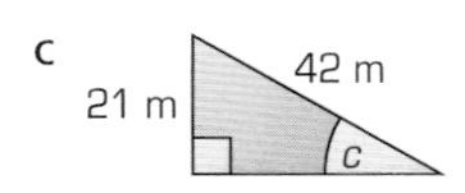

c

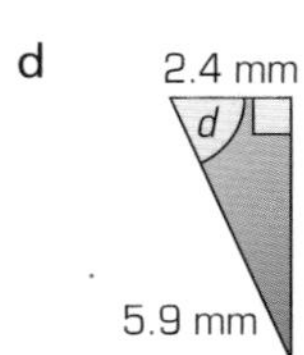

d

2.4 mm

d

5.9 mm

2 The measurements of this biscuit tin are given to the nearest centimetre.
Calculate its maximum and minimum possible volumes.

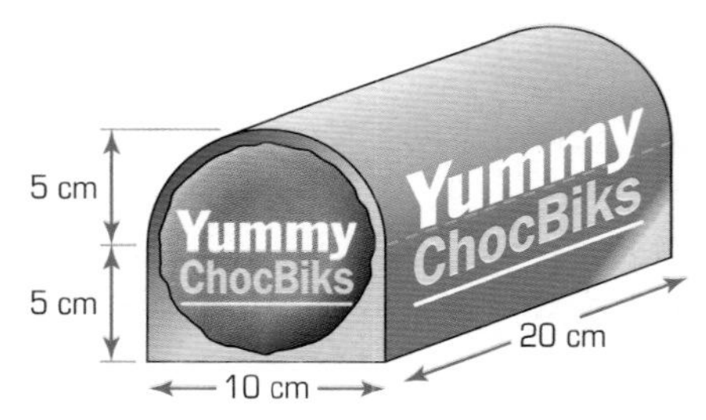

3 The surface area of a cylinder is 50 cm^2.
Find its volume if its radius is equal to its height.

4 Theoretical and experimental probability

This unit will show you how to:

- Use the vocabulary of probability in interpreting results involving uncertainty and prediction.
- Identify all the mutually exclusive outcomes of an experiment.
- Know that the sum of all mutually exclusive outcomes is 1 and use this when solving problems.
- Compare experimental and theoretical probabilities in a range of contexts.
- Estimate probabilities from experimental data.
- Understand relative frequency as an estimate of probability and use this to compare outcomes of experiments.
- Appreciate the difference between mathematical explanation and experimental evidence.
- Generate fuller solutions to problems.
- Recognise limitations on accuracy of data and measurements.
- Pose extra constraints and investigate whether particular cases can be generalised further.

You can work out strategies for lots of games using probability.

Before you start

You should know how to ...

1 Convert between percentages, decimals and fractions.

2 Draw a tree diagram.

Check in

1 Copy and complete the table.

Fraction	$\frac{2}{5}$			
Decimal			0.35	
Percentage		99		8

2 A dice and a coin are thrown.
Draw a tree diagram to represent all possible outcomes.

D4.1 Probability ideas

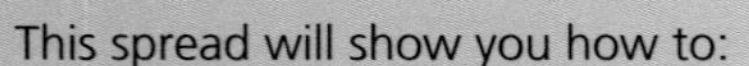

This spread will show you how to:

- Use the vocabulary of probability in interpreting results involving uncertainty and prediction.
- Identify all the mutually exclusive outcomes of an experiment.
- Know that the sum of probabilities of all mutually exclusive outcomes is 1 and use this when solving problems.

KEYWORDS

Event
Outcome
Independent
Mutually exclusive
Exhaustive

These events relate to this spinner being spun:

R = you spin a red shape
Y = you spin a yellow shape
G = you spin a green shape
C = you spin a circle
S = you spin a square

Remember:

- Events are **exhaustive** if one of them is bound to occur. Events C and S are exhaustive.
- Events are **mutually exclusive** if they cannot occur at the same time. Events R, Y and G are mutually exclusive.

example

Joely spins the spinner once.
Find the probability of it landing on either the red or the green shape.

$p(R) = \frac{1}{2}$ and $p(G) = \frac{1}{3}$ so $p(R \text{ or } G) = \frac{1}{2} + \frac{1}{3} = \frac{3}{6} + \frac{2}{6} = \frac{5}{6}$

The probability of either red or green is $\frac{5}{6}$.

Remember:

- You add the probabilities of mutually exclusive events.

The **OR** rule:
Add probabilities.

- Events are independent if the outcome of one event does not affect the outcome of the other.

example

Joely spins the spinner twice.
Find the probability that she gets red both times.

When the spinner is spun twice, the outcomes of each event are **independent**.

So $p(R) = \frac{1}{2} \Longrightarrow \quad p(R \text{ and } R) = \frac{1}{2} \times \frac{1}{2} = \frac{1}{4}$

The probability that Joely gets red both times is $\frac{1}{4}$.

- You multiply the probabilities of independent events.

The **AND** rule:
Multiply probabilities.

Exercise D4.1

1 Work out the probabilities when Joely's spinner is spun twice.

- **a** Probability of yellow and green in that order.
- **b** Probability of yellow and green (any order).
- **c** Probability of red and green.
- **d** Probability that the two colours are the same.
- **e** Probability that the two colours are different.

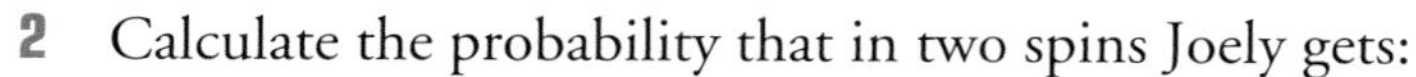

2 Calculate the probability that in two spins Joely gets:

- **a** two squares
- **b** the same shape each time.

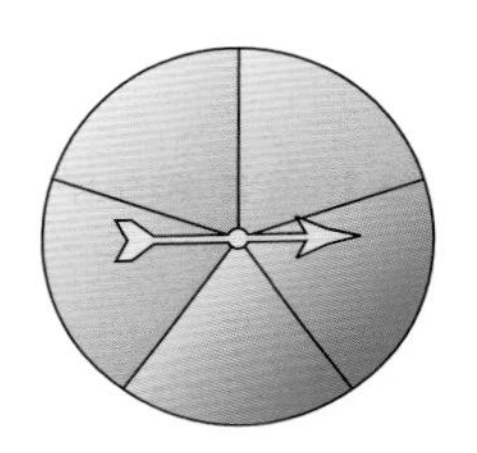

3 Kieran spins this spinner twice.
Calculate the probability that Kieran gets:

- **a** two blues
- **b** two different colours.

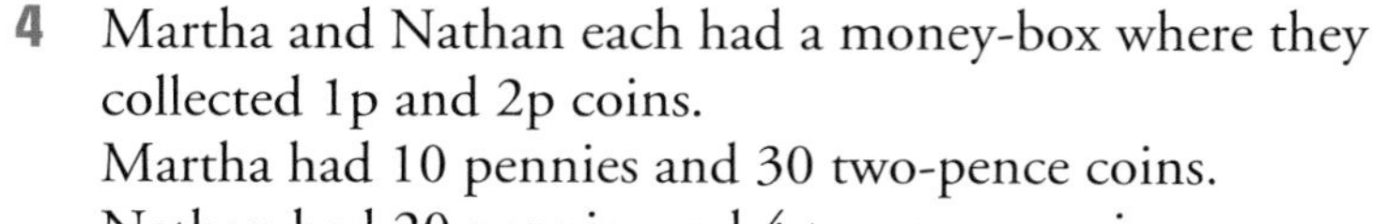

4 Martha and Nathan each had a money-box where they collected 1p and 2p coins.
Martha had 10 pennies and 30 two-pence coins.
Nathan had 20 pennies and 4 two-pence coins.
Martha and Nathan each chose a coin from their money-box.
Calculate the probability that they both chose:

- **a** pennies
- **b** the same value coin.
- **c** What is the probability that the total value of the coins was 3p?

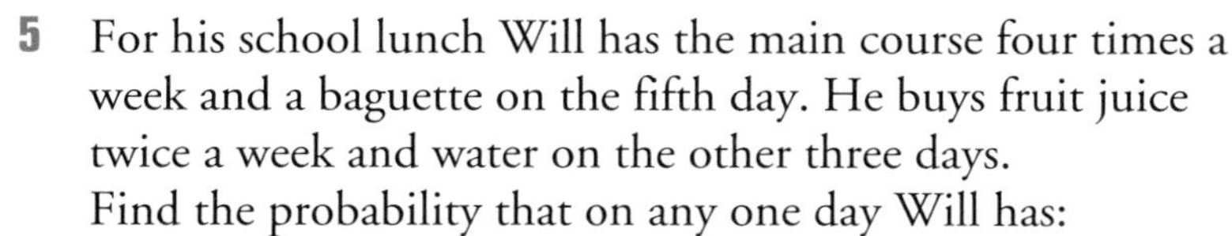

5 For his school lunch Will has the main course four times a week and a baguette on the fifth day. He buys fruit juice twice a week and water on the other three days.
Find the probability that on any one day Will has:

- **a** the main course and fruit juice
- **b** a baguette and fruit juice.

6 A spinner has 10 equal sections, 5 are red, 3 are green and 2 are blue.
The spinner is spun twice.
Find the probability that the spinner lands on the same colour both times.

D4.2 Sampling with and without replacement

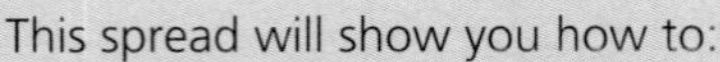

This spread will show you how to:

- Identify all the mutually exclusive outcomes of an experiment.
- Know that the sum of probabilities of all mutually exclusive outcomes is 1 and use this when solving problems.

KEYWORDS
Tree diagram
Sampling

Reuben has a bag of marbles, of which six are blue and four are green.
He wants to choose two marbles.

Reuben picks a marble at random, replaces it and then picks a second marble.

This is called **sampling with replacement**.

▶ You can use a **tree diagram** to illustrate combined events.

This tree diagram shows all the different possibilities of selecting marbles.

Write the probabilities as fractions.

first choice: B $\frac{6}{10}$, G $\frac{4}{10}$
second choice: after B: B $\frac{6}{10}$, G $\frac{4}{10}$; after G: B $\frac{6}{10}$, G $\frac{4}{10}$

(B, B)
(B, G)
(G, B)
(G, G)

The probability that Reuben picks two blue marbles:
$p(B, B) = \frac{6}{10} \times \frac{6}{10} = \frac{36}{100}$

The probability that Reuben picks two green marbles:
$p(G, G) = \frac{4}{10} \times \frac{4}{10} = \frac{16}{100}$

Reuben changes his method of selecting marbles.
Now he picks a marble at random, does **not** replace it and then picks a second marble.

This is called **sampling without replacement**.

If Reuben's first choice is blue, then there will be **five** blue marbles left and **nine** marbles in total.

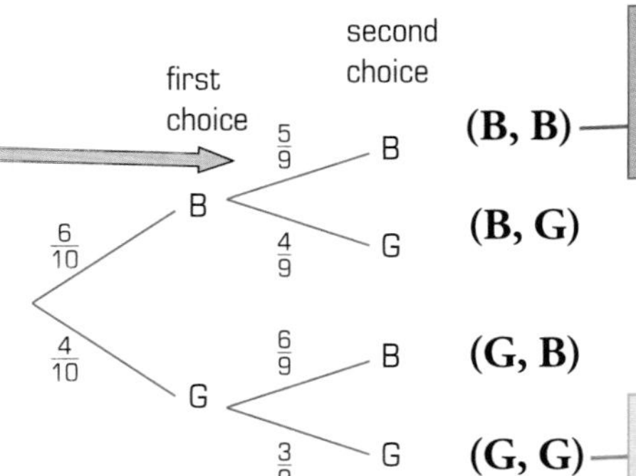

Now the probability that Reuben picks two blue marbles is:
$p(B, B) = \frac{6}{10} \times \frac{5}{9} = \frac{30}{90} = \frac{1}{3}$

Now the probability that Reuben picks two green marbles is:
$p(G, G) = \frac{4}{10} \times \frac{3}{9} = \frac{12}{90} = \frac{2}{15}$

▶ For sampling without replacement, the probability of an outcome changes for the second event.

Exercise D4.2

1 A bag contains 3 red and 5 blue counters.
Alfie chooses two counters.
He draws a tree diagram to show all the possible outcomes.

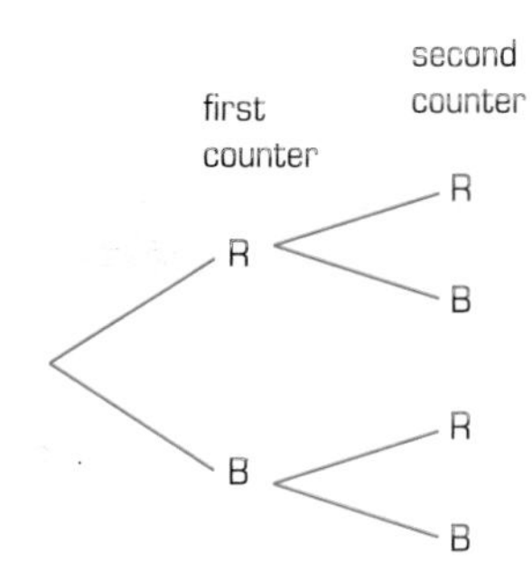

Remember:
Write the probabilities as fractions.

a Copy and complete the tree diagram, filling in all probabilities if the counters are replaced between each choice.
Calculate the probability of choosing two counters of the same colour.

b Copy the diagram and fill in the probabilities if the counters are **not** replaced between each choice.
Calculate the probability of choosing two counters of the same colour.

2 Jenna has a box of chocolates that contain 6 hard centres and 10 soft centres.
Jenna chooses a chocolate, eats it, then chooses a second chocolate.

a Draw a tree diagram to show the probabilities of Jenna choosing hard- and soft-centred chocolates.

b Calculate the probability that Jenna eats:

i two soft centres

ii one soft centre and one hard centre.

3 Roger has a bag of 8 lemon and 12 strawberry sweets.
He chooses two sweets to eat.

a Draw a tree diagram to show the probabilities of choosing lemon and strawberry sweets.

b Find the probability that Roger chooses:

i two strawberry sweets

ii one sweet of each flavour.

4 A bag contains 3 purple, 4 green and 5 blue marbles.
Ria chooses two marbles.

a Draw a tree diagram to show all the possible outcomes.

b Calculate the probability that the marbles Ria chose were:

i both purple

ii both the same colour.

Harder tree diagram problems

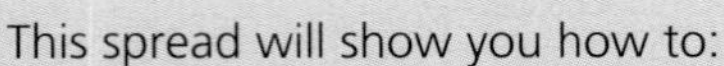

This spread will show you how to:
- Identify all the mutually exclusive outcomes of an experiment.
- Know that the sum of probabilities of all mutually exclusive outcomes is 1 and use this when solving problems.

KEYWORDS
Event
Tree diagram

For any event A,

- p(not A) = 1 − p(A) and p(A) = 1 − p(not A)

They say there's a 40% chance of rain tomorrow.

Look on the bright side – there's a 60% chance it'll be dry!

example

Reuben chooses two marbles without replacement from a bag containing six blue and four green marbles.
Find the probability that Reuben chooses at least one blue marble.

There are three possibilities that contain at least one blue marble.

First draw a tree diagram:

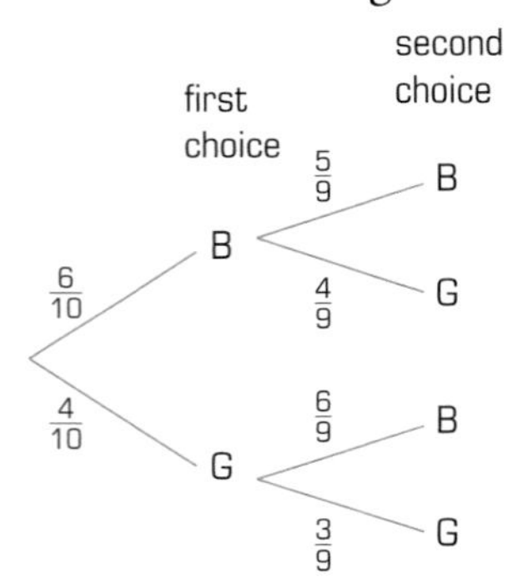

$p(\mathbf{B}, \mathbf{B}) = \frac{6}{10} \times \frac{5}{9} = \frac{30}{90} = \frac{1}{3}$

or

$p(\mathbf{B}, G) = \frac{6}{10} \times \frac{4}{9} = \frac{24}{90} = \frac{4}{15}$

or

$p(G, \mathbf{B}) = \frac{4}{10} \times \frac{6}{9} = \frac{24}{90} = \frac{4}{15}$

$$\begin{aligned} p(\text{at least one blue}) &= p(B, B) + p(B, G) + p(G, B) \\ &= \frac{6}{10} \times \frac{5}{9} + \frac{6}{10} \times \frac{4}{9} + \frac{4}{10} \times \frac{6}{9} \\ &= \frac{30}{90} + \frac{24}{90} + \frac{24}{90} \\ &= \frac{78}{90} \\ &= \frac{13}{15} \end{aligned}$$

Remember:
Or means +

A quicker method is to calculate the probability of **not** choosing a blue marble:

$p(G, G) = \frac{4}{10} \times \frac{3}{9} = \frac{12}{90} = \frac{2}{15}$

$p(\text{at least one blue}) = 1 - p(\text{no blue})$

$= 1 - \frac{2}{15} = \frac{13}{15}$ as before.

p(G, G) = p(no blue)

For a repeated event A,
- p(A occurs at least once) = 1 – p(A does not occur)

Exercise D4.3

1 Look at your diagram from question 1 on page 221.
When two counters are chosen, find the probability that Alfie chooses at least one red counter if:

a the counter is replaced after the first choice

b the counter is **not** replaced after the first choice.

2 Look at your diagram from question 2 on page 221.
When Jenna eats two chocolates, find the probability that she chooses:

a at least one soft-centred chocolate

b at least one hard-centred chocolate.

3 Look at your diagram from question 3 on page 221.
When Roger chooses two sweets, find the probability that he eats:

a at least one strawberry sweet

b at least one lemon sweet.

4 Look at your diagram from question 4 on page 221.
When Ria chooses two marbles, find the probability that she picks:

a at least one green or blue marble

b at least one purple marble.

5 A dice is thrown twice.
Find the probability that:

a at least one of the throws gives a number less than six

b at least one six is thrown.

6 Alison has a small box of sweets.
There are 12 sweets in the box, of which 2 are orange.
Alison will only eat orange sweets.
Alison chooses a sweet. If it is orange she eats it.
If it is not orange she replaces it in the box.

a Copy and complete the tree diagram to show the different probabilities for Alison's first two choices of sweet.

b Calculate the probability that Alison picks at least one orange sweet on her first two choices.

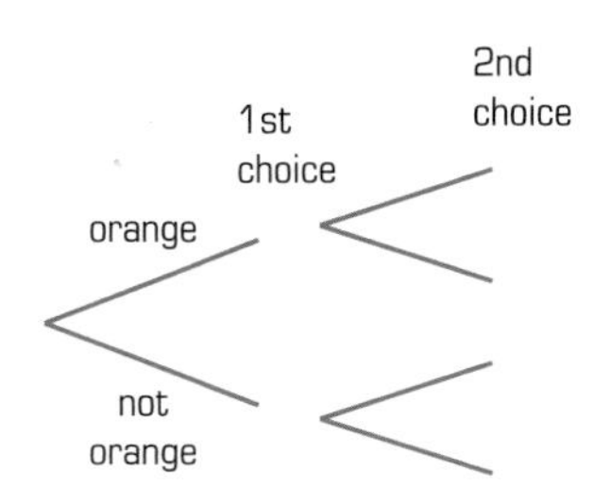

D4.4 More than two events

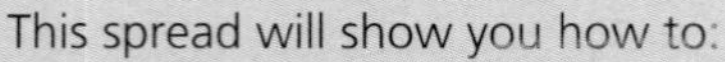

This spread will show you how to:

- Identify all the mutually exclusive outcomes of an experiment.
- Know that the sum of probabilities of all mutually exclusive outcomes is 1 and use this when solving problems.
- Compare experimental and theoretical probabilities.

KEYWORDS

Event

Tree diagram

You can use tree diagrams and all probability ideas for more than two events.

Reuben has a bag of marbles.
Six are blue and four are green.

Reuben chooses three marbles from the bag without replacement.
Find the probability that Reuben chooses two blue marbles and one green marble.

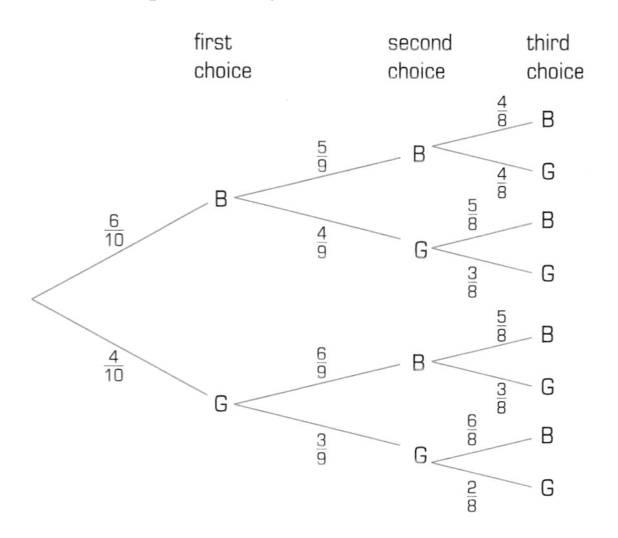

There are three possibilities: B,B,G or B,G,B or G,B,B.

$$
\begin{aligned}
p(B, B, G) + p(B, G, B) + p(G, B, B) &= \left(\tfrac{6}{10} \times \tfrac{5}{9} \times \tfrac{4}{8}\right) + \left(\tfrac{6}{10} \times \tfrac{4}{9} \times \tfrac{5}{8}\right) + \left(\tfrac{4}{10} \times \tfrac{6}{9} \times \tfrac{5}{8}\right) \\
&= 3 \times \left(\tfrac{6}{10} \times \tfrac{5}{9} \times \tfrac{4}{8}\right) \\
&= \tfrac{1}{2}
\end{aligned}
$$

For situations involving more than three events, a tree diagram can become too large. It may be better just to list the possibilities.

Exercise D4.4

1 A bag contains five toffees and seven chocolates.
Sara chooses three.

a Draw a tree diagram to show all the possible outcomes.

b Calculate the probability that Sara chooses:

i three toffees **ii** at least one chocolate

iii two toffees and one chocolate.

2 A bag contains nine orange and three lemon sweets.

a Draw a tree diagram to show the outcomes when three sweets are chosen.

b Calculate the probability that:

i all the sweets are orange

ii all the sweets are the same colour

iii two lemon sweets and one orange sweet are chosen.

3 In the game of dominoes there are 28 pieces, seven of which are 'doubles'.
Tom chooses three dominoes.

a Draw a tree diagram to show the possible outcomes of Tom choosing a 'double'.

b Calculate the probability that Tom chooses three 'doubles'.

c Calculate the probability that Tom chooses all seven 'doubles' when he chooses seven dominoes.

4 A coin is spun three times.

a Draw a tree diagram to show all the possible outcomes.

b Calculate the probability of the coin landing on:

i heads three times

ii two heads and a tail

iii one head and two tails

iv no heads.

c This is repeated 40 times.
Use your results from part **b** to calculate the expected number of times you would get:

i 3 heads **ii** 2 heads **iii** 1 head **iv** no head.

d **Experiment**
Spin a coin three times, and write down the number of heads in the three spins.
Repeat so that you have a total of 40 results.
Compare your experimental results with the expected results from part **c**.
Comment on what you find.

D4.5 Estimating probability

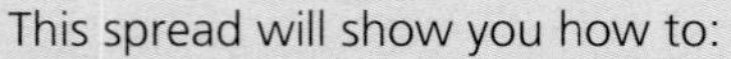

This spread will show you how to:

- Estimate probabilities from experimental data.
- Understand relative frequency as an estimate of probability and use this to compare outcomes of experiments.

KEYWORDS
Relative frequency
Experiment

Ursula is making a game to play at a children's party.
The object of the game is to push a coin to land inside a circle.

Ursula draws a circle on a large piece of paper.
She places the paper on a table.

Two players each place a coin partly over the edge of the table and push their coin over the paper, using their palms.

Whoever gets their coin completely inside the circle wins.
If neither coin lands in the circle, or if both do, they play the game again.

Ursula wonders what coins to use.

She decides to use two-pence coins because they have a large diameter (25 mm).

Ursula knows that the circle needs to be bigger than 25 mm.

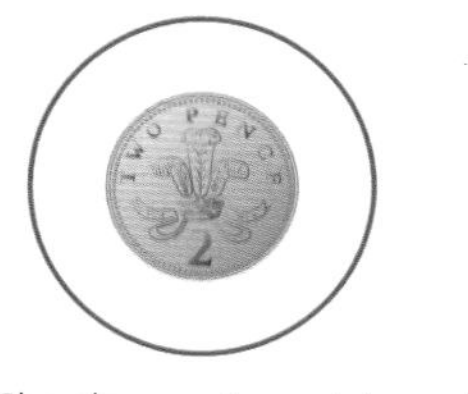

She does not want to make it too easy nor too hard.

Ursula decides to carry out an experiment to find the relative frequency of a coin ending inside a circle.

Exercise D4.5

Help Ursula estimate the probability of a coin landing inside a circle.

Decide:

- The size of coin you are going to use.
  Measure its diameter.
- The size of circle you are going to use in your experiment.
- How far from the edge of the paper you are going to draw the circle.
  Draw the circle.
- How many trials you are going to do.

Remember:

- Be methodical.
- Design and use a data collection sheet.

Carry out the experiment.

Use relative frequency to estimate the probability of a coin landing in the circle.

Remember: relative frequency $= \frac{\text{number of successful trials}}{\text{total number of trials}}$

Repeat the experiment and find a better estimate of the probability.

Compare the outcomes of your experiment with the outcomes of the experiment from other members of your class.

D4.6 Probability investigations

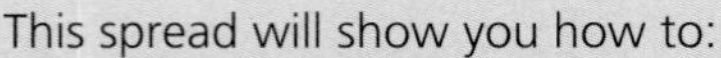

This spread will show you how to:

- Compare experimental and theoretical probabilities in a range of contexts.
- Appreciate the difference between mathematical explanation and experimental evidence.

KEYWORDS
Probability
Relative frequency
Estimate

Ursula is asked to use her game on a stall at a charity fair.

She wants her stall to make a profit to give to charity.

Ursula needs to draw circles large enough to encourage people to play, but not so large that players would win every time.

Ursula thinks that more children would play if ...

they win a sweet if their coin lies partly over the circumference of the circle

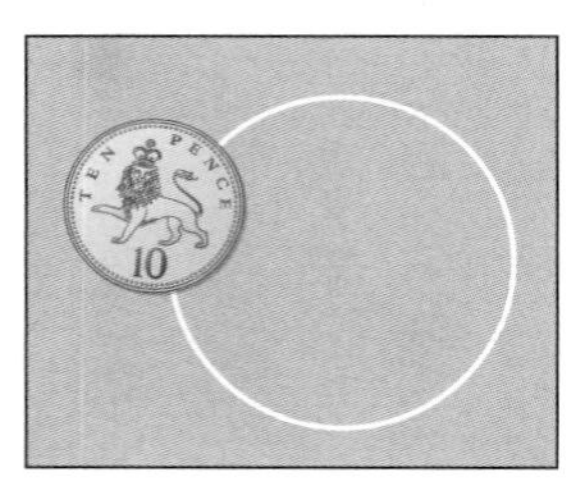

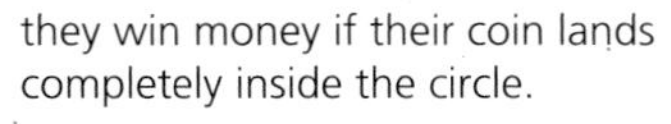

they win money if their coin lands completely inside the circle.

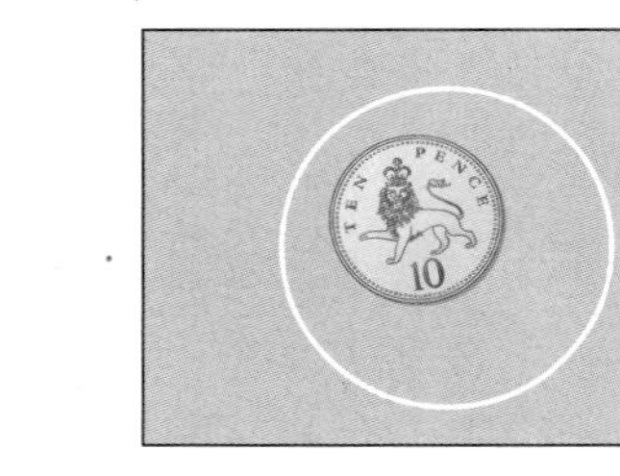

Ursula uses relative frequency to estimate the probabilities of a coin:

- landing completely inside a circle
- partly covering the circumference, for differently sized circles.

Ursula asks her friends to play the game 40 times using a 10p coin and circle diameter 40 mm.
The money prize is £1.
The sweets cost one penny.
It costs 10p to play the game (the 10p coin that you use!)

Coins landing inside the circle	4
Coins partly covering circumference	12

Here are their results:
So p(win money) = $\frac{4}{40} = \frac{1}{10}$ p(win sweet) = $\frac{12}{40} = \frac{3}{10}$

For the 40 games played by Ursula's friends, Ursula receives 40×10p = £4
But four friends win £1 each and 12 receive penny sweets.
Ursula makes a loss of 12p!

Exercise D4.6

Help Ursula investigate how she can make a profit for charity!

You could:

- ▶ Change the coin used to play.
- ▶ Change the size of the circle.
- ▶ Increase the number of trials.
- ▶ Change the amount it cost to play the game.
- ▶ Change the prize money.
- ▶ Design your own investigation to help Ursula.

Remember:

- ▶ Only change one thing at a time.
- ▶ Be methodical.
- ▶ Design and use a table for your results.

D4 Summary

You should know how to ...

1 Generate fuller solutions to problems.

Check out

1 A bag is said to contain an equal number of 2 pence and 10 pence coins.

Each student in a class is asked to choose two coins at random, note the total value of their coins and replace them.
The total value could be 4 pence, 12 pence or 20 pence.

You are investigating whether or not the bag does in fact contain equal amounts of 2 pence and 10 pence coins.

Describe an experiment you could carry out to test this investigation.
(Assume that there are too many coins to count.)

2 Recognise limitations on accuracy of data and measurements.

2 Describe any bias and limitations that may occur if you collect data for your experiment in question 1.

3 Know that the sum of all mutually exclusive outcomes is 1 and use this when solving problems.

3 A fair dice is thrown twice.
Find the probability that at least one square number is thown.

31 Trigonometry and Pythagoras' theorem

This unit will show you how to:

- Understand and apply Pythagoras' theorem.
- Begin to use sine, cosine and tangent in right-angled triangles to solve problems in two dimensions.
- Use bearings to specify direction.
- Link a graphical representation of an equation to the algebraic solution.
- Generate fuller solutions to problems.
- Recognise limitations on accuracy of data and measurements.
- Represent problems and synthesise information in geometric and graphical form.
- Move from one form to another to gain a different perspective on the problem.

Ships' crews navigate across huge distances using bearings.

Before you start

You should know how to ...

1 Use angle facts.

2 Find the volume of a cylinder.

Check in

1 Find the angles marked x.

a

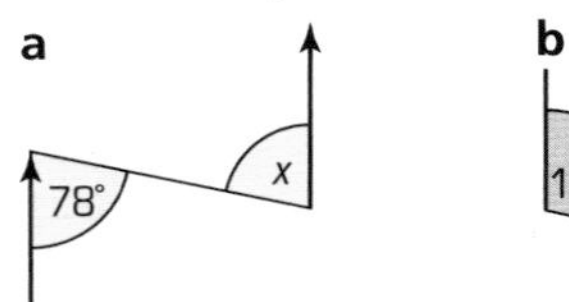

b

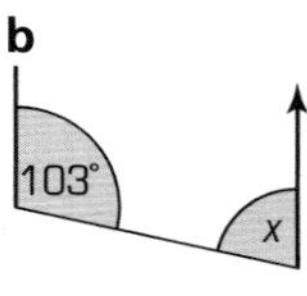

2 Calculate the volume of a cylinder with base radius 6 cm and height 10 cm.

B1.1 Pythagorean triples

This spread will show you how to:

- Understand and apply Pythagoras' theorem.

KEYWORDS
Integer
Hypotenuse
Pythagoras' theorem
Pythagorean triple

You can easily construct a triangle with integer sides.

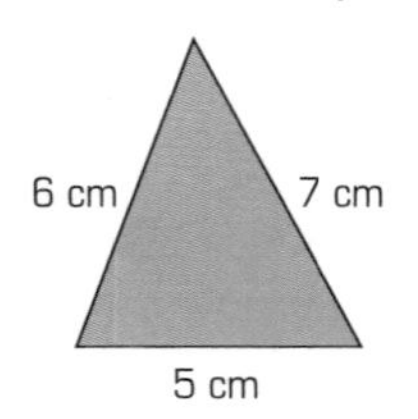

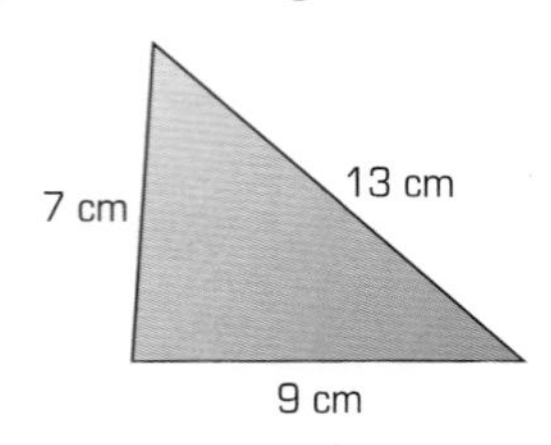

The sum of the two shorter sides must be greater than the longest side.

A triangle measuring 1 cm, 2 cm and 10 cm is impossible.

However there are fewer right-angled triangles with integer sides.

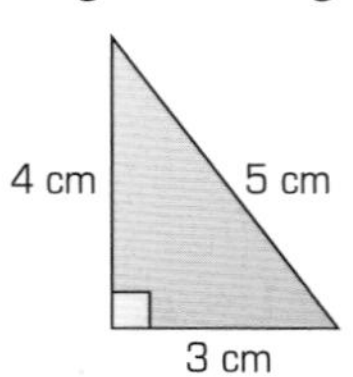

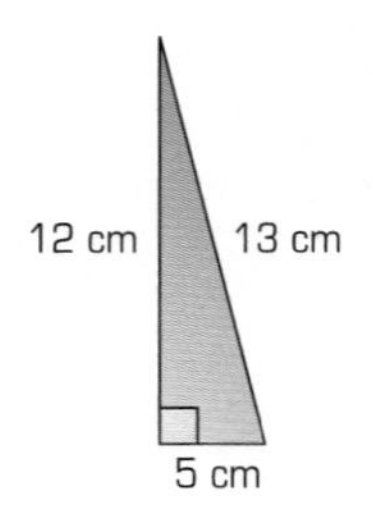

- A **Pythagorean triple** is a set of three integers (a, b, c) that form the measurements of a right-angled triangle.
- The numbers in a Pythagorean triple must obey Pythagoras' theorem: $a^2 = b^2 + c^2$, where a represents the hypotenuse.

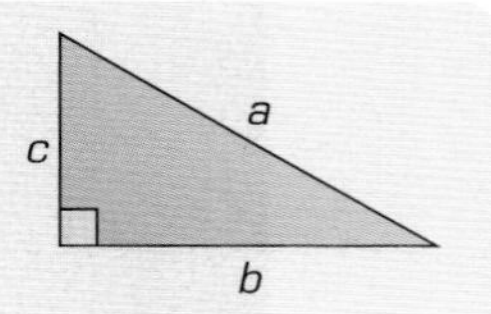

(3, 4, 5) and (5, 12, 13) are Pythagorean triples.

$3^2 + 4^2 = 5^2$ $\qquad$ $5^2 + 12^2 = 13^2$

There is a simple rule that helps you find Pythagorean triples.

1 Choose two positive integers m and n ($m > n$).

2 The three numbers (a, b, c) are found by:

- Multiply m and n and double the answer. $(b = 2mn)$
- Square both m and n and find their difference. $(c = m^2 - n^2)$
- Square both m and n and find their sum. $(a = m^2 + n^2)$

Katie chooses the numbers 3 and 8.

She writes:

$3 \times 8 = 24$ $\qquad$ $2 \times 24 = 48$

$3^2 = 9$ $\qquad$ $8^2 = 64$ $\qquad$ $64 - 9 = 55$

$3^2 = 9$ $\qquad$ $8^2 = 64$ $\qquad$ $64 + 9 = 73$

The numbers 48, 55 and 73 form a Pythagorean triple.

You can check Katie's numbers:

$48^2 = 2304$
$55^2 = 3025$
$73^2 = 5329$

$a^2 + b^2 = c^2$
$2304 + 3025 = 5329$

Exercise B1.1

1 Here is a Pythagorean triple.

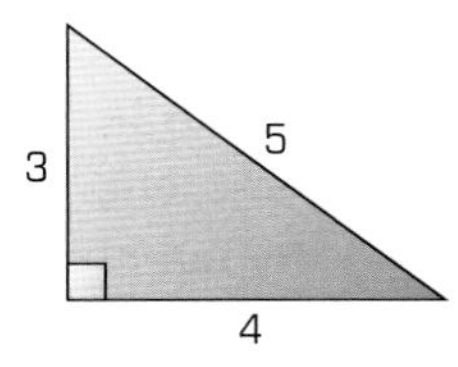

Page 232 describes two others.
Use the rule on page 232 to find four more sets of Pythagorean triples.
Check that your answers obey Pythagoras' theorem.

2 A Pythagorean triple is **primitive** if the only common factor is 1.

- 3, 4, 5 is primitive
- 6, 8, 10 is not primitive (2 is a common factor)

Find four sets of primitive Pythagorean triples.
Ensure that they are different from your answers in question 1.

3 Every Pythagorean triple can be represented by this diagram.

The green L-shape has area c^2.
The smaller square has area b^2.
The larger square has area a^2.

a Find the Pythagorean triple that the diagram represents, given these dimensions:

i $b = 99$, width of green strip is 2.
ii $b = 60$, width of green strip is 1.
iii $b = 21$, width of green strip is 8.
iv $b = 20$, width of green strip is 9.

Hint:
Find the value of c first, and check that your values fit with the diagram.

b Comment on your answers to **a** parts **iii** and **iv**.

B1.2 Solving problems with trigonometry

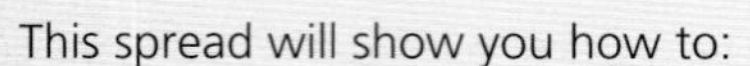

This spread will show you how to:

- Begin to use sine, cosine and tangent in right-angled triangles to solve problems in two dimensions.

KEYWORDS

Sine	Opposite
Cosine	Adjacent
Tangent	Hypotenuse
Trigonometry	

Trigonometry means 'triangle measure'.
Sine, cosine and tangent are ratios that relate sides in a right-angled triangle.

opposite hypotenuse x adjacent

$\sin x = \frac{\text{opposite}}{\text{hypotenuse}}$ $\quad \cos x = \frac{\text{adjacent}}{\text{hypotenuse}}$ $\quad \tan x = \frac{\text{opposite}}{\text{adjacent}}$

sin, cos and tan are abbreviations

You can use these ratios to solve problems.

example

A 6 m ladder rests against a wall.
The foot of the ladder is 1.7m from the bottom of the wall.

Calculate the angle that the top of the ladder makes with the wall.
Give your answer correct to 1 decimal place.

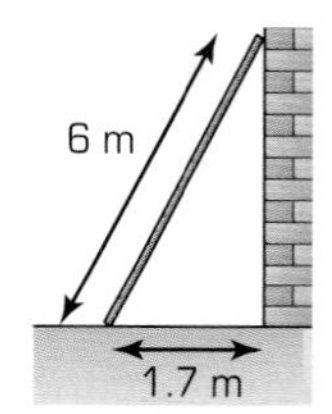

First draw a diagram, labelling the angle and sides.

Label the **angle** x.

x is **opposite** 1.7 m.

The length of the ladder (6 m) is the **hypotenuse**.

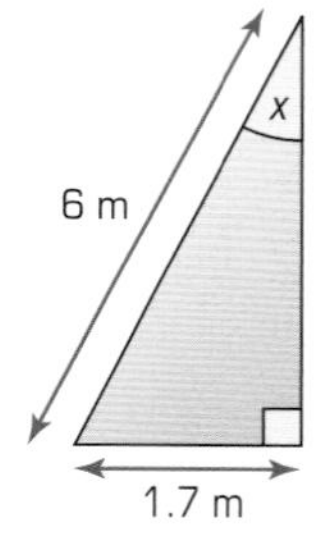

You know opposite and hypotenuse, so use **sine**.

$$\sin x = \frac{\text{opp}}{\text{hyp}}$$
$$= \frac{1.7}{6}$$
$$= 0.283$$

$$x = 16.46°$$

On this calculator, you would press the keys:

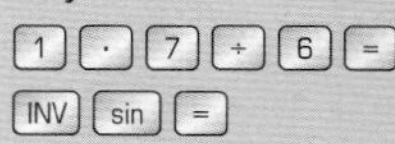

The angle that the top of the ladder makes with the wall is 16.5° (to 1 dp).

Note: When you find an angle in a right-angled triangle, the answer should be between 0° and 90°.

- To solve problems using trigonometry:
  - Draw a diagram
  - Decide which ratio, sin, cos or tan, you need to use
  - Use the ratio to work out the angle or side you have to find.

You may need to rearrange your equation.

Exercise B1.2

Solve these problems using trigonometry.
Draw a diagram for each question.

1 A 5 m ladder rests against a wall.
The foot of the ladder is 1.4 m from the bottom of the wall.
Find the angle the ladder makes with the ground.

2 An 8 m ladder leans against a wall making an angle of 49° with the ground.
How far up the wall is the top of the ladder?

3 A rectangle has side lengths 8 cm and 15 cm.
Find the angle that a diagonal makes with a longer side.

4 A triangle is drawn in a semicircle.
The shorter side of the triangle is 3 cm.
It makes an angle of 42° with the diameter.
Find the diameter of the circle.

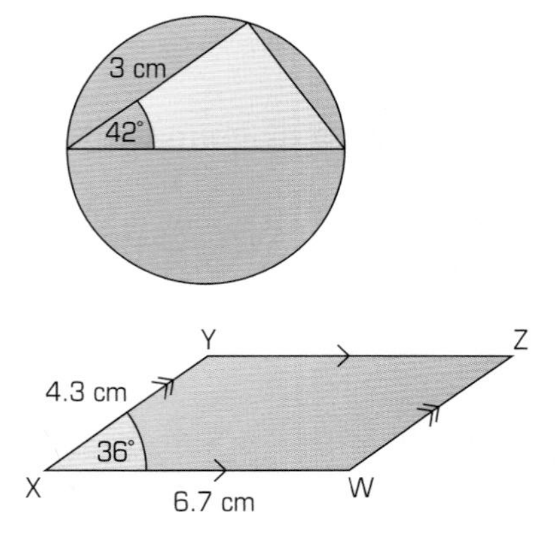

5 The parallelogram WXYZ has sides WX = 6.7 cm and XY = 4.3 cm.
The angle WXY = 36°.

a Calculate the perpendicular distance between the sides WX and YZ.

b Calculate the area of the parallelogram.

6 An equilateral triangle is drawn inside a circle with radius 5 cm.

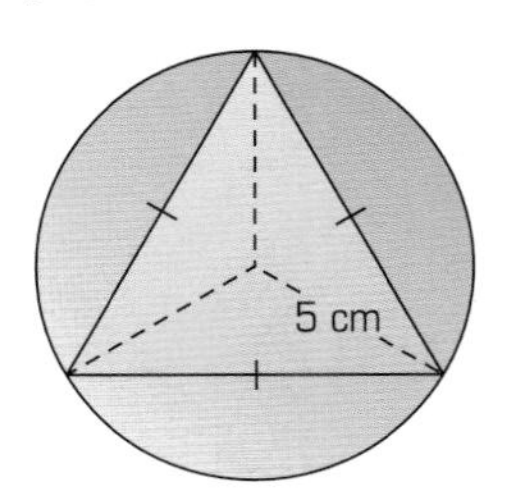

Calculate the perimeter of the triangle.

B1.3 Bearings and trigonometry

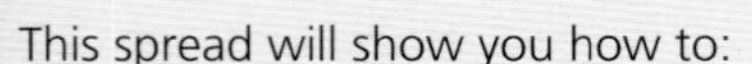

This spread will show you how to:

- Understand and use Pythagoras' theorem.
- Begin to use sine, cosine and tangent in right-angled triangles to solve problems in two dimensions.
- Use bearings.

KEYWORDS
Pythagoras' theorem
Trigonometry
Bearing

A **bearing** is an angle measured clockwise from north.

You can use Pythagoras' theorem and trigonometry to solve problems involving bearings.

example

A boat sails from port for 48 km on a bearing of 070°.
It then changes course and travels for 29 km on a bearing of 160°.

Bearings are always given as three digits.

Find the distance and bearing that the boat is from port.
Give your answers to 3 significant figures.

First draw a diagram with the facts that you know:

A boat sails from port for 48 km on a bearing of 070° ...

... It then changes course and travels for 29 km on a bearing of 160°.

N
N
70°
48 km
160°
Port
29 km
Boat

Now deduce some new facts:

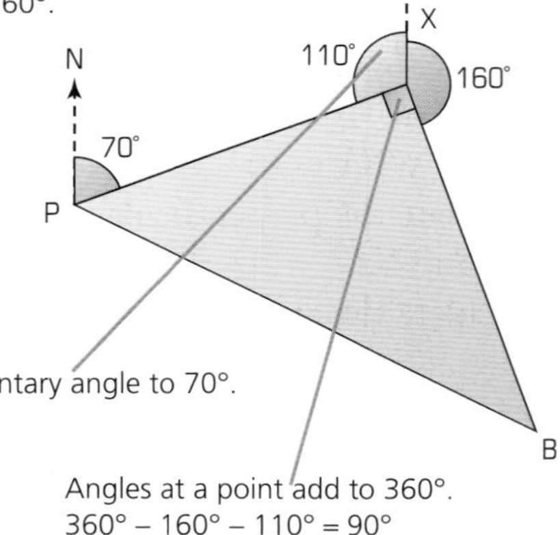

This is the supplementary angle to 70°.
180° – 70° = 110°

Angles at a point add to 360°.
360° – 160° – 110° = 90°

This is a right-angled triangle, so use Pythagoras' theorem:
$PB^2 = 48^2 + 29^2 = 3145 \quad \Longrightarrow \quad PB = 56.1$ km (to 3 sf)

The distance of the boat from port is 56.1 km.

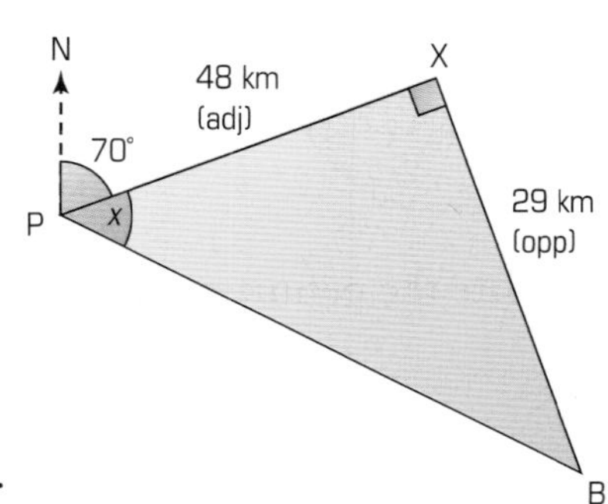

To find the bearing you need to find the angle, x.
You are given the opposite and the adjacent, so use tan:

$\tan x = \frac{29}{48} = 0.604\,16\ldots \qquad x = 31.1°$ (to 3 sf)

The bearing from port to the boat is 70° + 31.1° = 101.1°.

Exercise B1.3

Solve these problems using Pythagoras' theorem and trigonometry.
Draw a diagram for each question.

1 Rachael walks 5 km east and 8 km south.

a How far is Rachael from her starting point?

b On what bearing should Rachael walk to return to her starting point?

2 Stuart runs 10 km on a bearing of 315°.
He changes direction and runs a further 6 km on a bearing of 045°.
Lawrence is travelling to meet Stuart.
He is starting at the same place that Stuart began his run.

a Calculate the shortest possible distance that Lawrence can travel.

b Calculate the bearing that Lawrence has to travel to reach Stuart.

3 A ship is due east of a lighthouse.
It sails 100 km on a bearing of 324° until it is due north of the lighthouse.
How far away is the ship now from the lighthouse?

4 A ship is 64 km due south of a lighthouse.
It sails 80 km until it is due east of the lighthouse.
Calculate the bearing on which the ship sails.

5 A ship is 92 km due west of a lighthouse.
It sails on a bearing of 130° until it is due south of the lighthouse.
How far does the ship sail?

6 Sisters Emelia and Freya set out from home.
Emelia walks 4 km on a bearing of 045°.
Freya runs 7.5 km on a bearing of 135°.

a How far apart are Emelia and Freya?

Emelia decides to run and join Freya.

b On what bearing should Emelia run?

B1.4 Angles of elevation and depression

This spread will show you how to:

- Begin to use sine, cosine and tangent in right-angled triangles to solve problems in two dimensions.

KEYWORDS
Angle of elevation
Angle of depression

Sheree is looking up at a pigeon on the school roof.

The angle between Sheree's line of sight and the horizontal is called the **angle of elevation**.

Shaka is looking down at a boat on the river.

The angle between Shaka's line of sight and the horizontal is called the **angle of depression**.

You can use Pythagoras' theorem and trigonometry to solve problems involving angles of elevation and depression.

example

Josh is 1.6 m tall. He is standing 37 m from the foot of a tree.
He measures the angle of elevation of the top of the tree as 24°.

Calculate the height of the tree, giving your answer to 1 decimal place.

First draw a diagram.

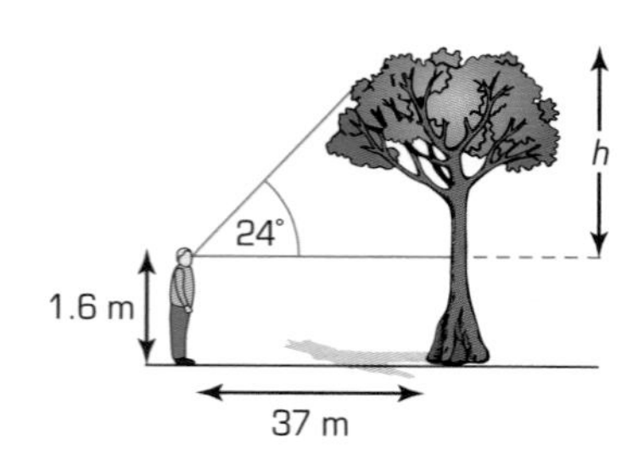

You know the adjacent, and you want to find the opposite $\Rightarrow$ use **tangent**.

$\tan 24° = \frac{\text{opposite}}{\text{adjacent}}$

$= \frac{h}{37}$

$h = 37 \times \tan 24°$

$h = 16.5$ m (to 1 decimal place)

But Josh is 1.6 m tall:

$16.5 + 1.6 = 18.1$

$\Rightarrow$ The tree is 18.1 m tall.

Sometimes the height of the person is so small compared with the other distances that you can ignore it.

Remember:

- Always draw a diagram.
- Give your answers to the degree of accuracy stated.

Exercise B1.4

Adam and Peter want to estimate the volume of wood contained in Foxglove Woods.

They choose an area of 10 m^2 and find that there are four trees.

Peter measures the circumference of each tree.

Adam, who is 1.8 m tall, stands 4 m from the foot of each tree and measures the angle of elevation to the top.

Here are their results.

Tree	A	B	C	D
Circumference, cm	110	98	125	107
Radius				
Angle of elevation, degrees	64	56	70	61
Height				
Volume				

1 Copy the table.

a Use the circumference to find the radius of each tree.

b Use the angle of elevation to find the height of each tree.

> **Hint:**
> Draw a diagram in each case and use trigonometry.

c Treating each of the trees as if it were a cylinder, calculate the volume of wood in each tree.

In Foxglove Woods there is a quarry containing a single tree.
The area of the quarry is 200 m^2.

The boys assume that the tree's circumference is half the average of the other four trees (it is visibly smaller).

They stand at the edge of the quarry and measure the angle of depression to both the top and the bottom of the tree.

Here are their results:

Angle of depression to the top of the tree	12°
Angle of depression to the bottom of the tree	78°
Estimated distance of tree from edge of the quarry	0.4 m

2 **a** Find the height of the tree.

b Find the volume of wood in the tree.

Foxglove Woods occupy a total area of 1 km^2.

3 Estimate the total volume of wood in Foxglove Woods.
State any assumptions that you have made in your estimate.

B1.5 Triangles in a circle

This spread will show you how to:

- Represent problems and synthesise information in geometric and graphical form.
- Move from one form to another to gain a different perspective on the problem.

KEYWORDS
Sine
Ratio
Circumference
Vertical

Jon wants to investigate the sine ratio.
He is going to see how the opposite side changes as the angle changes.

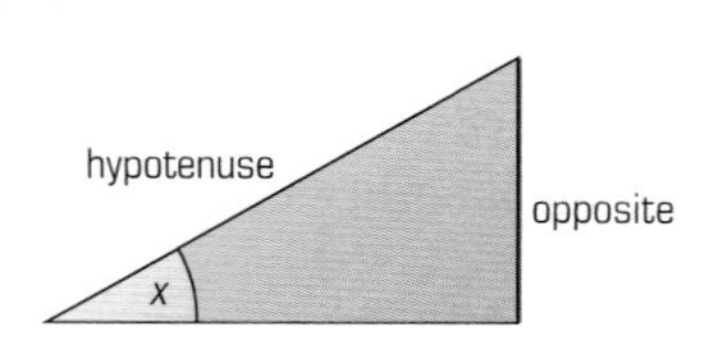

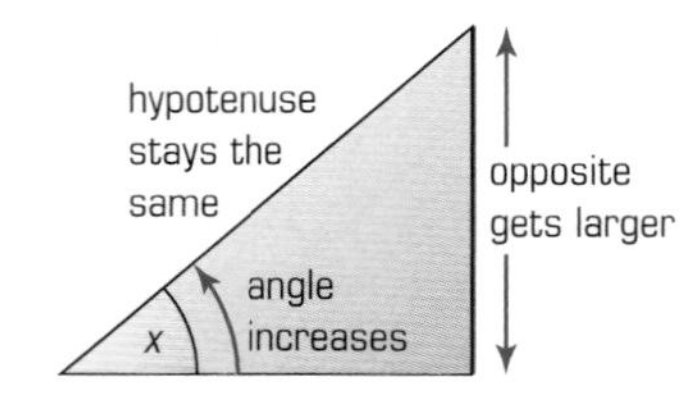

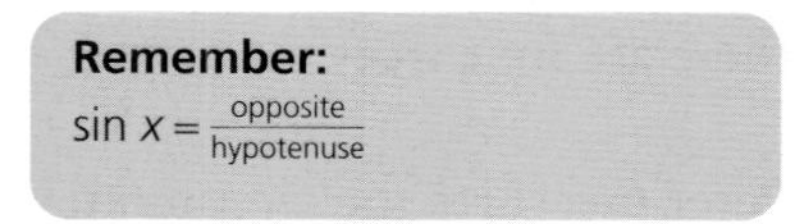

Jon draws a circle radius 4 cm on plain paper, with a zero line marked.

He marks points on the circumference every 15°, moving around anticlockwise.

He draws a right-angled triangle inside the circle every 15°.

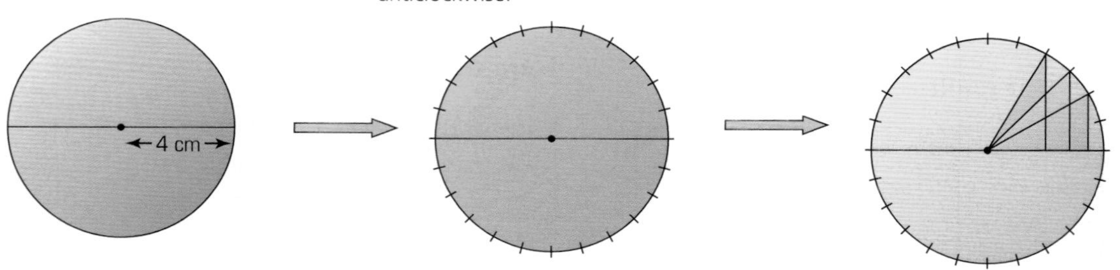

Jon measures the vertical side of the triangle (**opposite** the angle), and records his results in a table.

Angle, x	15°	30°	45°	60°	…	330°	345°	360°
Length of opposite (cm)	1.0	2.0	2.8	3.5		⁻2.0	⁻1.0	0

Jon takes lengths below the zero line to be negative.

Next to the circle Jon draws a pair of axes with angle on the *x*-axis, and length of opposite on the *y*-axis.
He plots the values that he has measured.

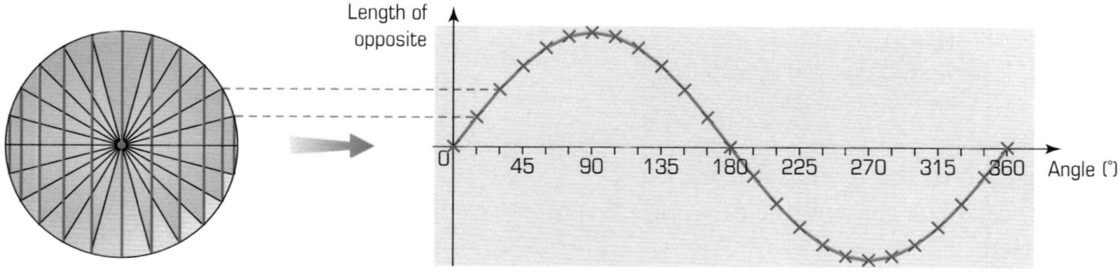

Jon joins all his points with a smooth curve.

Exercise B1.5

You are going to draw a sine curve.

You will need:
plain paper, pencil, ruler, compasses, protractor, string, spaghetti, glue

- Draw a circle radius 4 cm, and mark points on the circumference every 15°.
- Draw a pair of axes next to your circle.
  The x-axis should be 26 cm long, and the y-axis should range from $^-4$ cm to $^+4$ cm.
- Place string around the circumference of the circle and transfer the 15° marks from the circle to the string.
  Include marks for 0° and 360°.

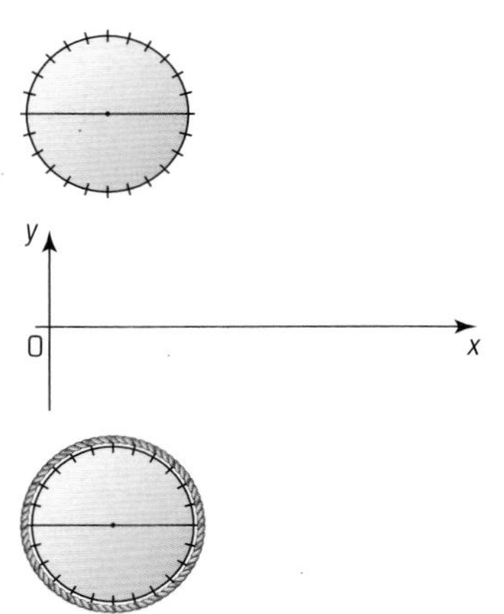

- Place the taut string along the x-axis and transfer the marks.
  Label the x-axis 'Angle'.

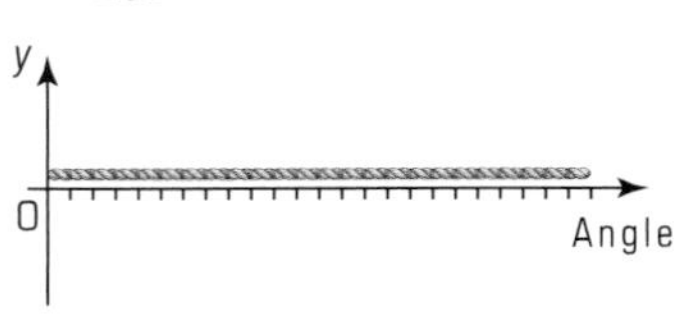

- Draw right-angled triangles all the way around the circle starting at 15° and ending at 345°.

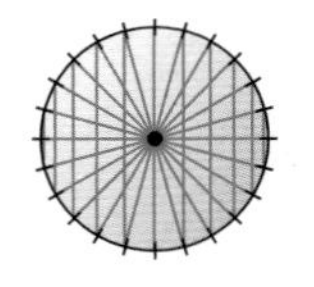

- Use pieces of dried spaghetti to determine the vertical lengths of each triangle.
  You can break pieces to the correct length.

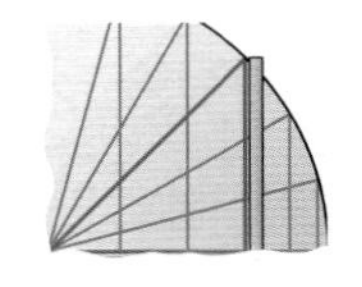

- Mark on the graph the spaghetti lengths for each triangle.
  If you want to, you can glue the pieces of spaghetti to your graph.

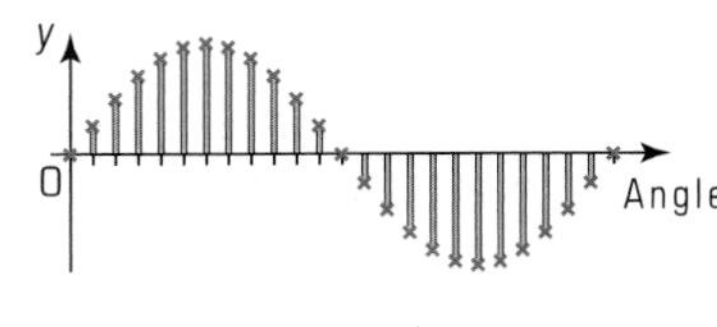

- Join the points, or lengths of spaghetti, with a smooth curve.
  This is a sine curve.

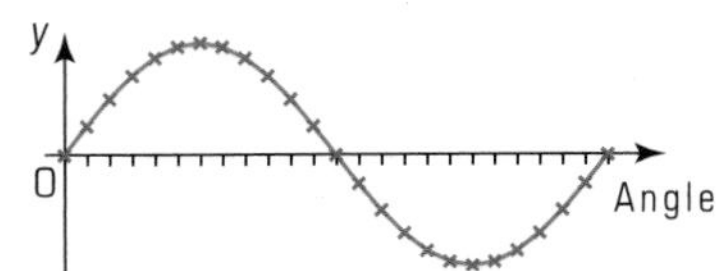

B1.6 Graph of $y = \sin x$

This spread will show you how to:

- Link a graphical representation of an equation to the algebraic solution.
- Recognise limitations on accuracy of measurements.

KEYWORDS

Sine	Opposite
Hypotenuse	Equation
Radius	Solution

Jon is investigating triangles in a circle.

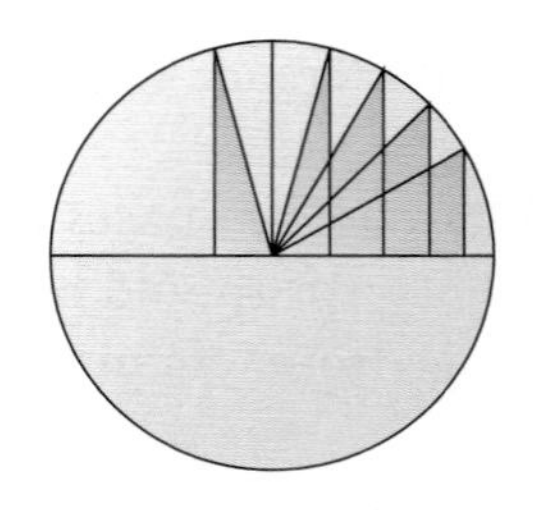

He drew this graph to show how the **opposite** side changes with angle.

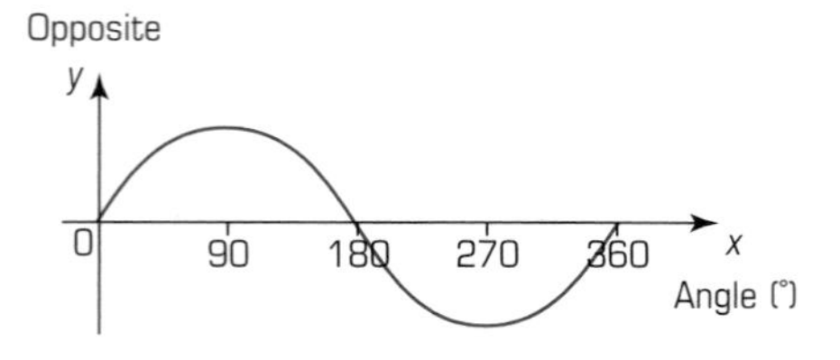

You can abbreviate **sine** to **sin**.

Jon wants to scale the y-axis to show how $\sin x$ changes with angle.

$$\sin x = \frac{\text{opposite}}{\text{hypotenuse}} = \frac{\text{opposite}}{4\text{ cm}}$$

The hypotenuse of each of Jon's triangles is 4 cm (equal to the radius).

Jon extends his original table (from page 240).

Jon calculates $\sin x$ to 2 sf because the opposite is measured to 2 sf.

Angle, x	15°	30°	45°	60°	...	330°	345°	360°
Length of opposite (cm)	1.0	2.0	2.8	3.5	...	$^{-}2.0$	$^{-}1.0$	0
$y = \sin x$ (= opposite ÷ hypotenuse)	0.25	0.50	0.70	0.88	...	$^{-}0.50$	$^{-}0.25$	0

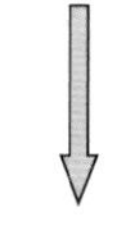

Jon's graph has limited accuracy because it is based on his measurements.

Jon uses his table to draw a graph of $y = \sin x$ (see opposite page).

You can use the graph of $y = \sin x$ to solve problems.

example

Solve the equation $\sin x = 0.5$ by using the graph of $y = \sin x$.

Draw a horizontal line from 0.5 on the y-axis across to the graph.

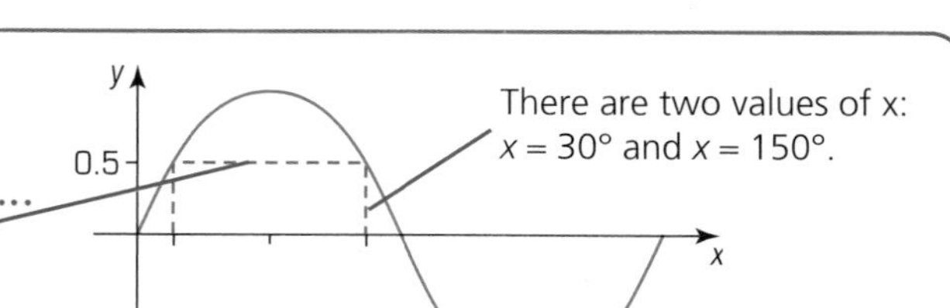

There are two values of x: $x = 30°$ and $x = 150°$.

So $x = 30°$ and $x = 150°$ are both solutions to the equation.

Exercise B1.6

Do not use a calculator for this exercise.

Here is the graph of $y = \sin x$.

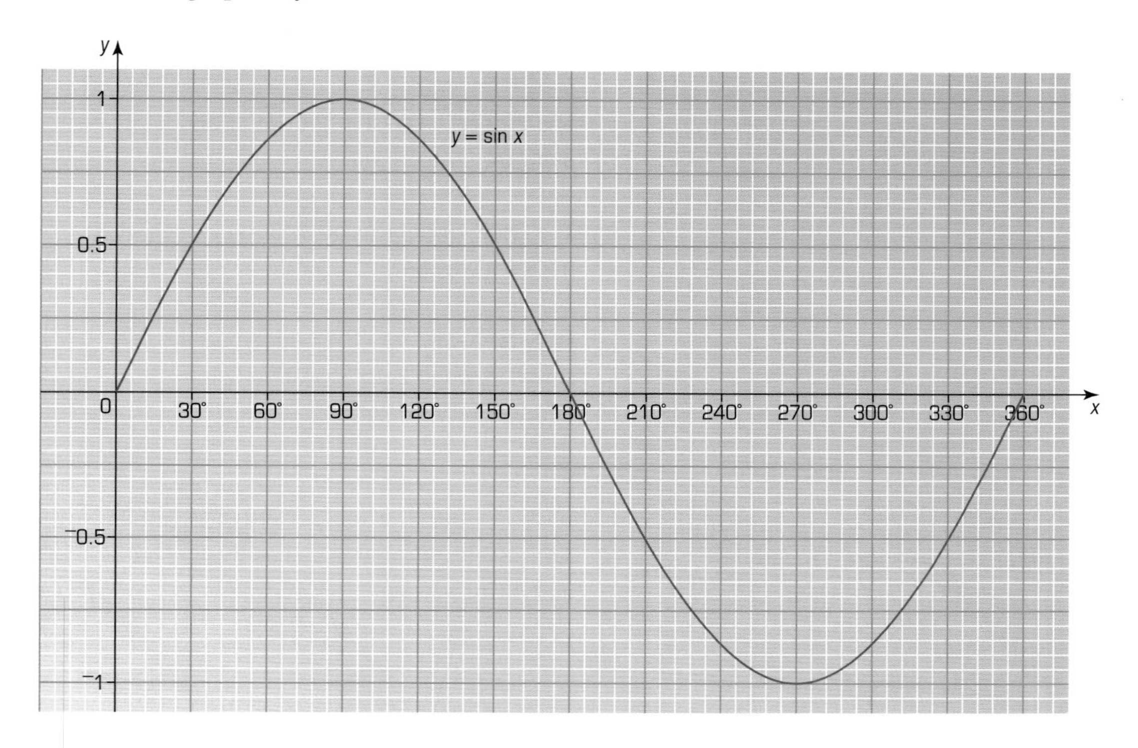

1 Use the sine curve to solve these equations for angles between 0° and 360°.

a $\sin x = 0.25$ **b** $\sin x = 0.625$

c $\sin x = 0.4$ **d** $\sin x = {}^{-}0.5$

e $\sin x = {}^{-}0.75$ **f** $\sin x = {}^{-}0.2$

Hint: You should find two solutions to each equation.

2 Sort these quantities into pairs of equal value.

sin 30°, sin 16°, sin 142°, sin 280°, sin 187, sin 290°, sin 205, sin 38°, sin 260°, sin 150°, sin 250°, sin 164°, sin 353°, sin 335°

Hint: Use the symmetry of the graph to help sort the pairs of values.

Summary

You should know how to ...

1 Begin to use sine, cosine and tangent in right-angled triangles to solve problems in two dimensions.

2 Understand and apply Pythagoras' theorem.

Check out

1 Betsy stood on a bridge that was 2 m above the water level of a river.

She dropped a stick in the water and let it float away from her.

The angle of depression as Betsy looked at her stick was 54°.

How far away from the bridge had the stick travelled when Betsy saw it?
(Ignore Betsy's height.)

2 An isosceles triangle has sides 7 cm, 7 cm and 4 cm.

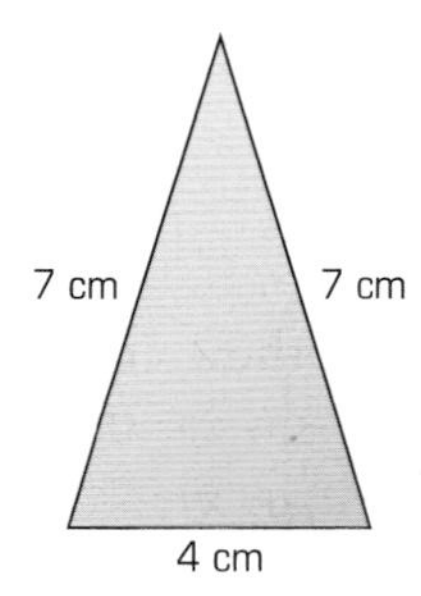

a Find the angle between the equal sides.

b Find the area of the triangle.

accuracy
N2.3, P1.1, P1.6

Accuracy describes the degree to which a number is rounded.

adjacent
S4.4, S4.5, B1.2

Adjacent means next to.

algebra
A4.3

Algebra is the branch of mathematics where symbols or letters are used to represent numbers.

algebraic expression
A4.3

An algebraic expression is a term, or several terms connected by plus and minus signs.

algebraic fraction
N1.1, N1.2, A5.4

A fraction containing letters, for example $\frac{2x}{y}$.

alternate
S1.3, S4.1

A pair of alternate angles is formed when a straight line crosses a pair of parallel lines. Alternate angles are equal.

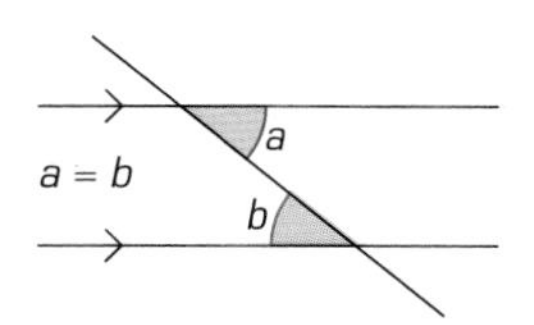

angle: acute, obtuse, right, reflex

An angle is formed when two straight lines cross or meet each other at a point.
The size of an angle is measured by the amount one line has been turned in relation to the other.

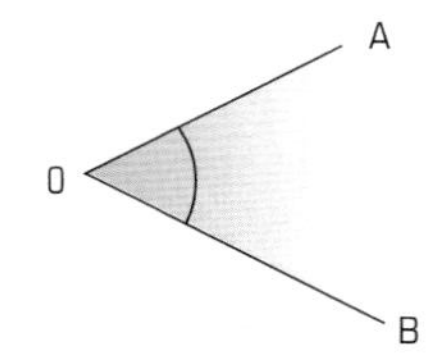

An acute angle is less than 90°.

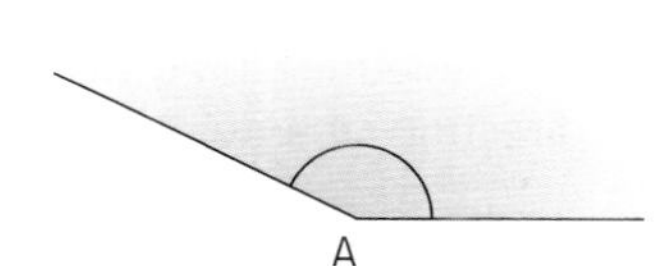

An obtuse angle is more than 90° but less than 180°.

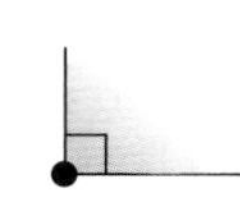
A right angle is a quarter of a turn, or 90°.

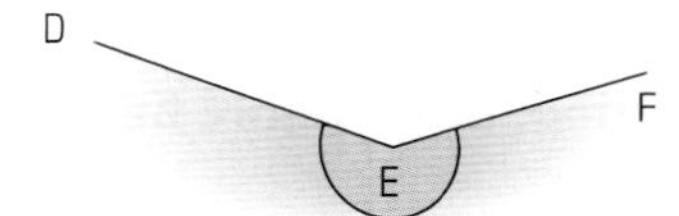

A reflex angle is more than 180° but less than 360°.

angle of elevation/depression
B1.4

When you look up, the angle between your line of sight and the horizontal is the angle of elevation. When you look down, the angle between your line of sight and the horizontal is the angle of depression.

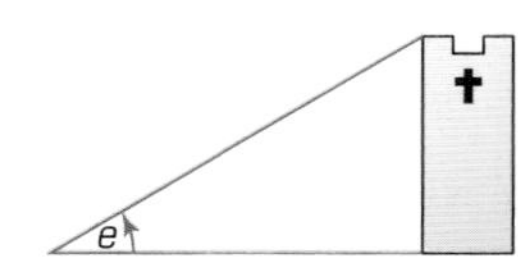

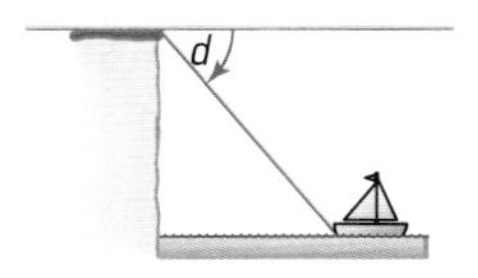

angle bisector
S1.8

An angle bisector cuts an angle in half.

angles on a straight line
P1.5

Angles on a straight line add up to 180°.

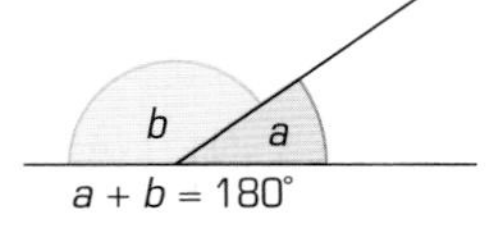

anomaly
D1.5

An anomaly is a value that does not fit a pattern.

approximate
S2.2, N2.6, P1.6

To approximate an answer is to work out a rough answer using easier figures. An approximate value is close to the real value.

arc
S2.3

An arc is part of a curve.

area
S2.1

The area of a surface is a measure of its size.
Square millimetre, square centimetre, square metre, square kilometre are all units of area.

ASA
S1.7, S3.3

A triangle given ASA ('angle side angle') is unique. Two triangles are congruent if two angles and the side between them in one triangle are equal to two angles and the corresponding side in the other triangle.

average rate
S2.5

An average rate is the mean value of a varying quantity, such as speed.

axes
A2.1

Axes are lines scaled to locate a point by its coordinates.

axis of rotation symmetry

When a shape is rotated about its axis of rotation it has rotation symmetry, that is, it fits onto itself more than once during a full turn.

bar chart

The heights of the bars on a bar chart represent the frequencies of the data.

base (number)
A4.1

The base is the number which is raised to a power.
For example in 2^3, 2 is the base.

bearing, three-figure bearing
B1.3

A bearing is a clockwise angle measured from the North line giving the direction of a point from a reference point.
A bearing should always have three digits.

N
120°
A
B

The bearing of B from A is 120°.

bias
D2.3, D3.1, D3.2

An experiment or selection is biased if not all outcomes are equally likely.

BIDMAS

BIDMAS is a mnemonic to remind you of the correct order of operations: **b**rackets, **i**ndices, **d**ivision or **m**ultiplication, **a**ddition or **s**ubtraction.

bisect, bisector
S1.6, S4.1

To bisect is to cut in half.
A bisector is a line that cuts something in half.

box-and-whisker plot
D3.4, D3.5

A box-and-whisker plot shows the median and the spread of a set of data.

brackets
A3.1

Brackets show you what part of a calculation to do first.

cancel
N1.1, N1.2, N1.3, A5.4

You cancel a fraction by dividing the numerator and denominator by a common factor.

capacity
S2.1

Capacity is a measure of how much liquid a hollow 3-D shape can hold.

census
D3.2

In a census, everyone in the group being surveyed is included.

centre
S1.6

The centre of a circle is the point from which all points on the circumference are equidistant.

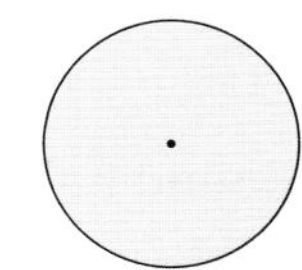

centre of enlargement
S3.5

The centre of enlargement is the point from which an enlargement is measured.

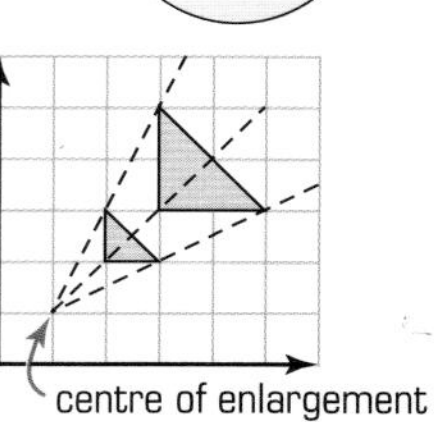

centre of rotation
S3.1

The centre of rotation is the fixed point about which a rotation takes place.

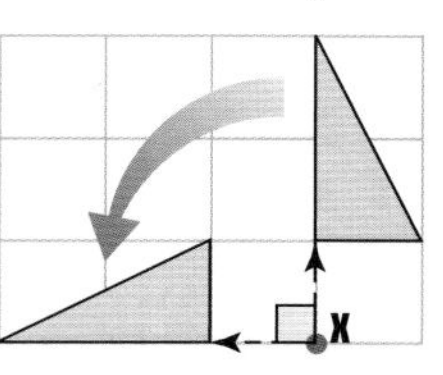

chord
S1.6

A chord is a straight line joining two points on a curve, or a circle.

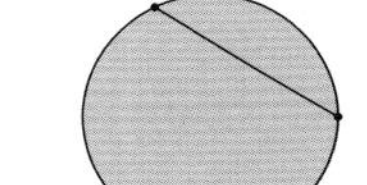

circumcentre
S1.6

The circumcentre is the centre of a circumcircle.

circumcircle
S1.6

A circumcircle is a circle drawn around another shape, so that all the vertices of the shape lie on the circle.

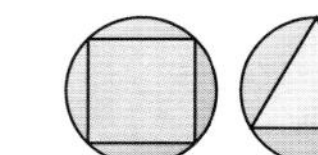
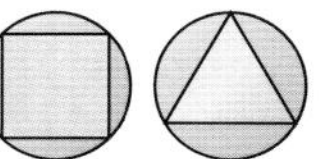

circumference
S2.2, S2.3, B1.5

The circumference is the distance around the edge of a circle.

circumscribed
S1.6

A circumscribed shape is a shape drawn around another shape, for example a circumcircle.

coefficient
A1.1, A1.2, A4.3

The coefficient is the number part of an algebraic term. For example in $3n^5$ the coefficient of n^5 is 3.

collect like terms
A2.3, A5.3

To collect like terms, put together terms with the same letter parts. For example $5x + 3x = 8x$ and $4y^2 - y^2 = 3y^2$.

common denominator
N1.1, N2.4

To add or subtract fractions, change them into equivalent fractions with the same denominator, that is a common denominator.

common factor
A4.4

A common factor is a factor of two or more numbers or terms. For example $2p$ is a common factor of $2p^2$ and $6p$.
You factorise an expression by taking out a common factor.
For example

$$15a^2b + 10bc = 5b(3a^2 + 2c)$$

commutative

An operation is commutative if the order of combining two terms does not matter.

compasses
S1.7

A pair of compasses is a geometrical instrument used to draw circles or arcs.

compensating
N2.4

The method of compensation is used to make calculations easier. For example to add 99, add 100 and then compensate by subtracting 1.

compound interest
N1.4, N1.6

With compound interest you are paid interest on the original amount and the previous years' interest.

conclude, conclusion
D3.6

To conclude is to formulate a result or conclusion based on evidence.

congruent
S3.1, S3.3

Congruent shapes are exactly the same shape and size.

consecutive
A1.3

Numbers that follow one another, for example 15 and 16, are consecutive numbers.

constant
A1.1, S2.5

Constant means unchanging. A constant is an algebraic term that remains unchanged. For example, in the expression $5x + 3$ the constant is 3.

construct, construction
S1.7, S1.9

To construct is to draw accurately. A construction is an accurate drawing.

construction lines
S1.7

Construction lines are the arcs drawn when making an accurate diagram.

continuous
N2.3

Continuous data can take any value between given limits, for example height.

convention
S1.1

A convention is the accepted way of describing something in maths.

convert, conversion
S2.1

To convert is to change from one unit to another, for example 1 m = 100 cm.

coordinates
S4.6

The coordinates of a point give its position in terms of its distance from the origin along the x and y axes.

correlation
D1.5, D3.3

Correlation is a measure of the relationship between two variables.

corresponding
S1.3, S3.4, S4.1

A pair of corresponding angles is formed when a straight line crosses a pair of parallel lines. Corresponding angles are equal.

$a = b$

cosine
S4.4, S4.5, B1.2

In a right-angled triangle, the cosine of an angle is the ratio of the length of the adjacent side to the length of the hypotenuse.

$$\cos = \frac{\text{adjacent}}{\text{hypotenuse}}$$

cost price
N1.4

Cost price is the price that a retailer pays for an article.

counter-example
A5.1

A counter-example is an example that shows that a rule does not work.

cross-multiply
A3.1

Cross-multiplying is a method for removing fractions from equations.

cross-section
S4.3
The cross-section of a solid is the shape of its transverse section, that is a section cut parallel to the end of the shape.

cube root
The cube root of x is the number that when cubed gives you x. For example $\sqrt[3]{64} = 4$, because $4 \times 4 \times 4 = 64$.

cubic
A2.3, A4.7
A cubic expression, equation or function contains a term in x^3 as the highest power.

cumulative frequency
D1.3, D1.4
The cumulative frequency is the sum of the frequencies.

curve
A2.3, A4.7
Graphs of quadratic, cubic and reciprocal functions are curves.

cylinder
S2.6, S4.6
A cylinder is a prism with a circular cross-section.

data
D3.1
Data are pieces of information.

data collection sheet
D3.1
A data collection sheet is a form designed for the systematic collection of data.

data logging
Data logging is the automatic collection of data.

decagon
A decagon has ten sides.

decimal
N2.5
A decimal number is a number written using base ten notation.

decimal multiplier
N1.6
A decimal multiplier is used in calculating percentages, for example to increase by $4\frac{1}{2}\%$ multiply by 1.045.

decimal place
N2.3, N2.6, N2.7
Each column after the decimal point is called a decimal place.

definition
S1.1
A definition explains the exact meaning of a word.

degree (°)
S1.1
Angles are measured in degrees. There are 360° in a full turn.

degree of accuracy
N2.3, P1.1, P1.6
The degree of accuracy of an answer depends on the accuracy of the figures used in the calculation.

demonstrate
A5.1
You can demonstrate that a statement is true for some values by giving examples.

denominator
N1.2, A5.4
The denominator is the bottom number in a fraction. It shows how many parts there are in the whole.

density
S2.5
Density is the mass (weight) of a unit volume (for example, per 1 cm^3) of a substance.

derive
A5.6
You can derive a formula from information given in a problem.

derived property
S1.1
In geometry, a derived property arises from the basic properties of a shape.

diagonal
S4.1

A diagonal line is one which is neither horizontal nor vertical.

diameter
S2.2

The diameter is a chord that passes through the centre of a circle.

difference of two squares
A5.2

The identity $(a + b)(a - b) \equiv a^2 - b^2$ is called the difference of two squares.

difference pattern
A1.1

You can find a general rule for a sequence by looking at the pattern of differences between consecutive terms.

digit
N2.8

A digit is any of the numbers 0, 1, 2, 3, 4, 5, 6, 7, 8, 9.

dimension

A dimension is a length, width or height of a shape or solid.

direct proportion
S2.5

Two quantities are in direct proportion if one quantity increases at the same rate as the other.

discrete
N2.3

Discrete data are data that can be counted.

distance–time graph

A distance–time graph is a graph of distance travelled against time taken. Time is plotted on the horizontal axis.

distribution
D3.5, D3.6

A distribution is a set of observations of a variable.

divisible, divisibility

A whole number is divisible by another if there is no remainder after division.

divisor
N1.3, N2.7

The divisor is the number that does the dividing.
For example, in $14 \div 2 = 7$ the divisor is 2.

edge (of solid)

An edge is a line along which two faces of a solid meet.

edge

elevation
S4.3

An elevation is an accurate drawing of the side or front of a solid.

elimination
A3.3, A3.4, A3.5

A method for solving simultaneous equations by removing (eliminating) one of the variables (letters).

enlargement
N1.7, S3.2, S3.5

An enlargement is a transformation that multiplies all the sides of a shape by the same scale factor.

equation
A3.1, A3.2, A5.3

An equation is a statement showing that two expressions have the same value.

equation (of a graph)
B1.6

An equation is a statement showing the relationship between the variables on the axes.

equidistant
S1.8

Equidistant means the same distance apart.

equivalent, equivalence
N1.1, N2.6
Two quantities, such as fractions which are equal, but are expressed differently, are equivalent.

estimate
N1.9, N2.6, D2.3, D2.4, D4.6
An estimate is an approximate answer. You can estimate a probability by carrying out an experiment.

estimate of the mean/median
D1.3
The exact mean and median cannot be calculated if the data are grouped. They can be estimated from a cumulative frequency curve.

evaluate
A4.2
Evaluate means to find the value of an expression.

event
D2.1, D2.2, D4.1, D4.3, D4.4
In probability, an event is a trial or experiment.

exhaustive
D2.1, D4.1
In probability, two events are exhaustive if they include all possible outcomes.

expand
A4.4, A4.5, A5.2, A5.3
To expand an expression you remove all the brackets.

expected frequency
D2.1
In probability, the expected frequency of an event is equal to the number of trials × the probability of the event.

experiment
D4.5
An experiment is a test or investigation to gather evidence for or against a theory.

experimental probability
D1.4, D1.5, D1.6
You can find the experimental probability of an event by conducting trials.

explain
P1.2
Explain means to give a reason for your answer.

exponent
A4.1
The exponent, or index, tells you how many of a number or variable to multiply together. For example, x^4 means $x \times x \times x \times x$.

expression
A3.2, A4.3, A4.4, A4.5, A5.2
An expression is a collection of terms linked with operations but with no equals sign.

exterior angle
S1.2, S1.3
An exterior angle is made by extending one side of a shape.

face
A face is a flat surface of a solid.

face

factor
N2.5
A factor is a number that divides exactly into another number. For example, 3 and 7 are factors of 21.

factorise
A4.4, A5.2
You factorise an expression by writing it with a common factor outside brackets.

formula, formulae
A3.2, A5.5, A5.6
A formula is a statement that links variables.

frequency
D4.2
The frequency is the number of times an event occurs.

frequency diagram
D3.3

A frequency diagram uses bars to display data.
The heights of the bars correspond to the frequencies.

function, linear function
A2.2, A4.7

A function is a rule.
The graph of a linear function is a straight line.

generalise
A5.1

Generalise means find a statement or rule that applies to all cases.

general term
A1.1, A1.2, A1.3

The general term in a sequence allows you to evaluate unknown terms. It is sometimes called the *n*th term.

gradient
A2.1, A4.6

Gradient is a measure of the steepness of a line.

graph
A2.1, A2.3

A graph is a diagram that shows a relationship between variables.

greater than or equal to (⩾)
A4.9

The symbol ⩾ means that the term on the left-hand side is greater than or equal to the term on the right-hand side.

hectare
S2.1, N4.6

A hectare is a unit of area equal to 10 000 (100×100) square metres.

hexagon
S1.2

A hexagon has six sides.

highest common factor (HCF)
A4.4

The highest common factor is the largest factor that is common to two or more numbers.
For example the HCF of 12 and 8 is 4.

horizontal
A3.1, S4.9

A horizontal line is parallel to the bottom edge of the page.

hyperbola
A4.7

The graph of a reciprocal function is a hyperbola.

hypotenuse
S1.4, S1.5, S2.4, S3.3, S4.2, S4.4, S4.5, B1.1, B1.2

The hypotenuse is the side opposite the right angle in a right-angled triangle.

hypothesis
D3.1

A hypothesis is a statement used as a starting point for a statistical investigation.

identically equal to (≡)
A5.2

One expression is identically equal to another if they are mathematically equivalent.

identity
A5.2

An identity is an equation which is true for all possible values.
For example $3x + 6 \equiv 3(x + 2)$ for all values of x.

identity function
A2.2

$x \longrightarrow x$ is called the identity function, because it maps any number on to itself.

image
S3.6

An image is an object after it has been transformed.

implicit
A2.1, A3.6

An equation in x and y is in implicit form if y is not the subject of the equation.

improper fraction
N1.2

In an improper fraction the numerator is bigger than the denominator.

independent
D2.1, D2.2, D4.1

In probability, events are independent if the outcome of one event does not affect the outcome of the other event.

index, indices
N2.1, A4.1, A4.2

The index tells you how many of a quantity must be multiplied together. For example x^3 means $x \times x \times x$.

index laws
N2.1, A4.2

To multiply powers of the same base add the indices, for example $2^5 \times 2^3 = 2^8$.
To divide powers of the same base subtract the indices.
For example $5^6 \div 5^3 = 5^2$.
To raise a power to a power, multiply the indices. For example $(4^3) = 4^6$.

index notation
N2.2

A number written as a power of a base number is expressed in index notation, for example $\frac{1}{1000} = 10^{-3}$.

inequality
A4.8, A4.9

An inequality is a relationship between two numbers or terms that are comparable but not equal.
For example, $7 > 4$.

infer
D2.3

Infer means to conclude from evidence.

inscribe, inscribed
S1.6

An inscribed polygon has every vertex lying on the perimeter of a shape, such as a circle.
An inscribed circle is drawn inside a polygon so that every side of the polygon is a tangent to the circle.

integer
(including zero).
A1.3, B1.1

An integer is a positive or negative whole number

The integers are: ..., $^-3$, $^-2$, $^-1$, 0, 1, 2, 3, ...

intercept
A2.1

The intercept is the point at which a graph crosses an axis.

interior angle
S1.2

An interior angle is inside a shape, between two adjacent sides.

interpret
D2.3, D3.5

You interpret data or a problem when you make sense of it.

interquartile range
D1.2, D1.3

The interquartile range (IQR) is the difference between the upper and lower quartiles.

intersection, intersecting
A3.6, S1.3

The intersection of two lines is the point where they cross.

inverse
N1.6

Inverse means opposite.

inverse function, operation
A2.2, A5.5, A5.6

An inverse function or operation acts in reverse to a specified function or operation.

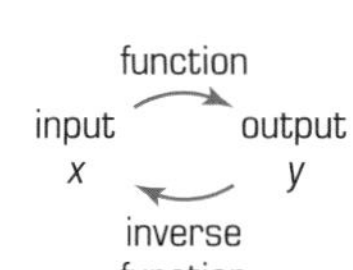

inverse mapping
A2.2

The inverse mapping reverses the direction of the mapping. For example, the inverse of $x \longrightarrow 3x$ is $x \longrightarrow \frac{x}{3}$.

investigation
D3.1
A investigation is research carried out to check a hypothesis.

isometric
S4.3
Isometric paper has three axes, ruled in equilateral triangles.

justify
P1.2
You justify a solution of a formula by explaining why it is correct.

less than or equal to (⩽)
A4.9
The symbol ⩽ means that the term on the left-hand side is less than or equal to the term on the right-hand side.

like terms
A4.3
Like terms are terms with the same letter parts, for example $3x^2$ and $^-5x^2$ are like terms.

limit, limiting value
A1.2, D2.4
If a sequence tends towards a certain value, that value is the limit .4or limiting value.

limitations
D3.6
Limitations are factors that restrict the usefulness of a statistical survey.

line graph
On a line graph, points are joined with straight lines.

line of best fit
D1.5, D3.3
A line of best fit passes through the points on a scatter graph, leaving roughly as many points above the line as below it.

line segment
S2.4
A line segment is the part of a line between two points.

linear equation, linear graph
A2.1, A3.1
A linear equation contains no squared or higher terms.
The graph of a linear equation is a straight line.

linear expression
A4.5
A linear expression contains no square or higher terms, for example $3x + 5$ is a linear expression.

linear sequence
A1.1
The terms of a linear sequence increase by the same amount each time.

locus, loci
S1.8, S1.9
A locus is a set of points (a line, a curve or a region) that satisfies certain conditions. Loci is the plural of locus.

lower bound
N2.3, P1.6
The lower bound is the lowest value a rounded figure could have.

lower quartile
D1.2, D1.3, D3.4
The lower quartile is the value $\frac{1}{4}$ of the way along a set of data arranged in ascending order.

lowest common denominator
A5.4
The lowest common denominator is the lowest number that is a multiple of the denominators of two unlike fractions. For example, 6 is the lowest common denominator of $\frac{1}{2}$ and $\frac{1}{3}$.

lowest common multiple (LCM)
N1.1
The lowest common multiple is the smallest multiple that is common to two or more numbers, for example the LCM of 4 and 6 is 12.

map
S3.6
A map is a representation of an area of land.

mapping
A2.2
A mapping is a rule that can be applied to a set of numbers to give another set of numbers.

mass
S2.5
The mass of an object is a measure of the quantity of matter in it.

maximum
A2.3
A maximum is the highest point on a curved graph, where the function has its highest value.

mean
D1.2
The mean is the average value found by adding the data and dividing by the number of data items.

median
D1.2, D1.4, D3.4
The median is the average which is the middle value when the data are arranged in order of size.

metric system
S2.1
In the metric system, units of measurement are related by powers of 10.

midpoint
S2.4
The midpoint of a line segment is the point that is halfway along.

minimum
A2.3
A minimum is the lowest point on a curved graph, where the function has its lowest value.

mirror line
S3.1
A mirror line is a line or axis of symmetry.

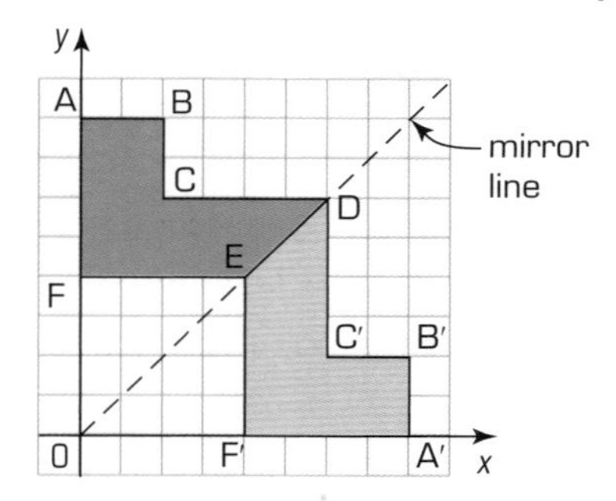

misleading
D1.6
Misleading means false. Statistics can be used to give a false impression.

mixed number
N1.2
A mixed number is a whole number with a fraction, for example $2\frac{1}{2}$.

modal class
N1.1
The modal class is the most commonly occurring class when the data is grouped.
It is the class with the highest frequency.

mode
The mode is an average.
It is the value that occurs most often.

multiple
N1.1
A multiple of an integer is the product of that integer and any other. For example 12, 18 and 30 are multiples of 6.

multiple bar chart
A multiple bar chart is a bar chart with two or more sets of bars.
It is used to compare two or more data sets.

multiplicative inverse
N1.3, N2.7
The multiplicative inverse of 3 is $\frac{1}{3}$. Multiplying by $\frac{1}{3}$ undoes the effect of multiplying by 3.

mutually exclusive
D2.1, D2.2, D4.1
Two events are mutually exclusive if they cannot occur at the same time.

negative — A negative number is a number less than zero.

net
S4.3
A net is a 2-D shape that can be folded to make a 3-D solid.

***n*th term**
A1.1
The *n*th term is the general term of a sequence.

numerator
N1.2, A5.4
The numerator is the top number in a fraction.
It tells you how many parts of the whole you have.

object, image
S3.1
The object is the original shape before a transformation.
An image is the shape after a transformation.

operation
A4.3
An operation is a rule for processing numbers.
The basic operations are addition, subtraction, multiplication and division.

opposite
S4.4, S4.5, B1.2, B1.6
The opposite side in a triangle is the side opposite the angle being considered.

order of operations
N1.9
The conventional order of operations is:
brackets first, then indices,
then division and multiplication,
then addition and subtraction (see BIDMAS).

order of rotational symmetry — The order of rotational symmetry is the number of times that a shape will fit on to itself during a full turn.

origin — The origin is the point where the x and y axes cross, that is (0, 0).

outcome
D2.2, D4.1
In probability, an outcome is the result of a trial.

outlier
D3.6
An outlier is an observation that is an exception.

p(*n*)
D2.1
p(n) stands for the probability of an event n.

parabola
A2.3, A4.7
The graph of a quadratic equation is a parabola.

parallel
A2.1, S1.3, A4.6
Parallel lines are always the same distance apart.

partitioning
N2.4
Partitioning means splitting a number into smaller parts.

pentagon
S1.2
A pentagon is a 5-sided polygon.

percentage change
N1.4, N1.5
Percentage change is an increase or decrease by a percentage of the original amount.

perimeter
S4.2

The perimeter is the distance round the edge of a shape.

perpendicular
S1.6, A4.6

A line or plane is perpendicular to another line or plane if they meet at a right angle.

perpendicular bisector
S1.8

The perpendicular bisector of a line is the line that divides it into two equal parts and is at right angles to it.

A M B
AM = MB

pi (π)
S2.2

The ratio $\frac{\text{circumference}}{\text{diameter}}$ is the same for all circles.
This ratio is denoted by the Greek letter π.

pie chart
D2.2

A pie chart is a circular diagram used to display data.
The angle in each sector is proportional to the frequency.

pilot survey
D3.1

A pilot survey is a small preliminary survey used to help plan an investigation.

place value

The place value is the value of a digit in a decimal number.
For example in 3.65 the digit 6 has a value of $\frac{6}{10}$.

plan
S4.3

The plan or plan view of a solid is an accurate drawing of the view from directly above.

plane
S4.3

A plane is a flat surface.

plane symmetry

A solid has plane symmetry if it can be divided into two identical halves.

plane of symmetry

A plane of symmetry divides a solid into two identical halves.

polygon
S1.2

A polygon is a shape with three or more straight sides.

population
D3.2

The population is the complete set of individuals from which a sample is drawn.

position-to-term rule

A position-to-term rule tells you how to calculate the value of a term if you know its position in the sequence.

positive

A positive number is greater than zero.

power
N1.9, N2.1, N2.2, A4.1

The power (or index) of a number or a term tells you how many of the number must be multiplied together.
For example 10 to the power 4 is 10 000.

practical demonstration
S1.1

A practical demonstration shows that a statement is true for some values.

predict
D1.5

Predict means to assign a value, such as a time or a length, by comparison with some known data.

pressure
S2.5
Pressure is weight per unit area (for example, per 1 cm^2).

primary data, primary source
Primary data is data you have collected yourself.

prime
A prime number is a number that has exactly two different factors.

prime factor
N1.1
A prime factor is a factor that is a prime number.

prime factor decomposition
Prime factor decomposition means splitting a number into its prime factors.

prism
S2.6, S4.6
A prism is a solid with a uniform cross-section.

probability
P1.2, D4.6
Probability is a measure of the likelihood of an event occurring.

problem
P1.1
A problem is a question requiring a solution.

properties
P1.5
The properties of a shape are its characteristic features.

product
A4.5
The product is the result of a multiplication.
For example, the product of 3 and 4 is 12.

projection
S4.3
A projection is a 2-D view of a solid.

proportion
N1.5, N1.7, D2.3, P1.4
A proportion compares the size of a part with the size of the whole.

proportional to (∝)
N1.8
When two quantities are in direct proportion, one quantity is proportional to the other.

proportionality
N1.8
The method of proportionality uses the proportion between quantities to solve problems.

prove, proof
S1.1, A5.1
You prove a statement is true by arguing from known facts.

Pythagoras' theorem
S1.4, S1.5, S2.4, S3.3, S4.2, S4.3, B1.1, B1.2
The area of the square drawn on the hypotenuse of a right-angled triangle is equal to the sum of the areas of the squares drawn on the other two sides.

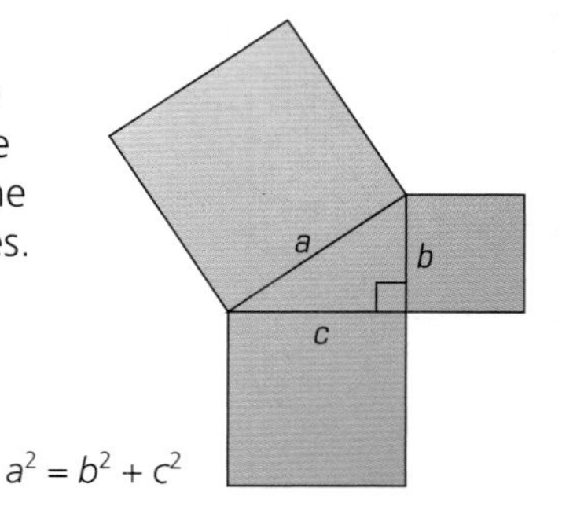

Pythagorean triple
B1.1
A Pythagorean triple is a set of three integers that form the sides of a right-angled triangle.

quadratic
A1.1, A1.2, A2.3, A4.5, A4.7

A quadratic expression, equation or function contains a square term, for example x^2.

quadratic sequence
A1.1

In a quadratic sequence the second difference is constant.

quadrilateral
S1.2, S4.1

A quadrilateral is a polygon with four sides.

rectangle

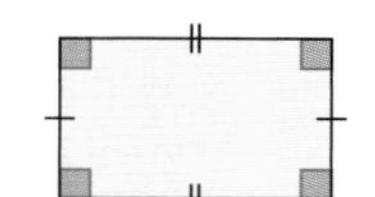

All angles are right angles. Opposite sides equal.

parallelogram

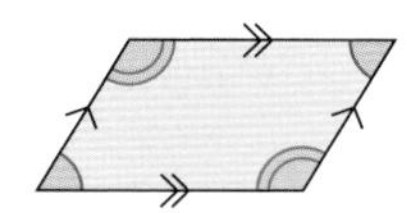

Two pairs of parallel sides.

kite

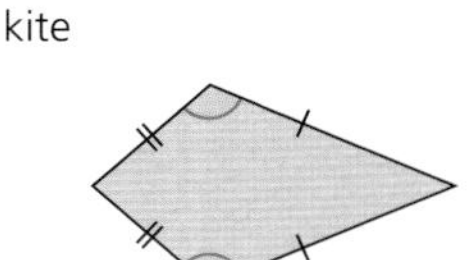

Two pairs of adjacent sides equal. No interior angle greater than 180°.

rhombus

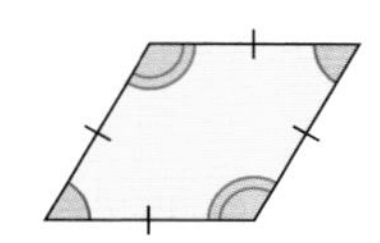

All sides the same length. Opposite angles equal.

square

All sides and angles equal.

trapezium

One pair of parallel sides.

quotient

A quotient is the result of a division.
For example, the quotient of 12 ÷ 5 is $2\frac{2}{5}$, or 2.4.

radius, radii
S1.6, S1.8, S1.9, S2.2, B1.6

The radius is the distance from the centre to the circumference of a circle. Radii is the plural of radius.

random process

The outcome of a random process cannot be predicted.

random sample
D3.2

In a random sample every item has an equal chance of being selected.

range
D1.2

The range is the difference between the largest and smallest values in a set of data.

rate
S2.5

A rate is the change per unit of time, area, volume, and so on.

ratio
N1.7, N1.8, S2.4, S3.2, S3.4, S3.5, S3.6, P1.4, B1.5

A ratio compares the size of one part with the size of another part.

raw data
D1.6

Raw data is data before it has been processed.

rearrange
A5.5

You rearrange an equation or formula by moving terms from one side to the other.

reciprocal
N1.3, N2.8, A4.2, A4.7

The reciprocal of a quantity k is $1 \div k$.
For example the reciprocal of 5 is $\frac{1}{5}$ or 0.2; the reciprocal of x^2 is $\frac{1}{x^2}$.

recurring
N2.8

A recurring decimal has a repeating pattern of digits after the decimal point, for example 0.33333 ...
You show the recurring digit with a dot: $0.3333 = 0.\dot{3}$

reflect, reflection
S3.1

A reflection is a transformation in which corresponding points in the object and the image are the same distance from the mirror line.

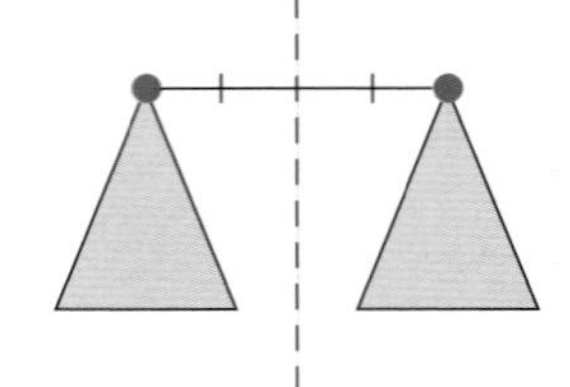

reflection symmetry

A shape has reflection symmetry if it has a line of symmetry.

region
S1.9, A4.9

A region is an area on a graph or a locus where certain rules hold.

regular
S1.3, S1.6

A regular polygon has equal sides and equal angles.

RHS
S1.7, S3.3

Two right-angled triangles are congruent if their hypotenuses and one other side are equal.
A triangle given RHS is unique.

relative frequency
D2.3, D4.5, D4.6

Relative frequency is the proportion of successful trials in an experiment.

relative frequency diagram
D2.4

A relative frequency diagram is a graph showing how relative frequency changes as the number of trials increases.

repeated subtraction
N2.7

Repeated subtraction is a method of long division in which multiples of the divisor are subtracted from the dividend (the number being divided).

representative
D3.2

A representative sample is a selection chosen rom a population to give an accurate indication of the population characteristic being studied.

rotate, rotation
S3.1

A rotation is a transformation in which every point in the object turns through the same angle relative to a fixed point.

rotation symmetry

A shape has rotation symmetry if it fits onto itself more than once during a full turn.

right prism
N2.1

Apart from its two end faces, the faces of a right prism are all rectangles.

root
N1.9, N2.1

The square root of 9 is 3, because $3 \times 3 = 3^2 = 9$.
The cube root of 8 is 2, because $2 \times 2 \times 2 = 2^3 = 8$.

rounding
N2.3

You round a number by expressing it to a given degree of accuracy.

rule
A1.1

A fixed procedure for finding a term in a sequence.

sample
D3.2

A sample is a set of individuals or items drawn from a population.

sampling
D4.2

Sampling is choosing some items from a set.

sample space, sample space diagram

In probability, the set of all possible outcomes in an experiment is called the sample space.
A sample space diagram is a diagram recording all the outcomes.

SAS
S1.7, S3.3

Two triangles are congruent if two sides and the angle between them in one triangle are equal to two sides and the corresponding angle in the other triangle.
A triangle given SAS is unique.

scale
S3.6

A scale gives the ratio between the size of an object and its diagram.

scale drawing
S1.9, S3.6

A scale drawing is an accurate drawing of a shape to a given scale.

scale factor
S3.2, S3.5, P1.4

A scale factor is a multiplier.

scatter diagram
D1.5, D3.3

Pairs of variables, for example age and height, can be plotted on a scatter diagram. The diagram shows whether there is a relationship between the two variables.

secondary data, secondary source

Secondary data is data that someone else has collected.
Common secondary sources include books, magazines and the internet.

sector
S1.6, S2.3

A sector is part of a circle bounded by an arc and two radii.

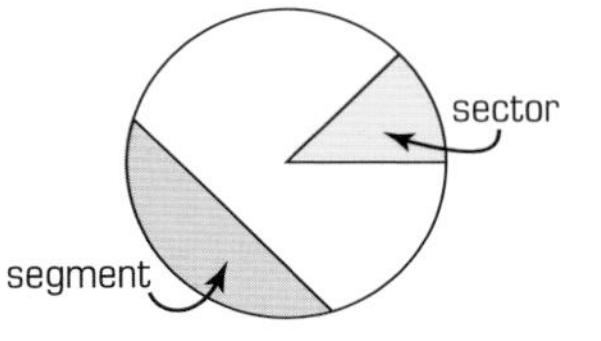

second difference
A1.2

The differences between the terms of a sequence are the *first* differences. The differences between the first differences are the *second* differences.

segment
S1.6

A segment is part of a circle bounded by an arc and a chord.

self-inverse
A2.2

The inverse of a self-inverse function is the function itself. For example, $x \longrightarrow 10 - x$ is the inverse of $x \longrightarrow 10 - x$. So $x \longrightarrow 10 - x$ is self-inverse.

selling price
A1.1

Selling price is the price an article is sold for. It is usually calculated by increasing the cost price by a percentage.

sequence
A1.1

A sequence is a set of numbers, objects or terms that follow a rule.

significant figures (sf)
N1.9, N2.3, S4.2, S4.6

The first non-zero figure in a number is its first significant figure.
For example, the first significant figure of 0.0308 is 3.

similar, similarity
S2.4, S3.2, S3.4, S3.5

Similar shapes have the same shape but are different sizes.

simplify
N1.2, A4.3, A4.5

You simplify a fraction by cancelling. You simplify an algebraic expression by collecting like terms or combining terms.

simulation

A simulation is an experiment designed to model a real-life situation.

simultaneous equations
A3.3, A3.4, A3.5, A3.6

Simultaneous equations are two or more equations whose unknowns have the same values.

sine
S4.4, S4.5, B1.2, B1.5, B1.6

In a right-angled triangle the sine of an angle is the ratio of the length of the opposite side to the length of the hypotenuse.

$$\sin = \frac{\text{opposite}}{\text{hypotenuse}}$$

skew
D3.5

A distribution is skewed if there are more values at either one end or the other.

slope

The slope of a line is measured by the angle it makes with the x-axis.

solid
S4.3

A solid is a shape formed in three-dimensional space.

cube

six square faces

cuboid

six rectangular faces

prism

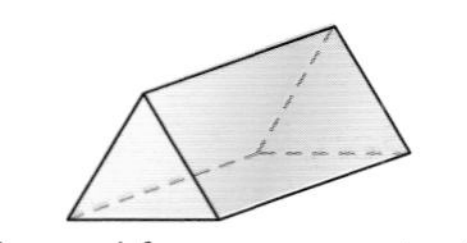

the end faces are constant

pyramid

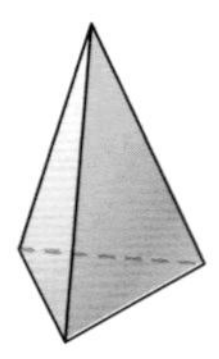

the faces meet at a common vertex

tetrahedron

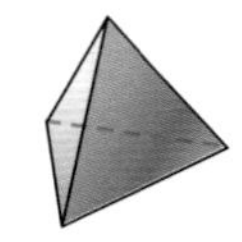

all the faces are equilateral triangles

square-based pyramid

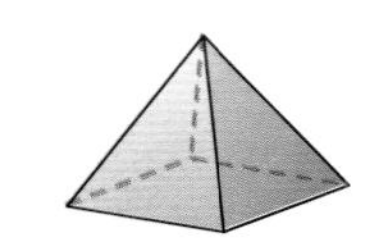

the base is a square

solution, solve
A3.1, A4.8, B1.6

The solution of an equation is the value that makes it true.

speed
S2.5

Speed is a measure of the rate at which distance is covered. It is often measured in miles per hour or metres per second.

sphere

A sphere is a 3-D shape in which every point on its surface is equidistant from the centre.

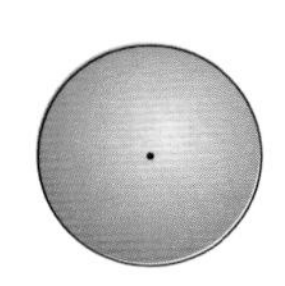

square
A4.5

To square an expression, you multiply it by itself.

square number
A1.3

If you multiply an integer by itself you get a square number.

square root
S1.4, N2.1

A square root is a number that when multiplied by itself is equal to a given number.
For example, $\sqrt{25} = 5$, because $5 \times 5 = 25$.

SSS
S1.7, S3.3

Two triangles are congruent if the three sides of one triangle are equal to the three sides in the other triangle.
A triangle given SSS is unique.

standard form
N2.2

A number in standard form is written as a number between 1 and 10 multiplied by a power of 10, for example $731 = 7.31 \times 10^2$.

statement
A5.1

A statement is a sentence giving a fact.

steepness

The steepness of a line depends on the angle the line makes with the x-axis. Gradient is a measure of steepness.

stem-and-leaf diagram
D3.4

A stem-and-leaf diagram is used to display raw data in numerical order.

straight-line graph
A2.1, A4.6

A straight-line graph is the graph of a linear equation.

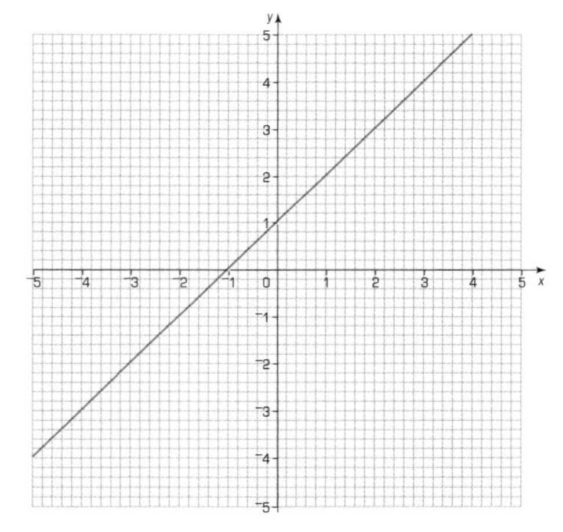

strategy
P1.1, P1.3, P1.5

A strategy is a plan for solving a problem.

stratified sample
D3.2

In a stratified sample, people are chosen from different groups so that the sample is representative of the population.

subject
A5.5, A5.6

The subject of an equation or formula is the term on its own in front of the equals sign. For example, the subject of $v = u + at$ is v.

substitute
A3.3, A3.4, A3.5, A5.5

To substitute is to replace a variable with a numerical value.

sum

A sum is the total of an addition.

supplementary
B1.3

Supplementary angles add up to 180°.
You can form a pair of supplementary angles on a straight line.

a
b
$a + b = 180°$

surd
N2.1

A surd is a square root that cannot be written as a decimal, for example $\sqrt{3} = 1.732\ 050\ 808 \ldots$

surface area
S2.6, S4.6

The surface area of a solid is the total area of its faces.

symmetry, symmetrical
S1.3, S4.1

A shape is symmetrical if it is unchanged after a rotation or reflection.

systematic sample
D3.2

A example of a systematic sample is to choose every tenth vehicle on a motorway.

T(*n*)
A1.1

T(*n*) stands for the general term in a sequence.

tangent
S1.6, S4.4, S4.5, B1.2

A tangent is a line that touches a circle or a curve at one point.

task
P1.3

A task can be a calculation that is part of a larger problem.

term
A3.1

A term is a number or object in a sequence.
It is also part of an expression.

terminating

A terminating decimal has a limited number of digits after the decimal point.

tessellation

A tessellation is a tiling pattern with no gaps.

theoretical probability

The theoretical probability of an event

$$= \frac{\text{number of favourable outcomes}}{\text{total possible number of outcomes}}$$

tonne

The tonne is a unit of mass, equal to 1000 kg.

transform
S3.1

You transform an expression by taking out single-term common factors.

transformation
S3.1

A transformation moves a shape from one place to another.

translate, translation
S3.1

A translation is a transformation in which every point in an object moves the same distance and direction.
It is a sliding movement.

tree diagram
D2.2, D2.3, D4.2, D4.3, D4.4

A tree diagram shows the possible outcomes of a probability experiment on branches.

trend

A trend is a general tendency.

trial
D2.1, D2.4, D4.1

In probability, a trial is an experiment.

trial and improvement
A3.2, P1.3

Square roots, cube roots and solutions to equations can be estimated by the method of trial and improvement.
An estimated solution is tried and refined by a better estimate until the required degree of accuracy is achieved.

triangle

A triangle is a polygon with three sides.

equilateral	isosceles	scalene	right-angled
three equal sides	two equal sides	no equal sides	one angle is 90°

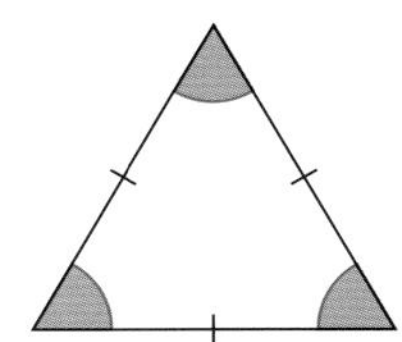

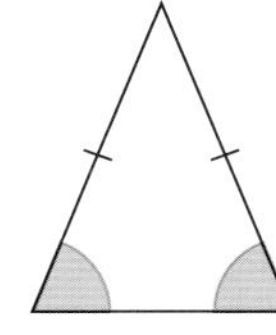
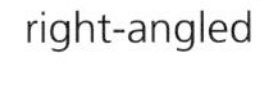
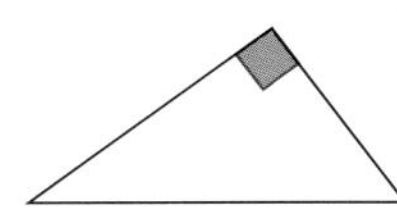

triangular number
A1.3

A triangular number is the number of dots in a triangular pattern.
The numbers form the sequence
1, 3, 6, 10, 15, 21, 28, ...

triangular prism
S4.3, S4.6

A triangular prism is a prism with a triangular cross-section.

trigonometry
B1.2, B1.3

Trigonometry is the relation between the sides and angles of a triangle.

unit fraction

A unit fraction has a numerator of 1.
For example, $\frac{1}{3}$ and $\frac{1}{7}$ are unit fractions.

unitary form
N1.7

A ratio is in unitary form when one of the numbers is 1.
For example the ratio 4 : 3 is 1 : 0.75 in unitary form.

unitary method
N1.6, P1.1, P1.4

In the unitary method, you calculate the value of one item or 1% first.

unknown
A3.1

In an equation, the unknown is a letter.

upper bound
N2.3, P1.6

The upper bound is the highest value a rounded figure could have.

upper quartile
D1.2, D1.3, D3.4

The upper quartile is the value $\frac{3}{4}$ of the way along a set of data arranged in ascending order.

variable
A3.4, A4.3

A variable is a quantity that can take a range of values.

vector
S3.1

A vector describes a translation by giving the *x*- and *y*-components of the translation.

vertex, vertices

A vertex of a shape is a point at which two or more edges meet.

vertex

vertical
B1.5

A vertical line is at right angles to the horizontal.

vertically opposite angles
S1.3

When two straight lines cross they form two pairs of equal angles called vertically opposite angles.

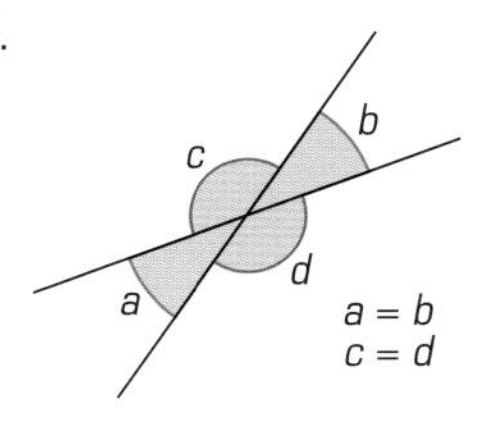

view

A view of a solid is an accurate drawing of the appearance of the solid above, in front or from the side.

volume
S2.1, S2.6, S4.6

Volume is a measure of the space occupied by a 3-D shape.
Cubic millimetres, cubic centimetres and cubic metres are all units of volume.

***x*-axis, *y*-axis**
A2.1, A4.7

On a coordinate grid, the *x*-axis is the horizontal axis and the *y*-axis is the vertical axis.

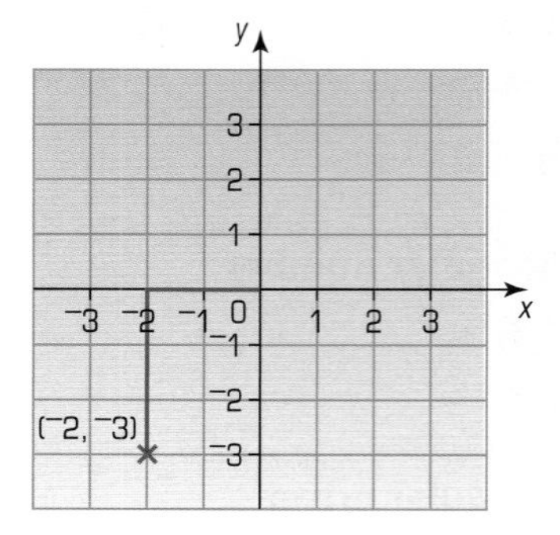

***x*-coordinate, *y*-coordinate**

The *x*-coordinate is the distance along the *x*-axis.
The *y*-coordinate is the distance along the *y*-axis.
For example, $(^-2, {}^-3)$ is $^-2$ along the *x*-axis and $^-3$ along the *y*-axis.

zero

Zero is nought or nothing.
A zero place holder is used to show the place value of other digits in a number.
For example, in 1056 the 0 allows the 1 to stand for 1 thousand. If it wasn't there the number would be 156 and the 1 would stand for 1 hundred.

Thousands	Hundreds	Tens	Units
1	0	5	6

A1 Check in

1 a 49, 60 **b** 79, 65 **c** 4, 9, 25

2 a

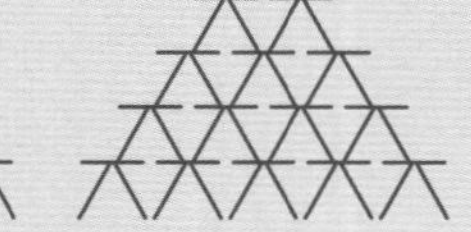

b 9, 18, 30, 45

c 63

A1 Check out

1 a i $T(n) = n^2 + 2$

ii $T(n) = 3n^2$

iii $T(n) = 2n^2 + n + 5$

b $\frac{1}{3}$

2 a $s = n^2$

b $s = \dfrac{n(n+1)}{2}$

c $s = n^2 + (n-1)^2$

3 At 65 he will receive £95.
In total he will receive £2304.

A2 Check in

1 a $abc = {}^{-}24$, $\frac{ab}{c} = {}^{-}6$, $3a - 2b = 1$, $10 - 2a = 4$, $2b - 3c = 14$, $2a^2 = 18$

b False, they are equal.

2 A T-shape:

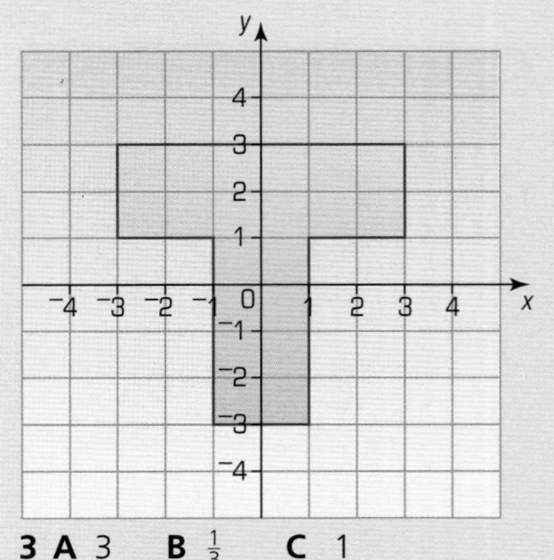

3 A 3 **B** $\frac{1}{3}$ **C** 1

D $^{-}2$ **E** $\frac{^{-}2}{3}$

A2 Check out

1 a i $y = \frac{x-1}{3}$ or $y = \frac{x}{3} - \frac{1}{3}$

ii $y = 2(x + 1)$ or $y = 2x + 2$

iii $y = \frac{x}{4} + 3$

b

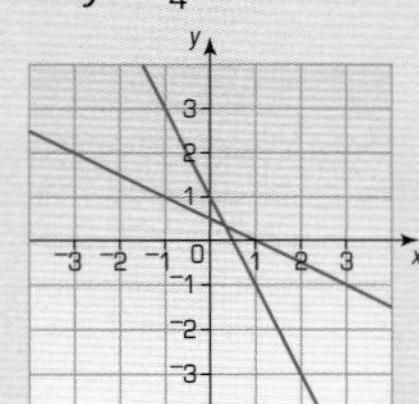

2 a $y = x^2 + 3$ has a minimum at (0, 3).
The line of symmetry is the y-axis ($x = 0$).

b

x	$^{-}3$	$^{-}2$	$^{-}1$	0	1	2	3
x^2	9	4	1	0	1	4	9
$^{-}3x$	9	6	3	0	$^{-}3$	$^{-}6$	$^{-}9$
8	8	8	8	8	8	8	8
y	26	18	12	8	6	6	8

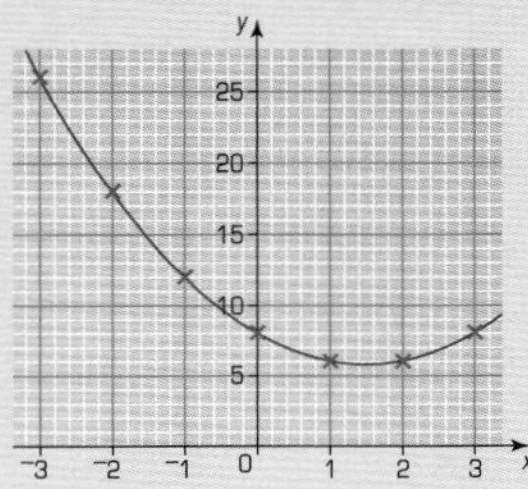

$y = 5.8$ when $x = 1.5$.

3 a Gradient $\frac{1}{2}$, y-intercept 1, equation $y = \frac{1}{2}x + 1$

b The cost is £1 (y-intercept) plus 50p per person (gradient).

N1 Check in

1 a 12

b 42

2 £149.76

3 a 13 : 19

b 1 : 3

c 31 : 10

4 a 32 150

b 2200

c 6.7

d 3.07

N1 Check out

1 a 35 : 4 : 6

b 78% water, 9% blackcurrant, 13% apple

c i 1.2 litres **ii** 0.3 litres

2 a 13.2 cm^3

b 1.6 cm

A3 Check in

1 $\frac{2x+1}{3} = 5$

2 $C = 50 - 3n$

3 $x^2 = 0.04$, $(x - 2y)^2 = 0.0196$, $xy^2 = 0.000\ 18$, $(xy)^2 = 0.000\ 036$, $y^3 = 0.000\ 027$

4

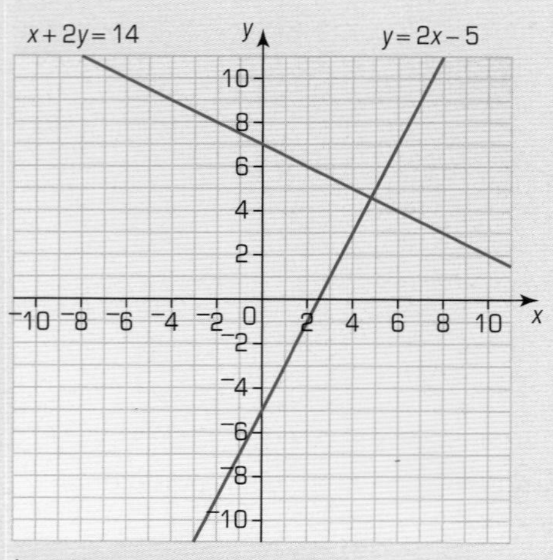

They intersect at (4.8, 4.6).

A3 Check out

1 a $x = 2, y = 4$

b $x = 1, y = {}^{-}3$

c $x = 5, y = 1$

2 a $x = 1.5, y = 1$

b $x = 1.5, y = 1$

3 Both companies charge £14 for 4 hours of calls. For fewer hours, A is better value. For more hours, B is better value.

S1 Check in

1 a 50°

b 97°

2

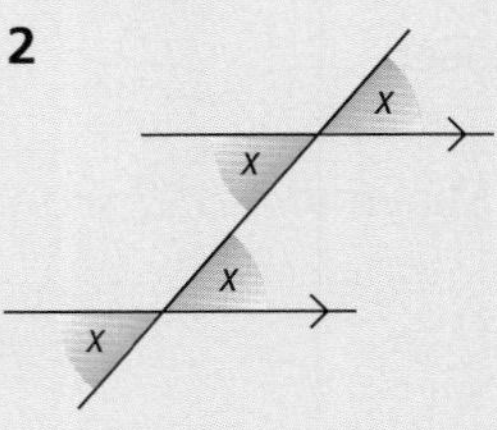

3 a 7.07 **b** 10.10 **c** 30.82

4

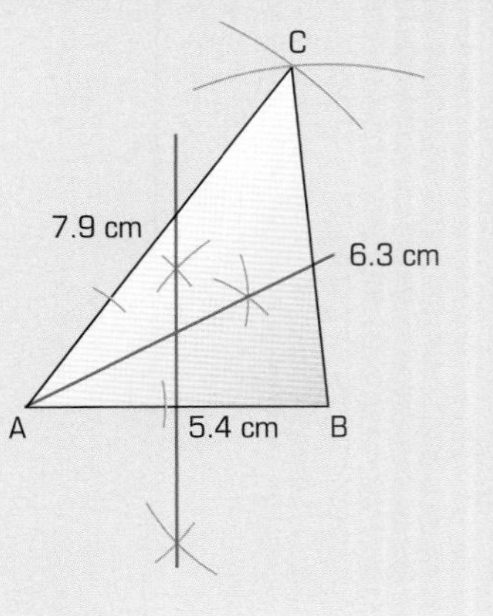

S1 Check out

1 a 7.21 cm

b 3.87 cm

c 4.90 cm

2 a Unique

b Not unique

c Unique

3 36.66 cm^2

D1 Check in

1 Mean = 26.5, Range = 6

2

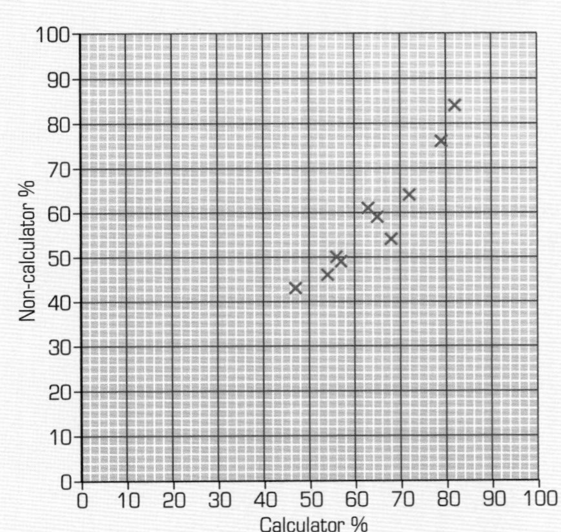

D1 Check out

1 Thao could time a simple random sample of students picked using random numbers. He should time them to the nearest tenth of a second.

2 Thao could compare alphabet and countdown times using a scatter diagram, stem-and-leaf diagram or box-and-whisker plots. He could compare means/medians to see whether, on average, students are quicker at reciting the alphabet or counting backwards. He could use the range or interquartile range to see how variable the times are.

3 The sample is too small.

4 Thao should use a larger sample. He could also collect separate data for male and female students.

S2 Check in

1 a 14.88 cm^2

b 34.56 mm^2

c 40 cm^2

2 a 40 mm

b 520 cm

c 2400 m

3 a 4.29

b 2.0

c 0.04

4 (2.5, 6)

5 a 7.8 m

b 7.8 m

S2 Check out

1 a 1.67 km/h

b 0.05 km/h

The man has the faster speed.

2 $6 \times 6 \times 6$

3 a Maximum = 202.5 g, Minimum = 197.5 g

b Largest = 162.75 cm^2, Smallest = 137.75 cm^2

N2 Check in

1 a 362, 3620, 36 200

b 21 500, 215 000, 2 150 000

c 0.063, 0.63, 6.3

2 a 0.62 m

b 6700 km

3 a 25.111

b 80.625

c 58.656

d 9.4

N2 Check out

1 a $(4^3)^2 = 4096$, $4^6 = 4096$

b $2^4 \times 2^3 = 128$, $2^7 = 128$

c $3^4 \div 3^2 = 9$, $3^2 = 9$

2 a $41 \times 32 = 1312$

b $da \times cb$ (where a, b, c, d are consecutive digits; a is the smallest)

c $da \times cb \times e$ (where a, b, c, d, e are consecutive digits, a is the smallest)

3 4.138 cm to 4.444 cm

A4 Check in

1 a $3^3 \times 5^6$

b x^4

c a^4b^2

d $6c^3d^3$

2 a $2(6x + 8)$

b i $4(x + 3)$

ii $5(2x - 1)$

iii $x(y + w)$

iv $x(x - 2)$

3 a $\frac{^-2}{5}$

b 2

A4 Check out

1 a $\frac{72x^4}{2x^{-3}}$, the other two are the same.

b $x^{\frac{3}{2}}$

2 a False

b i $x^2 + 7x + 12$

ii $x^2 - 3x - 28$

iii $x^2 - 10x + 25$

3 a B: $y = 3x + c(c > 0)$,

C: $y = {}^-\frac{1}{3}x + d(d < 0)$

b B: $4y = x + c(c > 0)$,

C: $y = -4x + d(d < 0)$

4 $(^-2, 4)$

D2 Check in

1 a $\frac{13}{15}$

b $\frac{2}{15}$

c 0.45

d 0.045

2 a 30

b 21

3

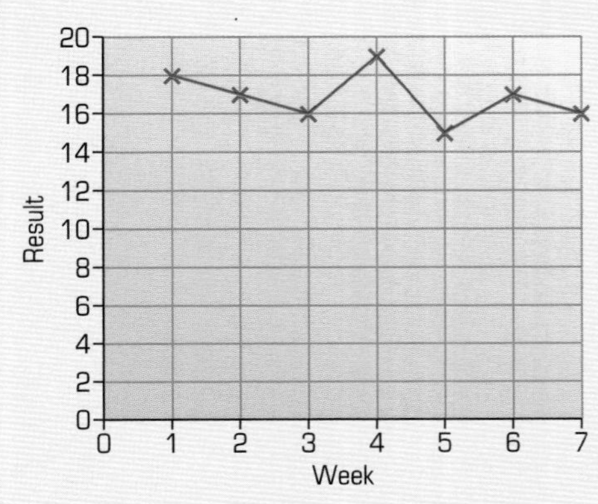

D2 Check out

1 The probability of Lesley making an error is $\frac{26}{500}$.

The probability of Duncan making an error is $\frac{31}{500}$.

Lesley is more reliable

2 a Dervla's relative frequency = 4.3

Emma's relative frequency = 3.6

b

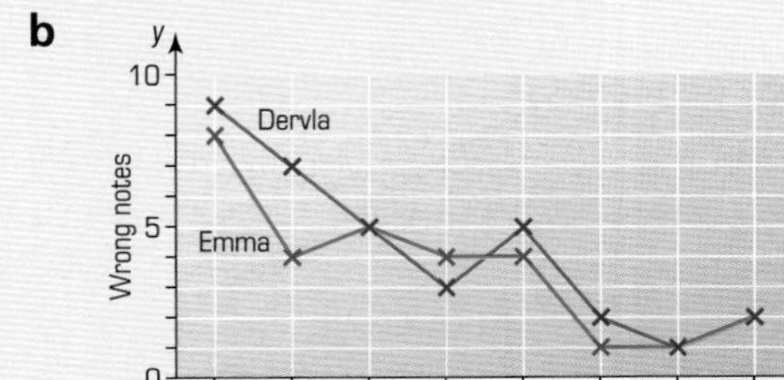

They may have been nervous at first, so made more mistakes.

Piece 5 may have been more difficult.

c i 4

ii 4

Leaving out value for Piece 1 gives estimates of:

i 3

ii 3

S3 Check in

1

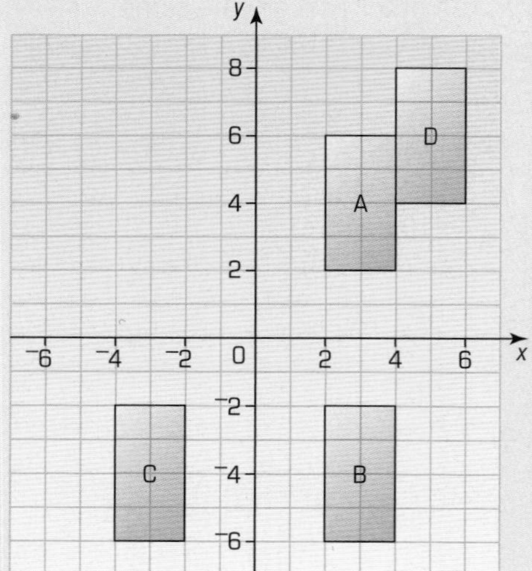

2 a 5 m

b 5.7 mm

3 a 5 : 1 : 2

b 30 : 1

c 1 : 5

S3 Check out

1 a Yes, all sides are the same.

b No, but they are similar.

2 a $x = 75°$, $y = 45°$, $z = 45°$, $w = 5$ cm

b $a = 18$ mm, $b = 5$ mm

3 a 1 : 4

b 1 : 2

P1 Check out

1 0.88 m/s (2 dp)

2 a The rods can be 3% of 1.62 m longer or shorter than 1.62 m.

b 1.5714 m $\leqslant x \leqslant$ 1.6686 m

c Costs could vary between £82 498.50 and £87 601.50

3 Show that any n-sided polygon can be divided into $(n - 2)$ triangles.

4 a

x	1	3	5	7	9	11
y	3	1	0.6	0.43	$0.\dot{6}$	0.27

Each triangle's angles total 180°.

b i y values 2 × previous

ii y values $\frac{1}{2}$ × previous

iii y values $\frac{1}{3}$ × previous

c For example y is the amount each person gets when £3 is shared between x people.

A5 Check in

1 $6(2x - 4) + 3(7 - 2x)$

2 **a** $x = 3\frac{1}{3}$

b $y = 1$

c $z = {}^{-}2\frac{1}{7}$

3 $x = \frac{41}{63}$ cm, $y = \frac{5}{28}$

A5 Check out

1 **a** **i** $10x^2 - 7x - 12y^2$

ii $36a^2 - 24ab + 4b^2$

b **i** $(x + 10)(x - 10)$ **ii** $(x + 5)(x - 5)$

iii $(2x + 3)(2x - 3)$ **iv** $(a + b)(a - b)$

c $\frac{a - b}{2}$

2 **a** $(x - 1)(x + 4) = (x + 1)^2$, $x = 5$

3 **a** $e = \frac{kp - z}{m}$ **b** $e = \sqrt{\frac{(m - k)}{wp}}$

c $e = 2k - f$ **d** $e = w - mR$

4 **a** For example: 1, 2, 3, 4, 5

b Let the smallest number be $x - 2$, then other four numbers are $x - 1$, x, $x + 1$, $x + 2$.

So the square of the average

$= (\frac{x - 2 + x - 1 + x + x + 1 + x + 2}{5})^2$

$= (\frac{5x}{5})^2 = x^2$

The average of the squares

$= \frac{(x - 2)^2 + (x - 1)^2 + x^2 + (x + 1)^2 + (x + 2)^2}{5}$

$= \frac{5x^2 + 10}{5} = x^2 + 2$

D3 Check in

1

Stem	Leaf
18	0 1 3
17	0 0 1 4 5 5 8 8 9
16	0 1 2 2 2 2 3 4 7 8
15	5 6 8 9 9

Key: 15 | 5 means 155 cm

D3 Check out

1 Only one weekday, only one time slot. Plan to include more than one day, weekend and weekdays and several time slots on each day.

2 Frequency diagrams for Week X and Y on the same axes. Statistical measure: the mean

3 Data is grouped and there is no maximum speed given for the last group, so answers will differ.

S4 Check in

1 a

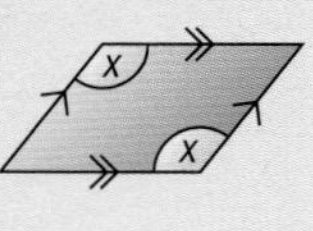

b

2 a 3.6 m **b** 5.5 m

3 a 430 **b** 0.48

c 4.3 **d** 0.000 48

e 600 **f** 250

4 a 8 m^2 **b** 16 m^2

c 12.6 m^2

S4 Check out

1 a 47 m

b 7.5 m

c 30°

d 66°

2 Min 1454 cm^3, max 2158 cm^3

3 24.9 cm^3

D4 Check in

1

Fraction	$\frac{2}{5}$	$\frac{99}{100}$	$\frac{7}{20}$	$\frac{2}{25}$
Decimal	0.4	0.99	0.35	0.08
Percentage	40%	99%	35%	8%

2

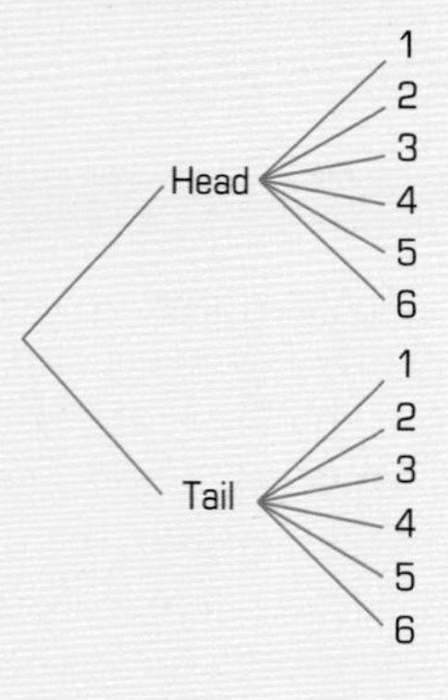

D4 Check out

1 If there are equal numbers of 2 pence and 10 pence coins, approximately half the class should have a total of 12p, approximately a quarter should have a total of 4p and approximately a quarter should have a total of 20p.

2 The coins are different sizes, so students may be able to feel which coin they are selecting.

3 $\frac{5}{9}$

B1 Check in

1 a 78°

b 77°

2 1131 cm^3

B1 Check out

1 1.45 m

2 a 33.2°

b 13.4 cm^2

Index

Index